FUNDAMENTALS OF APPLIED PROBABILITY THEORY

McGRAW-HILL SERIES IN PROBABILITY AND STATISTICS

DAVID BLACKWELL AND HERBERT SOLOMON, *Consulting Editors*

Bharucha-Reid Elements of the Theory of Markov Processes and Their Applications
Drake Fundamentals of Applied Probability Theory
Dubins and Savage How to Gamble If You Must
Ehrenfeld and Littauer Introduction to Statistical Methods
Graybill Introduction to Linear Statistical Models, Volume I
Jeffrey The Logic of Decision
Li Introduction to Experimental Statistics
Miller Simultaneous Statistical Inference
Mood and Graybill Introduction to the Theory of Statistics
Morrison Multivariate Statistical Methods
Pearce Biological Statistics: An Introduction
Pfeiffer Concepts of Probability Theory
Wadsworth and Bryan Introduction to Probability and Random Variables
Weiss Statistical Decision Theory
Wolf Elements of Probability and Statistics

ALVIN W. DRAKE

Operations Research Center and
Department of Electrical Engineering
Massachusetts Institute of Technology

fundamentals of applied probability theory

McGRAW-HILL BOOK COMPANY

New York St. Louis San Francisco Toronto London Sydney

fundamentals of applied probability theory

Library of Congress Catalog Card Number 67-11205

17815

34567890 MP 743210

To this Great Nation,
which has allowed
millions of refugees,
including my parents,
to enter a land of
freedom and opportunity

PREFACE

This is a first textbook in applied probability theory, assuming a background of one year of calculus. The material represents a one-semester subject taught at M.I.T. to about 250 students per year, most of whom are in the Schools of Engineering or Management. About two-thirds of these students are undergraduates. The subject, Probabilistic Systems Analysis, serves both as a terminal course and as a prerequisite for more advanced work in areas such as communication theory, control systems, decision theory, operations research, quantitative management, statistics, and stochastic processes.

My intention is to present a physically based introduction to applied probability theory, with emphasis on the continuity of fundamentals. A prime objective is to develop in the new student an understanding of the nature, formulation, and analysis of probabilistic situations. This text stresses the sample space of representation of probabilistic processes and (especially in the problems) the need for explicit modeling of nondeterministic processes.

In the attempt to achieve these goals, several traditional details have either been omitted or relegated to an appendix. Appreciable effort has been made to avoid the segmentation and listing of detailed applications which must appear in a truly comprehensive work in this area. Intended primarily as a student text, this book is not suitable for use as a general reference.

Scope and Organization

The fundamentals of probability theory, beginning with a discussion of the algebra of events and concluding with Bayes' theorem, are presented in Chapter 1. An axiomatic development of probability theory is used and, wherever possible, concepts are interpreted in the sample space representation of the model of an *experiment* (any nondeterministic process). The assignment of probability measure in the modeling of physical situations is not necessarily tied to a relative frequency interpretation. In the last section of this chapter, the use of sample and event spaces in problems of enumeration is demonstrated.

Chapter 2 is concerned with the extension of earlier results to deal with random variables. This introductory text emphasizes the local assignment of probability in sample space. For this reason, we work primarily with probability density functions rather than cumulative distribution functions. My experience is that this approach is much more intuitive for the beginning student. Random-variable concepts are first introduced for the discrete case, where things are particularly simple, and then extended to the continuous case. Chapter 2 concludes with the topic of derived probability distributions as obtained directly in sample space.

Discrete and continuous transform techniques are introduced in Chapter 3. Several applications to sums of independent random variables are included. Contour integration methods for obtaining inverse transforms are not discussed.

Chapters 4 and 5 investigate basic random processes involving, respectively, independent and dependent trials.

Chapter 4 studies in some detail the Bernoulli and Poisson processes and the resulting families of probability mass and density functions. Because of its significance in experimentation with physical systems, the phenomenon of random incidence is introduced in the last section of Chapter 4.

Discrete-state Markov models, including both discrete-transition and continuous-transition processes, are presented in Chapter 5. The describing equations and limiting state probabilities are treated, but closed form solutions for transient behavior in the general case are not discussed. Common applications are indicated in the text and in the problems, with most examples based on relatively simple birth-and-death processes.

Chapter 6 is concerned with some of the basic limit theorems, both for the manner in which they relate probabilities to physically observable phenomena and for their use as practical approximations. Only weak statistical convergence is considered in detail. A transform development of the central limit theorem is presented.

The final chapter introduces some common issues and techniques of statistics, both classical and Bayesian. My objectives in this obviously incomplete chapter are to indicate the nature of the transition from probability theory to statistical reasoning and to assist the student in developing a critical attitude towards matters of statistical inference.

Although many other arrangements are possible, the text is most effectively employed when the chapters are studied in the given order.

Examples and Home Problems

Many of the sections which present new material to the student contain very simple illustrative examples. More structured examples, usually integrating larger amounts of material, are solved and discussed in separate sections.

For the student, the home problems constitute a vital part of the subject matter. It is important that he develop the skill to formulate and solve problems with confidence. Passive agreement with other people's solutions offers little future return. Most of the home problems following the chapters are original, written by the author and other members of the teaching staff.

These problems were written with definite objectives. In particular, wherever possible, we have left for the student a considerable share in the formulation of physical situations. Occasionally, the probability assignments directly relevant to the problems must be derived from other given information.

It did not seem feasible to sample the very many possible fields of application with other than superficial problems. The interesting aspects of each such field often involve appreciable specialized structure and nomenclature. Most of our advanced problems are based on relatively simple operational situations. From these common, easily communicated situations, it seemed possible to develop compact representative problems which are challenging and instructive.

The order of the problems at the end of each chapter, by and large, follows the order of the presentation in the chapter. Although entries below are often not the most elementary problems, relatively comprehensive coverage of the material in this text is offered by the following skeleton set of home problems:

1.03	2.04	3.05	4.05	5.05	6.02	7.04
1.08	2.07	3.08	4.09	5.06	6.03	7.06
1.09	2.11	3.09	4.12	5.10	6.04	7.08
1.12	2.17	3.10	4.13	5.12	6.07	7.15
1.13	2.26	3.12	4.17	5.13	6.08	7.16
1.21	2.27	3.13	4.18	5.14	6.13	7.19
1.24	2.28	3.21	4.22	5.16	6.17	7.20
1.30	2.30					

Some Clerical Notes

I have taken some liberties with the usual details of presentation. Figures are not numbered but they do appear directly in context. Since there are few involved mathematical developments, equations are not numbered. Whenever it appeared advantageous, equations were repeated rather than cross-referenced.

Recommended further reading, including a few detailed references and referrals for topics such as the historical development of probability theory are given in Appendix 1. Appendix 2 consists of a listing of common probability mass and density functions and their expected values, variances, and transforms. Several of these probability functions do not appear in the body of the text. A brief table of the cumulative distribution for the unit normal probability density function appears in context in Chapter 6.

The general form of the notation used in this text seems to be gaining favor at the present time. To my taste, it is one of the simplest notations which allows for relatively explicit communication. My detailed notation is most similar to one introduced by Ronald A. Howard.

Acknowledgments

My interest in applied probability theory was originally sparked by the enthusiasm and ability of two of my teachers, Professors George P. Wadsworth and Ronald A. Howard. For better or worse, it is the interest they stimulated which led me to this book, rather than one on field theory, bicycle repair, or the larger African beetles.

Like all authors, I am indebted to a large number of earlier authors. In this case, my gratitude is especially due to Professors William Feller, Marek Fisz, and Emanuel Parzen for their excellent works.

Teaching this and related material during the past six years has been an exciting and rewarding experience, due to the intensity of our students and the interchange of ideas with my colleagues, especially Dr. Murray B. Sachs and Professor George Murray. The many teaching assistants associated with this subject contributed a great deal to its clarification. Some of the problems in this book represent their best educational (and Machiavellian) efforts.

During the preparation of this book, I had many productive discussions with Professor William Black. In return for his kindness, and also because he is a particularly close friend, I never asked him to look at the manuscript. Professor Alan V. Oppenheim and Dr. Ralph L. Miller were less fortunate friends; both read the manuscript with great care and offered many helpful suggestions. The publisher's review by Dr. John G. Truxal was most valuable.

Some award is certainly due Mrs. Richard Spargo who had the grim pleasure of typing and illustrating the entire manuscript—three times! My devoted wife, Elisabeth, checked all examples, proofread each revision of the manuscript, and provided unbounded patience and encouragement.

Finally, I express my gratitude to any kind readers who may forward to me corrections and suggestions for improvements in this text.

Alvin W. Drake

CONTENTS

CHAPTER ONE

events, sample space, and probability

Our study of probability theory begins with three closely related topics:

1 *The algebra of events* (*or sets or areas or spaces*) will provide us with a common language which is never ambiguous. This algebra will be defined by a set of seven axioms.

2 *Sample space* is the vital picture of a model of an "experiment" (any nondeterministic process) and all of its possible outcomes.

3 *Probability measure* is the relevant measure in sample space. This measure is established by three more axioms and it provides for the measurement of events in sample space.

1-1 A Brief Introduction to the Algebra of Events

We first introduce an explicit language for our discussion of probability theory. Such a language is provided by the algebra of events. We wish to develop an understanding of those introductory aspects of this algebra which are essential to our study of probability theory and its applications.

Let's begin with a brief tour of some of the definitions, concepts, and operations of the algebra of events.

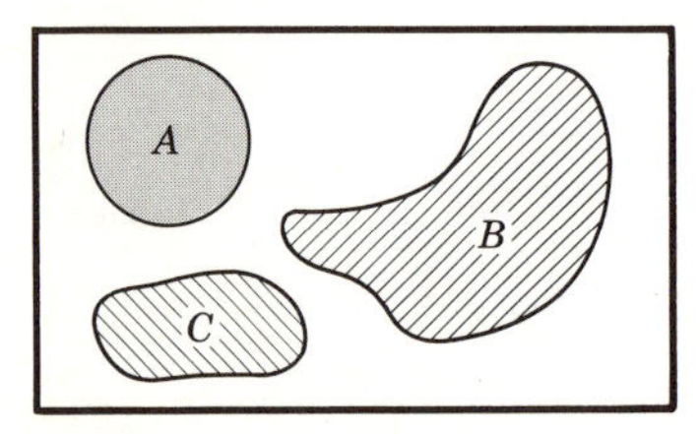

Events (or sets) are collections of points or areas in a space. The physical interpretation of this space will be provided in the following section.

The collection of all points in the entire space is called U, the *universal set* or the *universal event*.

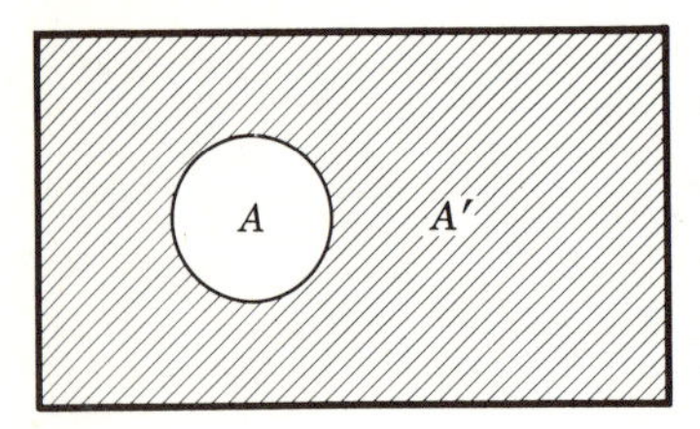

Event A', the *complement* of event A, is the collection of all points in the universal set which are not included in event A. The *null set* ϕ contains no points and is the complement of the universal set.

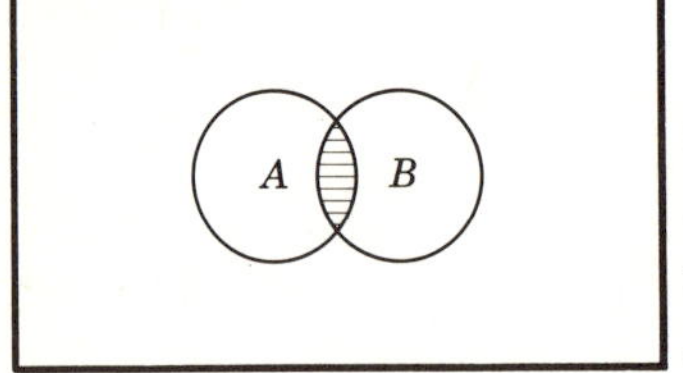

The *intersection* of two events A and B is the collection of all points which are contained both in A and in B. For the intersection of events A and B we shall use the simple notation AB.

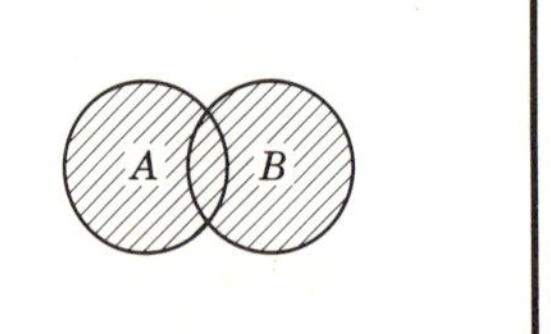

The *union* of two events A and B is the collection of all points which are either in A or in B or in both. For the union of events A and B we shall use the notation $A + B$.

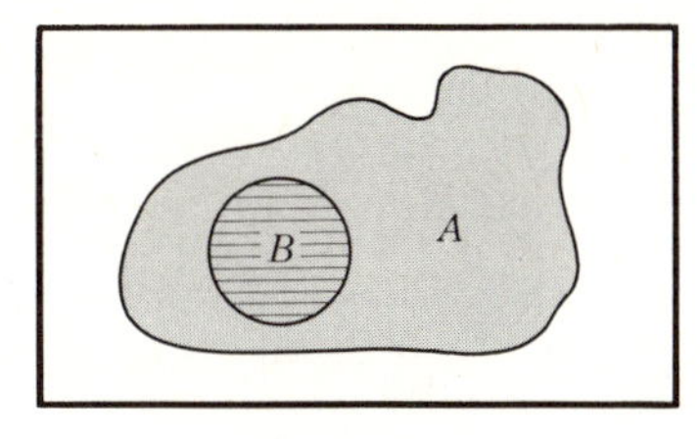

If all points of U which are in B are also in A, then event B is said to be *included* in event A.

Two events A and B are said to be *equal* if every point of U which is in A is also in B and every point of U which is in A' is also in B'. Another way to state the condition for the equality of two events would be to say that two events are equal if and only if each event is included in the other event.

We have sampled several of the notions of the algebra of events. Pictures of events in the universal set, such as those used above, are known as *Venn diagrams*. Formally, the following is a set of laws (*axioms*) which fully defines the algebra of events:

1 $A + B = B + A$ Commutative law
2 $A + (B + C) = (A + B) + C$ Associative law
3 $A(B + C) = AB + AC$ Distributive law
4 $(A')' = A$
5 $(AB)' = A' + B'$
6 $AA' = \phi$
7 $AU = A$

The Seven Axioms of the Algebra of Events

Technically, these seven axioms define everything there is to know about the algebra of events. The reader may consider visualizing each of these axioms in a Venn diagram. Our selection of a list of axioms is not a unique one. Alternative sets of axioms could be stated which would lead to the same results.

Any relation which is valid in the algebra of events is subject to proof by use of the seven axioms and with no additional information. Some representative relations, each of which may be interpreted easily on a Venn diagram, are

$A + A = A$ $\qquad$ $A + U = U$

$A + AB = A$ $\qquad$ $A + BC = (A + B)(A + C)$

$A + A'B = A + B$ $\qquad$ $A\phi = \phi$

$A + A' = U$ $\qquad$ $A(BC) = (AB)C$

If it were our intention to prove (or test) relations in the algebra of events, we would find it surprisingly taxing to do so using only the seven axioms in their given form. One could not take an "obvious" step such as using $CD = DC$ without proving the validity of this relation from the axioms. For instance, to show that $CD = DC$, we may proceed

$C' + D' = D' + C'$	Axiom (1) with $A = C'$ and $B = D'$
$(C' + D')' = (D' + C')'$	Take complement of both sides
$\therefore CD = DC$	By use of axioms (5) and (4)

The task of proving or testing relations becomes easier if we have relations such as the above available for direct use. Once such relations are proved via the seven axioms, we call them *theorems* and use them with the same validity as the original axioms.

We are already prepared to note those definitions and properties of the algebra of events which will be of value to us in our study of probability.

A list of events $A_1, A_2, \ldots, A_N$ is said to be composed of *mutually exclusive* events if and only if

$$A_iA_j = \begin{cases} A_i & \text{if } i = j \\ \phi & \text{if } i \neq j \end{cases} \qquad i = 1, 2, \ldots, N; \quad j = 1, 2, \ldots, N$$

A list of events is composed of mutually exclusive events if there is no point in the universal set which is included in more than one event in the list. A Venn diagram for three mutually exclusive events A, B, C could be

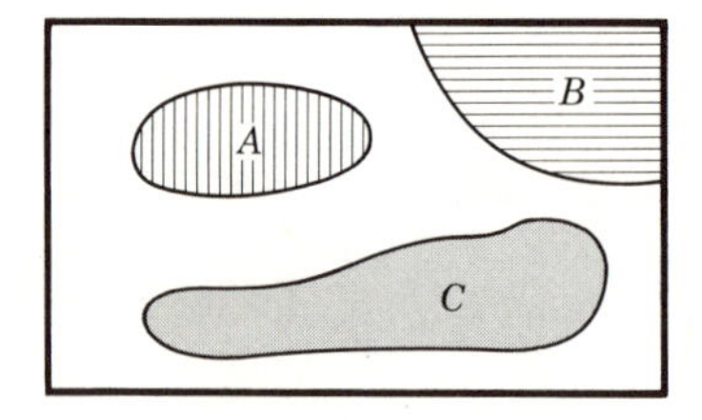

or

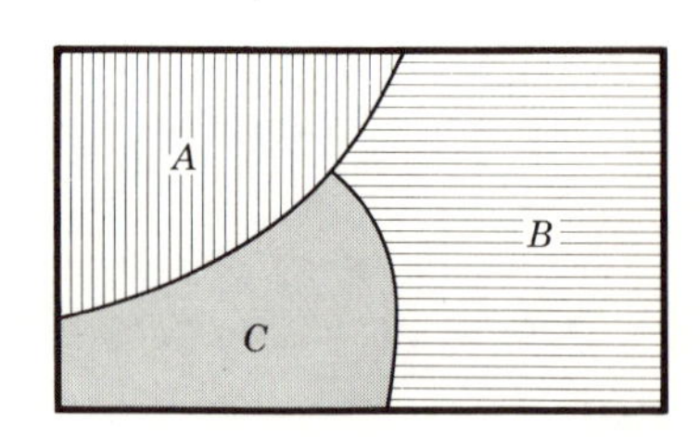

but in no case may there be any overlap of the events.

A list of events $A_1, A_2, \ldots, A_N$ is said to be *collectively exhaustive* if and only if

$$A_1 + A_2 + \cdots + A_N = U$$

A list of events is collectively exhaustive if each point in the universal set is included in at least one event in the list. A Venn diagram for three collectively exhaustive events A, B, C could be

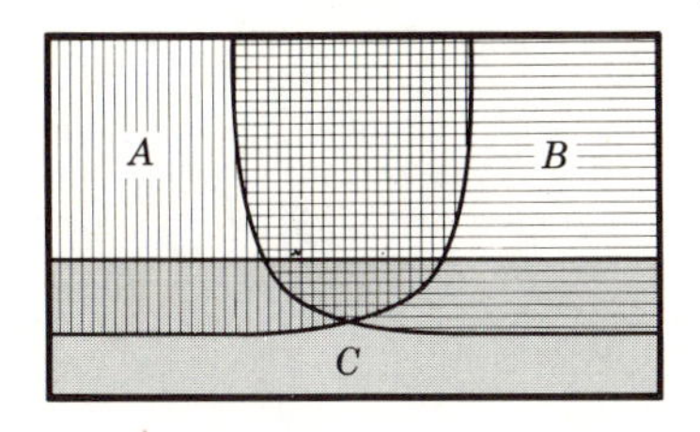

or

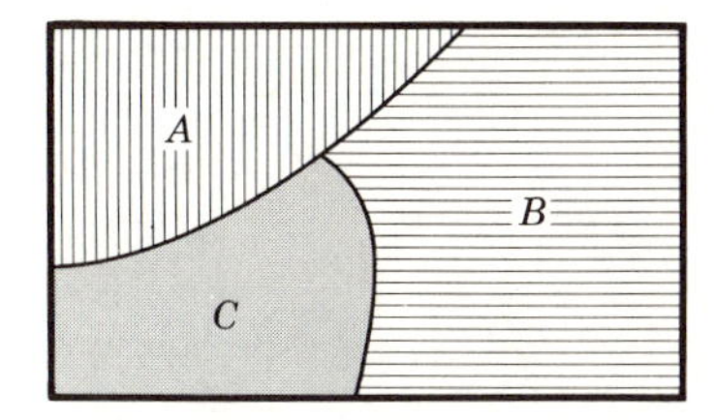

but in no case may there be any point in the universal set which is included in none of the events. A list of events may be mutually exclusive, collectively exhaustive, both, or neither. After the discussion of sample space in the next section, we shall have many opportunities to consider lists of events and familiarize ourselves with the use of these definitions.

We note two additional matters with regard to the algebra of events. One is that we do not make errors in the algebra of events if we happen to include the same term several times in a union. If we are trying to collect all the points in a rather structured event, we need only be sure to include every appropriate point at least once—multiple inclusion will do no harm.

Another consideration is of particular importance in dealing with actual problems. The algebra of events offers a language advantage in describing complex events, *if* we are careful in defining all relevant simple events. Since we shall be making conscious and subconscious use of the algebra of events in all our work, one necessary warning should be sounded. "He who would live with the algebra of events had better know exactly with which events he is living." The original defined events should be simple and clear. If this is the case, it is then an easy matter to assemble expressions for complex events from the original definitions. For instance:

NEVER Event A: Neither Tom nor Mary goes without Harry unless they see Fred accompanied by Phil or it is raining and . . . .

BETTER Event A: Tom goes. Event B: Mary goes.
Event C: Harry goes. Etc.

1-2 Sample Spaces for Models of Experiments

In this book, we use the word "experiment" to refer to any process which is, to some particular observer, nondeterministic. It makes no difference whether the observer's uncertainty is due to the nature of the process, the state of knowledge of the observer, or both.

The first six chapters of this book are concerned with the analysis of abstractions, or *models* of actual physical experiments. Our last chapter is concerned with the relation of the model to the actual experiment.

It is probably safe to state that there are more variables associated with the outcome of any physical experiment than anybody could ever care about. For instance, to describe the outcome of an actual coin toss, we could be concerned with the height of the toss, the number of bounces, and the heating due to impact, as well as the more usual consideration of which face is up after the coin settles. For most purposes, however, a reasonable model for this experiment would involve a simple nondeterministic choice between a head and a tail.

Many of the "trick" probability problems which plague students are based on some ambiguity in the problem statement or on an inexact formulation of the model of a physical situation. The precise statement of an appropriate *sample space*, resulting from a detailed description of the model of an experiment, will do much to resolve common difficulties. In this text, we shall literally live in sample space.

Sample space: The finest-grain, mutually exclusive, collectively exhaustive listing of all possible outcomes of a model of an experiment

The "finest-grain" property requires that all possible distinguishable outcomes allowed by the model be listed separately. If our model for the flip of a coin is simply a nondeterministic selection between possible outcomes of a head and a tail, the sample space for this model of the experiment would include only two items, one corresponding to each possible outcome.

To avoid unnecessarily cumbersome statements, we shall often use the word experiment in place of the phrase "model of an experiment." Until the final chapter, the reader is reminded that all our work is concerned with abstractions of actual physical situations. When we wish to refer to the real world, we shall speak of the "physical experiment."

Soon we shall consider several experiments and their sample spaces. But first we take note of two matters which account for our interest in these spaces. First, the universal set with which we deal in the study of probability theory will always be the sample space for an experiment. The second matter is that *any* event described in terms of the outcome of a performance of some experiment can be identified as a collection of events in sample space. In sample space one may collect the members of any event by taking a union of mutually exclusive points or areas from a collectively exhaustive space. The advantage of being able to collect any event as a union of mutually exclusive members will become clear as we learn the properties of probability measure in the next section.

A sample space may look like almost anything, from a simple listing to a multidimensional display of all possible distinguishable outcomes of an experiment. However, two types of sample spaces seem to be the most useful.

One common type of sample space is obtained from a sequential picture of an experiment in terms of its most convenient parameters. Normally, this type of sample space is not influenced by our particular interests in the experimental outcome. Consider the sequential sample space for a reasonable model of two flips of a coin. We use the notation

$$\text{Event } \begin{Bmatrix} H_n \\ T_n \end{Bmatrix} : \begin{Bmatrix} \text{Heads} \\ \text{Tails} \end{Bmatrix} \text{ on the } n\text{th toss of the coin}$$

This leads to the sequential sample space

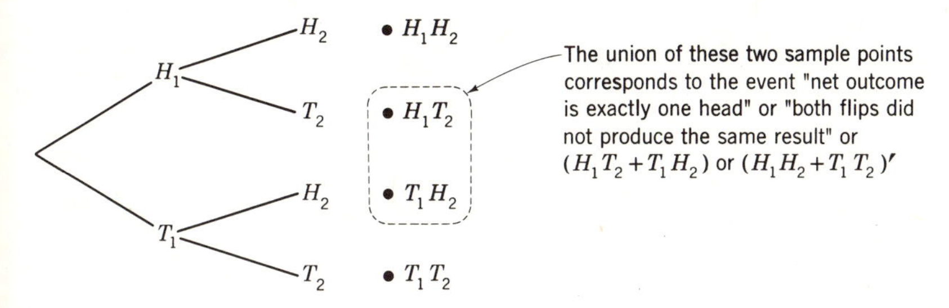

Above we picture the experiment proceeding rightward from the left origin. Each sample point, located at the end of a *terminal* tree branch, represents the event corresponding to the intersection of all events encountered in tracing a path from the left origin to that sample point. On the diagram, we have noted one example of how an event may be collected as a union of points in this sample space.

Formally, the four sample points and their labels constitute the sample space for the experiment. However, when one speaks of a *sequential* sample space, he normally pictures the entire generating tree as well as the resulting sample space.

For an experiment whose outcomes may be expressed numerically, another useful type of sample space is a coordinate system on which is displayed a finest-grain mutually exclusive collectively exhaustive set of points corresponding to every possible outcome. For instance, if we throw a six-sided die (with faces labeled 1, 2, 3, 4, 5, 6) twice and call the value of the up face on the nth toss x_n, we have a sample space with 36 points, one for each possible experimental outcome,

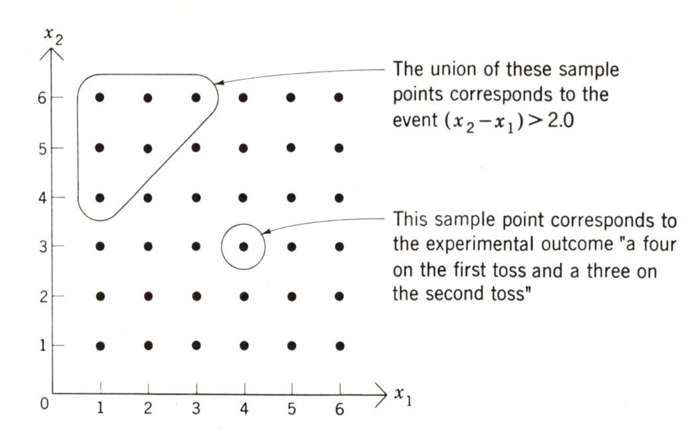

Often we shall have reason to abbreviate a sample space by working with a mutually exclusive collectively exhaustive listing of all possible outcomes which is not finest-grain. Such spaces, possessing all attributes of sample spaces other than that they may not list separately all possible distinguishable outcomes, are known as *event spaces*. If a list of events $A_1, A_2, \ldots, A_N$ forms an event space, each possible finest-grain experimental outcome is included in exactly one event in this list. However, more than one distinguishable outcome may be included in any event in the list.

We now present some sample and event spaces for several experiments.

1 Experiment: Flip a coin twice.

Notation: Let $\begin{Bmatrix} H_n \\ T_n \end{Bmatrix}$ be the event $\begin{Bmatrix} \text{heads} \\ \text{tails} \end{Bmatrix}$ on the nth toss.

We already have seen the sequential sample space for this experiment, but let's consider some event spaces and some other displays of the sample space. The sample space in a different form is

	H_2	T_2
H_1	•	•
T_1	•	•

The sample space displayed as a simple listing is

$\bullet H_1H_2$ $\quad\bullet H_1T_2$ $\quad\bullet T_1H_2$ $\quad\bullet T_1T_2$

An example of an event space but not a sample space for this experiment is

$\bullet H_1$ $\quad\bullet T_1$

Another example of an event space but not a sample space is

$\bullet H_1T_2$ $\quad\bullet (H_1T_2)'$

The following is neither an event space nor a sample space:

$\bullet H_1$ $\quad\bullet T_1$ $\quad\bullet H_2$ $\quad\bullet T_2$

2 Experiment: Flip a coin until we get our first head.

Notation: Let $\begin{Bmatrix} H_n \\ T_n \end{Bmatrix}$ be the event $\begin{Bmatrix} \text{heads} \\ \text{tails} \end{Bmatrix}$ on the nth toss.

A sample space generated by a sequential picture of the experiment is

$\bullet H_1$ $\quad\bullet T_1H_2$ $\quad\bullet T_1T_2H_3$ $\quad\bullet T_1T_2T_3H_4$

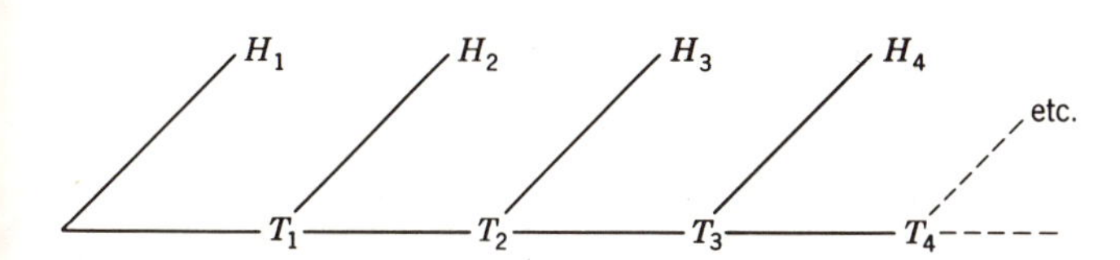

A sample space which is a simple listing for this experiment, with N representing the number of the toss on which the first head occurred, is

$\bullet N = 1$ $\quad\bullet N = 2$ $\quad\bullet N = 3$ $\quad\bullet N = 4$ $\quad\cdots$

Note that this is still a finest-grain description and that one can specify the exact sequence of flips which corresponds to each of these sample points.

An event space but not a sample space for this experiment is

$\bullet N > 4$ $\quad\bullet N \leq 4$

Neither a sample space nor an event space for this experiment is

$\bullet N \geq 1$ $\quad\bullet N \geq 2$ $\quad\bullet N \geq 3$ $\quad\bullet N \geq 4$ $\quad\cdots$

3 Experiment: Spin a wheel of fortune twice. The wheel is continuously calibrated, not necessarily uniformly, from zero to unity.

Notation: Let x_n be the exact reading of the wheel on the nth spin.

A sample space for this experiment is

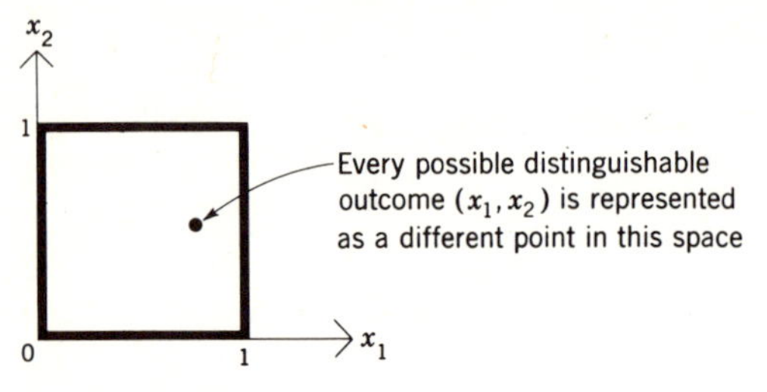

Let's collect several typical events in this sample space.

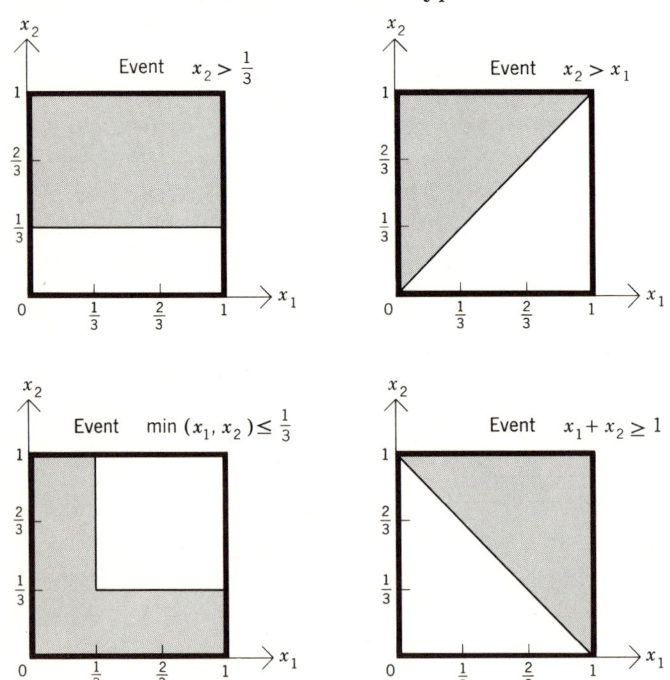

Any point in a sample space corresponds to one possible outcome of a performance of the experiment. For brevity, we shall often refer to the sample point as though it were the event to which it corresponds.

1-3 Probability Measure and the Relative Likelihood of Events

To complete our specification of the model for a physical experiment, we wish to assign *probabilities* to the events in the sample space of the experiment. The probability of an event is to be a number, representing the chance, or "relative likelihood," that a performance of the experiment will result in the occurrence of that event.

This measure of events in sample space is known as *probability measure.* By combining three new axioms with the seven axioms of the algebra of events, one obtains a consistent system for associating real nonnegative numbers (probabilities) with events in sample space and for computing the probabilities of more complex events. If we use the notation $P(A)$ for the probability measure associated with event A, then the three additional axioms required to establish probability measure are:

1 For any event A, $P(A) \geq 0$
2 $P(U) = 1$
3 If $AB = \phi$, then $P(A + B) = P(A) + P(B)$

The Three Axioms of Probability Measure

The first axiom states that all probabilities are to be nonnegative. By attributing a probability of unity to the universal event, the second axiom provides a normalization for probability measure. The third axiom states a property that most people would hold to be appropriate to a reasonable measure of the relative likelihood of events. If we accept these three axioms, all of conventional probability theory follows from them.

The original assignment of probability measure to a sample space for an experiment is a matter of how one chooses to model the physical experiment. Our axioms do not tell us how to compute the probability of any experimental outcome "from scratch." Before one can begin operating under the rules of probability theory, he must first provide an assignment of probability measure to the events in the sample space.

Given a physical experiment to be modeled and analyzed, there is no reason why we would expect two different people to agree as to what constitutes a reasonable probability assignment in the sample space. One would expect a person's modeling of a physical experiment to be influenced by his previous experiences in similar situations and by any other information which might be available to him. Probability theory will operate "correctly" on *any* assignment of probability to the sample space of an experiment which is consistent with the three axioms. Whether any such assignment and the results obtained from it are of any physical significance is another matter.

If we consider the definition of sample space and the third axiom of probability measure, we are led to the important conclusion,

Given an assignment of probability measure to the finest-grain events in a sample space, the probability of any event A may be computed by summing the probabilities of all finest-grain events included in event A.

One virtue of working in sample space is that, for any point in the sample space there must be a "yes" or "no" answer as to whether the point is included in event A. Were there some event in our space for which this was not the case, then either the space would not be a sample space (because it did not list separately certain distinguishable outcomes) or our model is inadequate for resolution between events A and A'. We shall return to this matter when we discuss the sample-space interpretation of conditional probability.

One can use the seven axioms of the algebra of events and the three axioms of probability measure to prove various relations such as

$$P(A') = 1 - P(A) \qquad P(A + B) = P(A) + P(B) - P(AB)$$
$$P(\phi) = 0 \qquad P(A + B + C) = 1 - P(A'B'C')$$

Because we shall live in sample space, few, if any, such relations will be required formally for our work. We shall always be able to employ directly the third axiom of probability measure to compute the probability of complex events, since we shall be expressing such events as unions of mutually exclusive members in a sample space. In computing the probability of any event in sample space, that axiom states that we must include the probability of every sample point in that event exactly *once*. Multiple counting caused no error in taking a union in the algebra of events to describe another event, but one must carefully confine himself to the axioms of probability measure when determining the probability of a complex event.

There are many ways to write out a "formula" for the probability of an event which is a union such as $P(A + B + C)$. Using the third axiom of probability measure and looking at a Venn diagram,

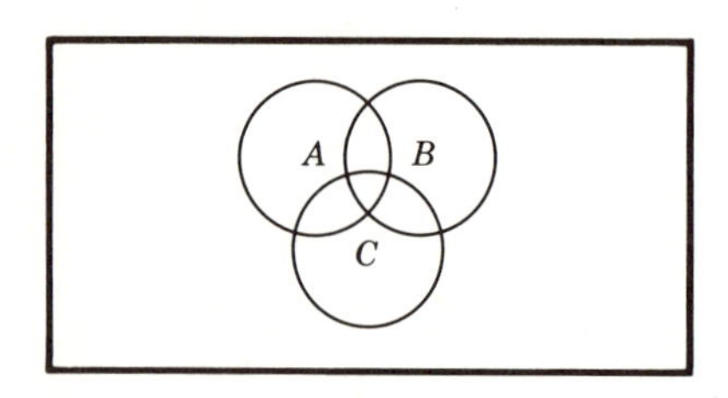

we may, for instance, write any of the following:

$$P(A + B + C) = P(A) + P(A'B) + P(A'B'C)$$
$$P(A + B + C) = P(A) + P(B) + P(C) - P(AB) - P(AC) - P(BC) + P(ABC)$$
$$P(A + B + C) = P(AB'C') + P(A'BC') + P(A'B'C) + P(AB) + P(AB'C) + P(A'BC)$$
$$P(A + B + C) = 1 - P(A'B'C')$$

Obviously, one could continue listing such relations for a very long time.

Now that we have considered both the algebra of events and probability measure, it is important to recall that:

1 Events are combined and operated upon only in accordance with the seven axioms of the algebra of events.
2 The probabilities of events are numbers and can be computed only in accordance with the three axioms of probability measure.
3 Arithmetic is something else.

$$P[\underbrace{A + CD + B(A + C'D)}]$$

The algebra of events applies inside the brackets. It operates upon events and has nothing to do with their measure.

The axioms of probability theory are used in obtaining the numerical value of this quantity.

In practice, we shall usually evaluate such probabilities by collecting the event as a union of mutually exclusive points in sample space and summing the probabilities of all points included in the union.

1-4 Conditional Probability and Its Interpretation in Sample Space

Assume that we have a fully defined experiment, its sample space, and an initial assignment of probability to each finest-grain event in the sample space. Let two events A and B be defined on the sample space of this experiment, with $P(B) \neq 0$.

We wish to consider the situation which results if the experiment is performed once and we are told only that the experimental outcome has attribute B. Thus, if B contains more than one sample point, we are considering the effect of "partial information" about the experimental outcome.

Let's look at a sample-space picture of this situation. Consider a sample space made up of sample points $S_1, S_2, \ldots, S_N$.

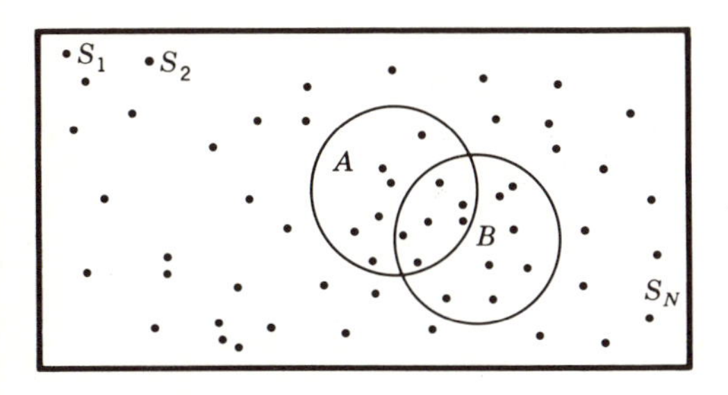

Given that event B has occurred, we know that the sample point representing the experimental outcome must be in B and cannot be in B'. We have no information which would lead us to alter the *relative* probabilities of the sample points in B. Since we know one of the sample points in B must represent the experimental outcome, we scale up their original probabilities by a constant, $1/P(B)$, such that they now add to unity, to obtain *conditional probabilities* which reflect the influence of our partial information.

We formalize these ideas by defining $P(S_j \mid B)$, the "conditional probability of S_j given B," to be

$$P(S_j \mid B) = \begin{Bmatrix} \dfrac{P(S_j)}{P(B)} & \text{if } S_j \text{ in } B \\ 0 & \text{if } S_j \text{ in } B' \end{Bmatrix} = \frac{P(S_jB)}{P(B)}$$

The conditional probability of any other event, such as A, is to be the sum of the conditional probabilities of the sample points included in A, leading to the common definition of the conditional probability of event A given B.

$$P(A \mid B) = \frac{P(AB)}{P(B)} \qquad \text{defined only for } P(B) \neq 0$$

which may be obtained from our previous statements via

$$P(A \mid B) = \sum_{\text{all } j \text{ in } A} P(S_j \mid B) = \sum_{\text{all } j \text{ in } A} \frac{P(S_jB)}{P(B)} = \frac{P(AB)}{P(B)}$$

We may conclude that one way to interpret conditional probability is to realize that a conditioning event (some partial information about the experimental outcome) allows one to move his analysis from the original sample space into a new conditional space. Only those finest-grain events from the original sample space which are included in the conditioning event appear in the new conditional space with a nonzero assignment of conditional probability measure. The original ("a priori") probability assigned to each of these finest-grain events is multiplied by the same constant such that the sum of the conditional ("a posteriori") probabilities in the conditional space is unity. *In the resulting conditional sample space, one uses and interprets these a posteriori probabilities exactly the same way he uses the a priori probabilities in the original sample space.* The conditional probabilities obtained by the use of some partial information will, in fact, serve as initial probabilities in the new sample space for any further work.

We present one simple example. A fair coin is flipped twice, and Joe, who saw the experimental outcome, reports that "at least one toss resulted in a head." Given this partial information, we wish to determine the conditional probability that both tosses resulted in heads. Using the notation $\begin{Bmatrix} H_n \\ T_n \end{Bmatrix}$ and the problem statement, we may draw a sample space for the experiment and indicate, in the $P(\bullet)$ column, our a priori probability for each sample point.

	$P(\bullet)$	A	B	AB
$\bullet H_1H_2$	0.25	✓	✓	✓
$\bullet H_1T_2$	0.25	✓		
$\bullet T_1H_2$	0.25	✓		
$\bullet T_1T_2$	0.25			
		$P(A) = 0.75$	$P(B) = 0.25$	$P(AB) = 0.25$

For the above display, we defined the events A and B to be "at least one head" and "two heads," respectively. In the A, B, and AB columns we check those sample points included in each of these events. The probability of each of these events is simply the sum of the probabilities of the sample points included in the event. (The only use of the B column in this example was to make it easier to identify the sample points associated with the complex event AB.)

The desired answer $P(B \mid A)$ may be found either directly from the definition of conditional probability

$$P(B \mid A) = \frac{P(AB)}{P(A)} = \frac{0.25}{0.75} = \frac{1}{3}$$

or by noting that, in the conditional sample space given our event A, there happen to be three equally likely sample points, only one of which has attribute B.

The solution to this problem could have been obtained in far less space. We have taken this opportunity to introduce and discuss a type of sample-space display which will be valuable in the investigation of more complex problems.

It is essential that the reader realize that certain operations (such as collecting events and conditioning the space) which are always simple in a sample space may not be directly applicable in an arbitrary event space. Because of the *finest-grain* property of a sample space, any sample point must be either wholly excluded from or wholly included in any arbitrary event defined within our model. However, in an event space, an event point A might be *partially* included in B, some other event of interest. Were this the case, the lack of detail in the event space would make it impossible to collect event B in this event space or to condition the event space by event B.

For instance, suppose that for the coin example above we were given only the event space

$\bullet H_1 \qquad \bullet T_1H_2 \qquad \bullet T_1T_2$

its probability assignment, and no other details of the experiment. Because this event space lacks adequate detail, it would be impossible to calculate the probability of event H_2 or to condition this space by event H_2.

When we are given a *sample* space for an experiment and the probability assigned to each sample point, we can answer all questions with regard to any event defined on the possible experimental outcomes.

1-5 Probability Trees for Sequential Experiments

Sequential sample and event spaces were introduced in Sec. 1-2. Such spaces, with all branches labeled to indicate the probability structure of an experiment, are often referred to as *probability trees*. Consider the following example:

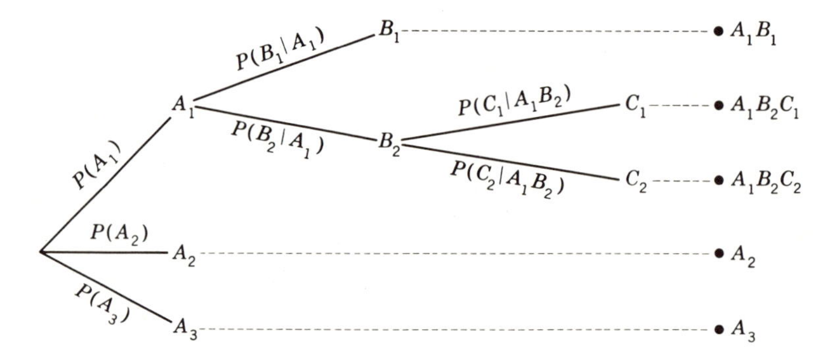

This would be a sample space for an experiment in which, for instance, the "C_1 or C_2" trial occurs only if A_1B_2 has resulted from the earlier stages of the sequential experiment. As before, one sample point appears for each terminal branch of the tree, representing the intersection of all events encountered in tracing a path from the left origin to a terminal node.

Each branch is labeled such that the product of all branch probabilities from the left origin to any node equals the probability that the event represented by that node will be the outcome on a particular performance of the experiment. Only the first set of branches leaving the origin is labeled with a priori probabilities; all other branches must be labeled with the appropriate conditional probabilities. The sum of the probabilities on the branches leaving any nonterminal node must sum to unity; otherwise the terminal nodes could not represent a collectively exhaustive listing of all possible outcomes of the experiment.

Of course, any tree sample space which contains the complete set of finest-grain events for an experiment is a "correct" sample space. In any physical situation, however, the model of the experiment will usually specify the sequential order of the tree if we wish to label all branches with the appropriate conditional probabilities without any calculations.

Sometimes, a complete picture of the sample space would be too large to be useful. But it might still be of value to use an "outline" of the actual sample space. These outlines may be "trimmed" probability trees for which we terminate uninteresting branches as soon as possible. But once we have substituted such an event space for the sample space, we must again realize that we may be unable to perform certain calculations in this event space.

1-6 The Independence of Events

Thus far, our structure for probability theory includes seven axioms for the algebra of events, three more for probability measure, and the definition and physical interpretation of the concept of conditional probability. We shall now formalize an intuitive notion of the *independence of events.* This definition and its later extensions will be of considerable utility in our work.

In an intuitive sense, if events A and B are defined on the sample space of a particular experiment, we might think them to be "independent" if knowledge as to whether or not the experimental outcome had attribute B would not affect our measure of the likelihood that the experimental outcome also had attribute A. We take a formal statement of this intuitive concept to be our definition of the independence of two events.

Two events A and B are defined to be independent if and only if

$P(A \mid B) = P(A)$

From the definition of conditional probability, as long as $P(A) \neq 0$ and $P(B) \neq 0$, we may write

$$P(AB) = P(A)P(B \mid A) = P(B)P(A \mid B)$$

When we substitute the condition for the independence of A and B into this equation, we learn both that $P(A \mid B) = P(A)$ requires that

$P(B \mid A) = P(B)$ and that an alternative statement of the condition for the (mutual) independence of two events is $P(AB) = P(A)P(B)$.

If A and B are other than trivial events, it will rarely be obvious whether or not they are independent. To test two events for independence, we collect the appropriate probabilities from a sample space to see whether or not the definition of independence is satisfied. Clearly, the result depends on the original assignment of probability measure to the sample space in the modeling of the physical experiment.

We have *defined* conditional probability such that (as long as none of the conditioning events is of probability zero) the following relations always hold:

$$P(AB) = P(A \mid B)P(B) = P(B \mid A)P(A)$$
$$P(ABC) = P(A)P(BC \mid A) = P(B)P(C \mid B)P(A \mid BC) = P(AC)P(B \mid AC) = \cdots$$

but only when two events are independent may we write

$$P(AB) = P(A)P(B).$$

We extend our notion of independence by defining the mutual independence of N events $A_1, A_2, \ldots, A_N$.

N events $A_1, A_2, \ldots, A_N$ are defined to be mutually independent if and only if

$$P(A_i \mid A_j A_k \cdots A_p) = P(A_i) \qquad \text{for all } i \neq j, k, \ldots p; \quad 1 \leq i, j, k, \ldots p \leq N$$

This is equivalent to requiring that the probabilities of *all* possible intersections of these (different) events taken any number at a time [such as $P(A_1A_3A_4A_9)$] be given by the products of the individual event probabilities [such as $P(A_1)P(A_3)P(A_4)P(A_9)$]. Pairwise independence of the events on a list, as defined at the beginning of this section, does not necessarily result in the mutual independence defined above.

One should note that there is no reason why the independence or dependence of events need be preserved in going from an a priori sample space to a conditional sample space. Similarly, events which are mutually independent in a particular conditional space may or may not be mutually independent in the original universal set or in another conditional sample space. Two events A and B are said to be *conditionally independent*, given C, if it is true that

$$P(AB \mid C) = P(A \mid C)P(B \mid C)$$

This relation does not require, and is not required by, the separate condition for the unconditional independence of events A and B,

$$P(AB) = P(A)P(B)$$

We close this section with a consideration of the definition of the independence of two events from a sample-space point of view. One statement of the condition for the independence of events A and B, with $P(B) \neq 0$, is

$$P(A \mid B) = P(A) \qquad \text{or, equivalently,} \qquad P(A) = \frac{P(AB)}{P(B)}$$

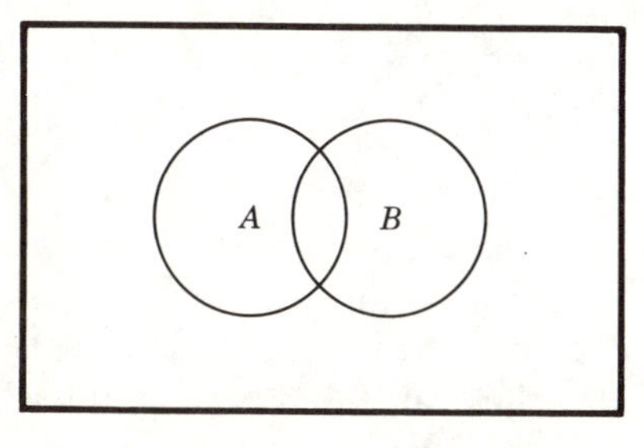

Thus, the independence requirement is that event AB is assigned a fraction of the probability measure of event B which is numerically equal to $P(A)$. As one would expect, we see (from the above diagram and the last equation) that, as long as $P(A) \neq 0$ and $P(B) \neq 0$, events A and B cannot be independent if they are mutually exclusive.

1-7 Examples

example 1 Our first example is simply an exercise to review the properties of probability measure, conditional probability, and some definitions from the algebra of events. The reader is encouraged to work out these examples for himself before reading through our discussion of the solutions.

Suppose that we are given three lists of events, called lists 1, 2, and 3. All the events in the three lists are defined on the same experiment, and none of the events is of probability zero.

List 1 contains events $A_1, A_2, \ldots, A_k$ and is mutually exclusive and collectively exhaustive.

List 2 contains events $B_1, B_2, \ldots, B_l$ and is mutually exclusive and collectively exhaustive.

List 3 contains events $C_1, C_2, \ldots, C_m$ and is mutually exclusive but *not* collectively exhaustive.

Evaluate each of the following quantities numerically. If you cannot evaluate them numerically, specify the *tightest upper and lower* numerical bounds you can find for each quantity.

(a) $\sum_{i=1}^{m} P(C_i)$ (b) $\sum_{j=1}^{k} P(A_2 \mid A_j)$

(c) $\sum_{i=1}^{k} \sum_{j=1}^{k} P(A_iA_j)$ (d) $\sum_{j=1}^{k} P(C_2 \mid A_jC_3)$

(e) $\sum_{j=1}^{k} P(A_j')$ (f) $\sum_{i=1}^{k} \sum_{j=1}^{l} P(A_i)P(B_j \mid A_i)$

(g) $\sum_{j=1}^{k} P(B_1 \mid A_j)$

Part (a) requires that we sum the probability measure associated with each member of a mutually exclusive but not collectively exhaustive list of events defined on a particular experiment. Since the events are mutually exclusive, we are calculating the probability of their union. According to the conditions of the problem, this union represents an event of nonzero probability which is not the universal set; so we obtain

$$0.0 < \sum_{i=1}^{m} P(C_i) < 1.0$$

For part (b), the A_i's are mutually exclusive so we note that $P(A_2 \mid A_j)$ is zero unless $j = 2$. When $j = 2$, we have $P(A_2 \mid A_2)$, which is equal to unity. Therefore, we may conclude that

$$\sum_{j=1}^{k} P(A_2 \mid A_j) = 1.0$$

Again in part (c), the mutually exclusive property of the A_i's will require $P(A_iA_j) = 0$ unless $j = i$. If $j = i$, we have $P(A_jA_j)$, which is equal to $P(A_j)$; so upon recalling that the A_i list is also collectively exhaustive, there follows

$$\sum_{i=1}^{k} \sum_{j=1}^{k} P(A_iA_j) = \sum_{i=1}^{k} \sum_{\substack{j=1 \\ i \neq j}}^{k} P(A_iA_j) + \sum_{i=1}^{k} P(A_i) = 0 + 1 = 1.0$$

In part (d), we know that C_2 and C_3 are mutually exclusive. Therefore, C_2 and C_3 can never describe the outcome of the same performance of the experiment on which they are defined. So, with no attention to the properties of the A_j's (assuming that we can neglect any pathological cases where the conditioning event A_jC_3 would be of probability zero and the conditional probability would be undefined), we have

$$\sum_{j=1}^{k} P(C_2 \mid A_j C_3) = 0.0$$

Part (e) is most easily done by direct substitution,

$$\sum_{j=1}^{k} P(A_j') = \sum_{j=1}^{k} [1 - P(A_j)] = k - \sum_{j=1}^{k} P(A_j)$$

and since we are told that the A_j's are mutually exclusive and collectively exhaustive, we have

$$\sum_{j=1}^{k} P(A_j') = k - 1.0$$

For part (f), we can use the definition of conditional probability and the given properties of lists 1 and 2 to write

$$\sum_{i=1}^{k} \sum_{j=1}^{l} P(A_i)P(B_j \mid A_i) = \sum_{i=1}^{k} \sum_{j=1}^{l} P(A_i B_j) = \sum_{i=1}^{k} P(A_i) = 1.0$$

The quantity to be evaluated in part (g) involves the summation of conditional probabilities each of which applies to a different conditioning event. The value of this sum need not be a probability. By decomposing the universal set into $A_1, A_2, \ldots, A_k$ and considering a few special cases (such as $B_1 = U$) consistent with the problem statement, we see

$$0 < \sum_{j=1}^{k} P(B_1 \mid A_j) \leq k$$

example 2 To the best of our knowledge, with probability 0.8 Al is guilty of the crime for which he is about to be tried. Bo and Ci, each of whom knows whether or not Al is guilty, have been called to testify.

Bo is a friend of Al's and will tell the truth if Al is innocent but will lie with probability 0.2 if Al is guilty. Ci hates everybody but the judge and will tell the truth if Al is guilty but will lie with probability 0.3 if Al is innocent.

Given this model of the physical situation:

(a) Determine the probability that the witnesses give conflicting testimony.
(b) Which witness is more likely to commit perjury?
(c) What is the conditional probability that Al is innocent, given that Bo and Ci gave conflicting testimony?
(d) Are the events "Bo tells a lie" and "Ci tells a lie" independent? Are these events conditionally independent to an observer who knows whether or not Al is guilty?

We begin by establishing our notation for the events of interest:

Event A: Al is innocent.
Event B: Bo testifies that Al is innocent.
Event C: Ci testifies that Al is innocent.
Event X: The witnesses give conflicting testimony.
Event Y: Bo commits perjury.
Event Z: Ci commits perjury.

Now we'll draw a sample space in the form of a probability tree and collect the sample points corresponding to the events of interest.

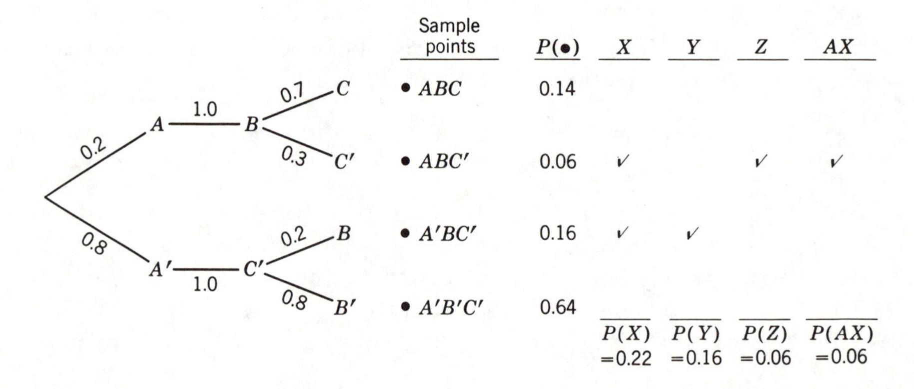

To find the probability of any event in sample space, we simply sum the probabilities of all sample points included in that event. (It is because of the *mutually exclusive* property of the sample space that we may follow this procedure.) Now, to answer our questions,

a $P(X) = P(BC' + B'C) = P(BC') + P(B'C) = 0.22$

b $P(Y) = P(AB' + A'B) = P(AB') + P(A'B) = 0.00 + 0.16 = 0.16$

$P(Z) = P(AC' + A'C) = P(AC') + P(A'C) = 0.06 + 0.00 = 0.06$

Therefore Bo is the more likely of the witnesses to commit perjury.

c $P(A \mid X) = \dfrac{P(AX)}{P(X)} = \dfrac{0.06}{0.06 + 0.16} = \dfrac{3}{11}$

Conflicting testimony is more likely to occur if Al is innocent than if he is guilty; so given X occurred, it should increase our probability that Al is innocent. This is the case, since $3/11 > 1/5$.

c (One other method of solution.) Given X occurred, we go to a conditional space containing only those sample points with attribute X. The conditional probabilities for these points are found by scaling up the original a priori probabilities by the same constant $[1/P(X)]$ so that they add to unity.

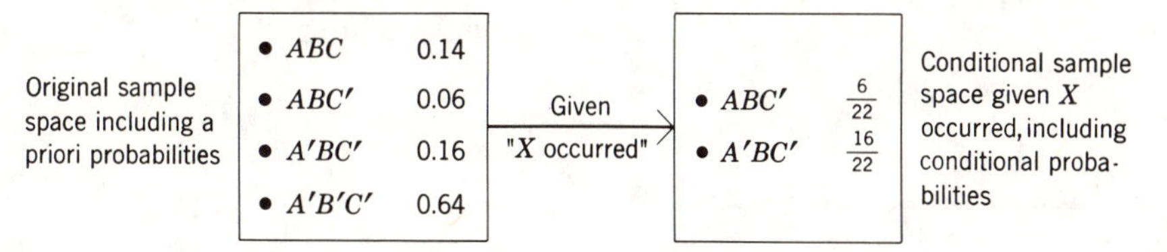

Now, in the conditional space we simply sum the conditional probabilities of all sample points included in any event to determine the conditional probability of that event.

$$P(A \mid X) = P(ABC' \mid X) = 3/11$$

d To determine whether "Bo tells a lie" and "Ci tells a lie" are independent in the original sample space, we need only test $P(YZ) \stackrel{?}{=} P(Y)P(Z)$. Since $P(YZ) = 0$ but $P(Y) > 0$ and $P(Z) > 0$, we see that events Y and Z are not independent, in fact they are mutually exclusive.

To determine whether Y and Z are conditionally independent given A or A', we must test $P(YZ \mid A) \stackrel{?}{=} P(Y \mid A)P(Z \mid A)$ and $P(YZ \mid A') \stackrel{?}{=} P(Y \mid A')P(Z \mid A')$. From the sample space, we find that the left-hand side and one term on the right-hand side of each of these tests is equal to zero, so events Y and Z are conditionally independent to one who knows whether or not Al is innocent.

Since the testimony of the two witnesses depends only on whether or not Al is innocent, this is a reasonable result. If we don't know whether or not Al is innocent, then Y and Z are dependent because the occurrence of one of these events would give us partial information about Al's innocence or guilt, which would, in turn, change our probability of the occurrence of the other event.

example 3 When we find it necessary to resort to games of chance for simple examples, we shall often use four-sided (tetrahedral) dice to keep our problems short. With a tetrahedral die, one reads the "down" face either by noticing which face *isn't* up or by looking up at the bottom of the die through a glass cocktail table. Suppose that a fair four-sided die (with faces labeled 1, 2, 3, 4) is tossed twice and we are told only that the product of the resulting two down-face values is less than 7.

(a) What is the probability that at least one of the face values is a two?

(b) What is the probability that the sum of the face values is less than 7?

(c) If we are told that both the product and the sum of the face values is less than 7 and that at least one of the face values is a two, determine the probability of each possible value of the other face.

We use the notation

Event F_i: Value of down face on first throw is equal to i
Event S_j: Value of down face on second throw is equal to j

to construct the sample space

	F_1	F_2	F_3	F_4
S_1	•	•	•	•
S_2	•	•	•	•
S_3	•	•	•	•
S_4	•	•	•	•

in which, from the statement of the problem ("fair die"), all 16 sample points are equally likely. Given that the product of the down faces is less than 7, we produce the appropriate conditional space by eliminating those sample points not included in the conditioning event and scaling up the probabilities of the remaining sample points. We obtain

	F_1	F_2	F_3	F_4
S_1	•	•	•	•
S_2	•	•	•	
S_3	•	•		
S_4	•			

in which each of the remaining 10 sample points is still an equally likely outcome in the conditional space.

a There are five sample points in the conditional space which are included in the event "At least one down face is a two"; so the conditional probability of this event is $5 \times 0.1 = 0.5$.

b All the outcomes in the conditional space represent experimental outcomes for which the sum of the down-face values is less than 7; so the conditional probability of this event is unity.

c Given all the conditioning events in the problem statement, the resulting conditional sample space is simply

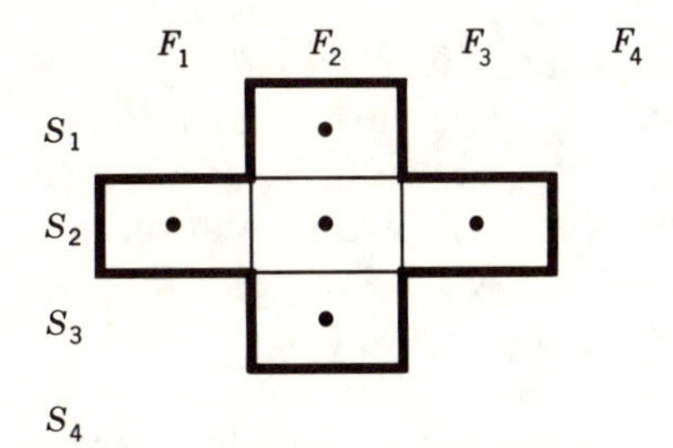

and the conditional probability that the other face value is unity is 2/5; the same applies for the event that the other face value is 3, and there is a conditional probability of 1/5 that the other face value is also a two.

For Example 3 we have considered one approach to a simple problem which may be solved by several equivalent methods. As we become familiar with random variables in the next chapter, we shall employ extensions of the above methods to develop effective techniques for the analysis of experiments whose possible outcomes may be expressed numerically.

1-8 Bayes' Theorem

The relation known as *Bayes' theorem* results from a particular application of the definition of conditional probability. As long as the conditioning events are not of probability zero, we have

$$P(AB) = P(A)P(B \mid A) = P(B)P(A \mid B)$$

We wish to apply this relation to the case where the events $A_1, A_2, \ldots, A_N$ form a mutually exclusive and collectively exhaustive list of events. For this case where the events $A_1, A_2, \ldots, A_N$ form an event space, let's consider the universal set and include some other event B.

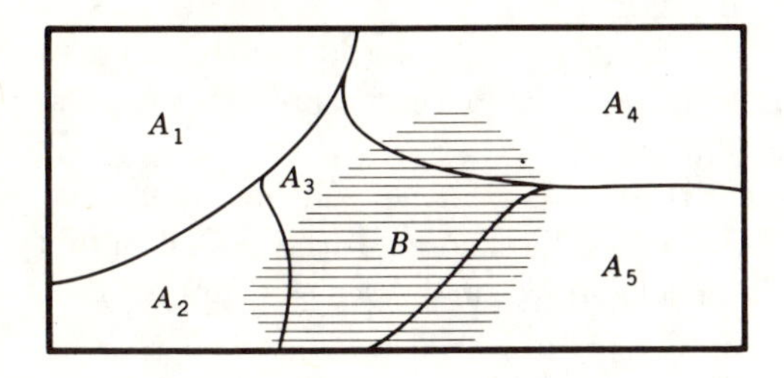

For the case of interest, assume that we know $P(A_i)$ and $P(B \mid A_i)$ for all $1 \leq i \leq N$ and we wish to determine the $P(A_i \mid B)$'s.

An example of this situation follows: Let A_i represent the event that a particular apple was grown on farm i. Let B be the event that an apple turns blue during the first month of storage. Then the quantity $P(B \mid A_i)$ is the probability that an apple will turn blue during the first month of storage given that it was grown on farm i. The question of interest: Given an apple which did turn blue during its first month of storage, what is the probability it came from farm i?

We return to the problem of determining the $P(A_i \mid B)$'s. As long as $P(B) \neq 0$ and $P(A_i) \neq 0$ for $i = 1, 2, \ldots, N$, we may substitute A_i for A in the above definition of conditional probability to write

$$P(A_i \mid B) = \frac{P(A_i)P(B \mid A_i)}{P(B)}$$

$$P(B) = P(UB) = P[(A_1 + A_2 + \cdots + A_N)B] = \sum_{i=1}^{N} P(A_iB) = \sum_{i=1}^{N} P(A_i)P(B \mid A_i)$$

and we substitute the above expression for $P(B)$ into the equation for $P(A_i \mid B)$ to get

Bayes' theorem:

$$P(A_i \mid B) = \frac{P(A_i)P(B \mid A_i)}{\sum_{i=1}^{N} P(A_i)P(B \mid A_i)}$$

with $P(B) \neq 0$ and $A_1, A_2, \ldots, A_N$ an event space

Bayes' theorem could also have been obtained directly from our sample space interpretation of conditional probability. Note that, using a sequential sample space (shown at the top of the following page) this interpretation would allow us to derive Bayes' theorem [the expression for $P(A_i \mid B)$] by inspection.

A knowledge of Bayes' theorem by name will not affect our approach to any problems. However, the theorem is the basis of much of the study of statistical inference, and we shall discuss its application for that purpose in the last chapter of this book.

In the literature, one finds appreciable controversy about the "validity" of certain uses of Bayes' theorem. We have learned that the theorem is a simple consequence of the definition of conditional probability. Its application, of course, must always be made with conscious knowledge of just what model of reality the calculations represent. The trouble is generated, not by Bayes' theorem, but by

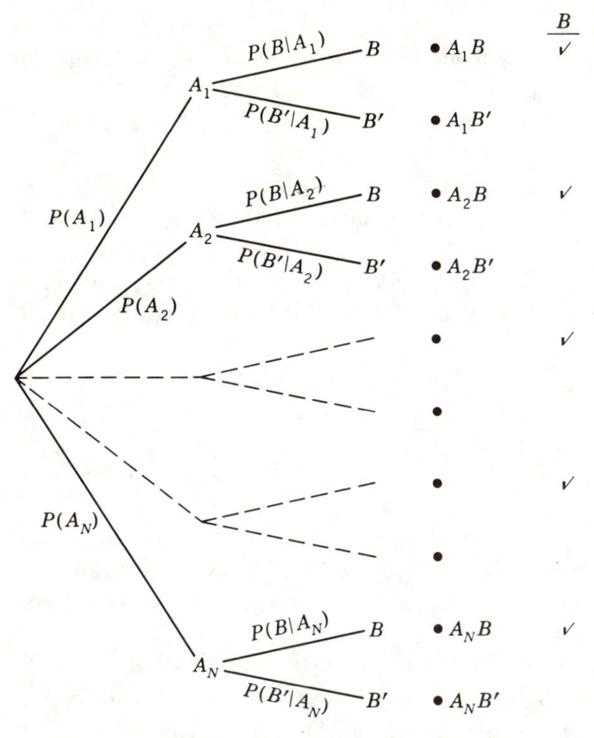

the attempt to assign a priori probabilities in the sample space for a model when one has very little information to assist in this assignment. We'll return to this matter in Chap. 7.

One last note regarding conditional probability is in order. No matter how close to unity the quantity $P(A \mid B)$ may be, it would be foolish to conclude from this evidence alone that event B is a "cause" of event A. *Association and physical causality may be different phenomena.* If the probability that any man who consults a surgeon about lung cancer will pass away soon is nearly unity, few of us would conclude "Aha! Lung surgeons are the cause of these deaths!" Sometimes, equally foolish conclusions are made in the name of "statistical reasoning." We wish to keep an open mind and develop a critical attitude toward physical conclusions based on the mathematical analyses of probabilistic models of reality.

1-9 Enumeration in Event Space: Permutations and Combinations

Sample spaces and event spaces play a key role in our presentation of introductory probability theory. The value of such spaces is that they enable one to display events in a mutually exclusive, collectively exhaustive form. Most problems of a combinatorial nature, whether probabilistic or not, also require a careful listing of events. It is the purpose of this section to demonstrate how combinatorial problems may be formulated effectively in event spaces.

Given a set of N distinguishable items, one might wonder how many distinguishable orderings (or arrangements) may be obtained by using these N items K at a time. For instance, if the items are the four events A, B, C, D, the different arrangements possible, subject to the requirement that each arrangement must include exactly two of the items, are

$$\begin{array}{cccc} AB & BA & CA & DA \\ AC & BC & CB & DB \\ AD & BD & CD & DC \end{array}$$

Such arrangements of N distinguishable items, taken K at a time, are known as *K-permutations* of the items. In the above example, we have found by enumeration that there are exactly 12 different 2-permutations of 4 distinguishable items.

To determine the number of K-permutations which may be formed from N distinguishable items, one may consider forming the permutations sequentially. One begins by choosing a first item, then a second item, etc. For instance, for the above example, this process may be shown on a sequential sample space,

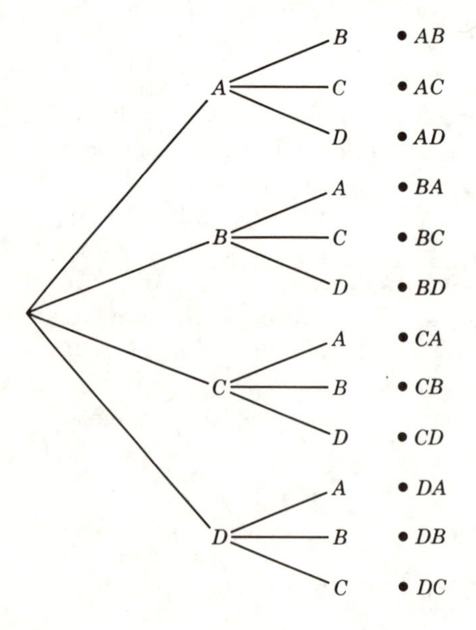

There were N choices for the first member, any such choice results in $N - 1$ possible choices for the second member, and this continues until there are $N - (K + 1)$ possible choices for the Kth (and last) member of the permutation. Thus we have

Number of K-permutations of N distinguishable items

$$= N(N-1)(N-2) \cdots (N-K+1) = \frac{N!}{(N-K)!} \qquad N \geq K$$

A selection of K out of N distinguishable items, without regard to ordering, is known as a K-*combination*. Two K-combinations are identical if they both select exactly the same set of K items out of the original list and no attention is paid to the order of the items. For the example considered earlier, AB and BA are different permutations, but both are included in the same combination.

By setting $N = K$ in the above formula, we note that any combination containing K distinguishable items includes $K!$ permutations of the members of the combination. To determine the number of K-combinations which may be formed from N distinguishable items, we need only divide the number of K-permutations by $K!$ to obtain

Number of K-combinations of N distinguishable items

$$= \frac{N!}{K!(N-K)!} = \binom{N}{K} \qquad N \geq K$$

Many enumeration and probability problems require the orderly collection of complex combinatorial events. When one attacks such problems in an appropriate sequential event space, the problem is reduced to several simple counting problems, each of which requires nothing more than minor (but careful) bookkeeping. The event-space approach provides the opportunity to deal only with the collection of mutually exclusive events in a collectively exhaustive space. This is a powerful technique and, to conclude this section, we note one simple example of its application.

Suppose that an unranked committee of four members is to be formed from a group of four males R, S, T, U and five females V, W, X, Y, Z. It is also specified that R and S cannot be on the same committee unless the committee contains at least one female. We first wish to determine the number of different such committees which may be formed.

We draw an event space for the enumeration of these committees. Our only methodology is to decompose the counting problem into smaller pieces, always branching out into a *mutually exclusive* decomposition which includes all members of interest. (We may omit items which are not relevant to our interests.)

NOTATION Event X: X is on the committee.
Event f_n: Exactly n females are on the committee.

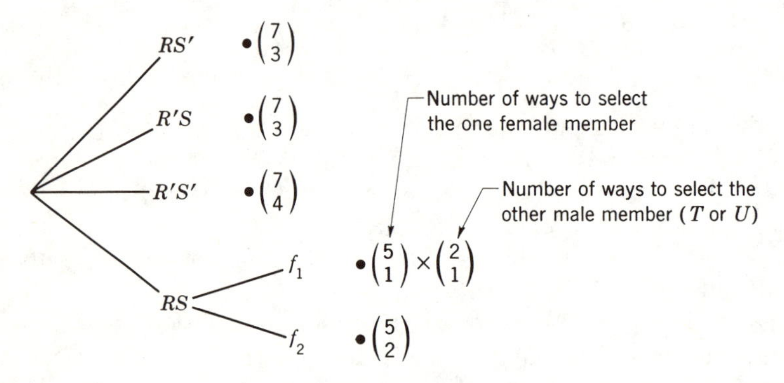

The quantity following each terminal node is the number of acceptable committees associated with the indicated event. For instance, for the RS' committees, note that there are as many acceptable committees containing R but not S as there are ways of selecting three additional members from T, U, V, W, X, Y, Z without regard to the order of selection. We have obtained, for the number of acceptable committees,

$$2\binom{7}{3} + \binom{7}{4} + \binom{5}{1}\binom{2}{1} + \binom{5}{2} = 125 \text{ possible acceptable committees}$$

Finally we determine, were the committees selected by lot, the probability that the first four names drawn would form an acceptable committee. Since any committee is as likely to be drawn as any other, we note that there are 125 acceptable committees out of $\binom{9}{4} = 126$ possible results of the draw. Thus there is a probability of 125/126 that a randomly drawn committee will meet the given constraint. Had we decided to solve by counting unacceptable committees and subtracting from 126, we would have noticed immediately that only one draw, $RSTU$, would be unacceptable.

The extension of our event-space approach to problems involving permutations, such as ranked committees, merely requires that we work in a more fine-grained event space.

PROBLEMS

1.01 Use the axioms of the algebra of events to prove the relations
a $U' = \phi$ **b** $A + B = A + A'B + ABC$ **c** $A + U = U$

1.02 Use Venn diagrams, the axioms of the algebra of events, or anything else to determine which of the following are valid relations in the algebra of events for arbitrary events A, B, and C:
a $(A + B + C)' = A' + B' + C'$
b $A + B + C = A + A'B + (A + A'B)'C$
c $(A + B)(A' + B') = AB' + A'B + A'BC'$
d $AB + AB' + A'B = (A'B')'$

1.03 Fully explain your answers to the following questions:
a If events A and B are mutually exclusive and collectively exhaustive, are A' and B' mutually exclusive?
b If events A and B are mutually exclusive but not collectively exhaustive, are A' and B' collectively exhaustive?
c If events A and B are collectively exhaustive but not mutually exclusive, are A' and B' collectively exhaustive?

1.04 We desire to investigate the validity of each of four proposed relations (called relations 1, 2, 3, and 4) in the algebra of events. Discuss what evidence about the validity of the appropriate relation may be obtained from each of the following observations. We observe that we can:
a Obtain a valid relation by taking the intersection of each side of relation 1 with some event E_1
b Obtain an invalid relation by taking the intersection of each side of relation 2 with some event E_2
c Obtain a valid relation by taking the union of each side of relation 3 with some event E_3
d Obtain an invalid relation by taking the complement of each side of relation 4

1.05 In a group of exactly 2,000 people, it has been reported that there are exactly:
612 people who smoke
670 people over 25 years of age
960 people who imbibe
86 smokers who imbibe
290 imbibers who are over 25 years of age

158 smokers who are over 25 years of age
44 people over 25 years of age each of whom both smokes and imbibes
250 people under 25 years of age who neither smoke nor imbibe

Determine whether or not this report is consistent.

1.06 Consider the experiment in which a four-sided die (with faces labeled 1, 2, 3, 4) is thrown twice. We use the notation

$$\text{Event } \begin{Bmatrix} F_n \\ S_n \end{Bmatrix}\text{: Down face value on } \begin{Bmatrix} \text{first} \\ \text{second} \end{Bmatrix} \text{ throw equals } n.$$

For each of the following lists of events, determine whether or not the list is (i) mutually exclusive, (ii) collectively exhaustive, (iii) a sample space, (iv) an event space:

a $F_1, F_2, (F_1 + F_2)'$
b $F_1S_1, F_1S_2, F_2, F_3, F_4$
c $F_1S_1, F_1S_2, (F_1)', (S_1 + S_2)'$
d $(F_1 + S_1), (F_2 + S_2), (F_3 + S_3), (F_4 + S_4)$
e $(F_1 + F_2)(S_1 + S_2 + S_3), (F_1 + F_2 + F_3)S_4,$
$(F_3 + F_4)(S_1 + S_2 + S_3), S_4F_4$

1.07 For three tosses of a fair coin, determine the probability of:
a The sequence HHH
b The sequence HTH
c A total result of two heads and one tail
d The outcome "More heads than tails"

Determine also the conditional probabilities for:
e "More heads than tails" given "At least one tail"
f "More heads than tails" given "Less than two tails"

1.08 Joe is a fool with probability of 0.6, a thief with probability 0.7, and neither with probability 0.25.
a Determine the probability that he is a fool or a thief but not both.
b Determine the conditional probability that he is a thief, given that he is not a fool.

1.09 Given $P(A) \neq 0, P(B) \neq 0, P(A + B) = P(A) + P(B) - 0.1$, and $P(A \mid B) = 0.2$. Either determine the exact values of each of the following quantities (if possible), or determine the tightest numerical bounds on the value of each:
a $P(ABC) + P(ABC')$ **b** $P(A' \mid B)$ **c** $P(B)$ **d** $P(A')$
e $P(A + B + A'B')$

1.10 If $P(A) = 0.4, P(B') = 0.7$, and $P(A + B) = 0.7$, determine:
a $P(B)$ **b** $P(AB)$ **c** $P(A' \mid B')$

1.11 A game begins by choosing between dice A and B in some manner such that the probability that A is selected is p. The die thus selected is then tossed until a white face appears, at which time the game is concluded.

Die A has 4 red and 2 white faces. Die B has 2 red and 4 white faces. After playing this game a great many times, it is observed that the probability that a game is concluded in exactly 3 tosses of the selected die is $\frac{7}{81}$. Determine the value of p.

1.12 Oscar has lost his dog in either forest A (with a priori probability 0.4) or in forest B (with a priori probability 0.6). If the dog is alive and not found by the Nth day of the search, it will die that evening with probability $N/(N+2)$.

If the dog is in A (either dead or alive) and Oscar spends a day searching for it in A, the conditional probability that he will find the dog that day is 0.25. Similarly, if the dog is in B and Oscar spends a day looking for it there, he will find the dog that day with probability 0.15.

The dog cannot go from one forest to the other. Oscar can search only in the daytime, and he can travel from one forest to the other only at night.

All parts of this problem are to be worked separately.

a In which forest should Oscar look to maximize the probability he finds his dog on the first day of the search?

b Given that Oscar looked in A on the first day but didn't find his dog, what is the probability that the dog is in A?

c If Oscar flips a fair coin to determine where to look on the first day and finds the dog on the first day, what is the probability that he looked in A?

d Oscar has decided to look in A for the first two days. What is the a priori probability that he will find a live dog for the first time on the second day?

e Oscar has decided to look in A for the first two days. Given the fact that he was unsuccessful on the first day, determine the probability that he does not find a dead dog on the second day.

f Oscar finally found his dog on the fourth day of the search. He looked in A for the first 3 days and in B on the fourth day. What is the probability he found his dog alive?

g Oscar finally found his dog late on the fourth day of the search. The only other thing we know is that he looked in A for 2 days and and in B for 2 days. What is the probability he found his dog alive?

1.13 Considering the statement of Prob. 1.12, suppose that Oscar has decided to search each day wherever he is most likely to find the dog on that day. He quits as soon as he finds the dog.

a What is the probability that he will find the dog in forest A?
b What is the probability that he never gets to look in forest B?
c Given that he does get to look in forest B, what is the probability that he finds the dog during his first day of searching there?
d Given that he gets to look in forest B at least once, what is the probability that he eventually finds the dog in forest A?

1.14 Joe is an astronaut for project Pluto. Mission success or failure depends only on the behavior of three major systems. Joe decides the following assumptions are valid and apply to the performance of an entire mission: (1) The mission is a failure only if two or more of the major systems fail. (2) System I, the Gronk system, will fail with probability 0.1. (3) If at least one other system fails, no matter how this comes about, System II, the Frab system, will fail with conditional probability 0.5. If no other system fails, the Frab system will fail with probability 0.1. (4) System III, the Beer Cooler, fails with probability 0.5 if the Gronk system fails. Otherwise, the Beer Cooler cannot fail.

a What is the probability that the mission succeeds but that the Beer Cooler fails?
b What is the probability that all three systems fail?
c Given that more than one system failed, determine the conditional probabilities that:
  i The Gronk did not fail.
  ii The Beer Cooler failed.
  iii Both the Gronk and the Frab failed.
d About the time when Joe was due back on Earth, you overhear a radio broadcast about Joe in a very noisy room. You are not positive what the announcer did say, but, based on all available information, you decide that it is twice as likely that he reported "Mission a success" as that he reported "Mission a failure." What now is the conditional probability (to you) that the Gronk failed?

1.15 A box contains two fair coins and one biased coin. For the biased coin, the probability that any flip will result in a head is $\frac{1}{3}$. Al draws two coins from the box, flips each of them once, observes an outcome of one head and one tail, and returns the coins to the box. Bo then draws one coin from the box and flips it. The result is a tail. Determine the probability that neither Al nor Bo removed the biased coin from the box.

1.16 Joe, the bookie, is attempting to establish the odds on an exhibition baseball game. From many years of experience, he has learned that his prime consideration should be the selection of the starting pitcher.

The Cardinals have only two pitchers, one of whom must start: C_1, their best pitcher; C_2, a worse pitcher. The other team, the Yankees, has only three pitchers, one of whom must start: Y_1, their best pitcher; Y_2, a worse one; Y_3, an even worse one.

By carefully weighting information from various sources, Joe has decided to make the following assumptions:
The Yankees will not start Y_1 if C_1 does not start. If C_1 does start, the Yankees will start Y_1 with probability $\frac{1}{3}$.
The Cardinals are equally likely to start C_1 or C_2, no matter what the Yankees do.
Y_2 will pitch for the Yankees with probability $\frac{3}{4}$ if Y_1 does not pitch, no matter what else occurs.
The probability the Cardinals will win given C_1 pitches is $m/(m+1)$, where m is the number (subscript) of the Yankee pitcher.
The probability the Yankees will win given C_2 pitches is $1/m$, where m is the number of the Yankee pitcher.

a What is the probability that C_2 starts?
b What is the probability that the Cardinals will win?
c Given that Y_2 does not start, what is the probability that the Cardinals will win?
d If C_2 and Y_2 start, what is the probability that the Yankees win?

1.17 Die A has five olive faces and one lavender face; die B has three faces of each of these colors. A fair coin is flipped once. If it falls heads, the game continues by throwing die A alone; if it falls tails, die B alone is used to continue the game. However awful their face colors may be, it is known that both dice are fair.

a Determine the probability that the nth throw of the die results in olive.
b Determine the probability that both the nth and $(n+1)$st throw of the die results in olive.
c If olive readings result from all the first n throws, determine the conditional probability of an olive outcome on the $(n+1)$st toss. Interpret your result for large values of n.

1.18 Consider events A, B, and C with $P(A) > P(B) > P(C) > 0$. Events A and B are mutually exclusive and collectively exhaustive. Events A and C are independent. Can C and B be mutually exclusive?

1.19 We are given that the following six relations hold:

$$P(A) \neq 0 \qquad P(B) \neq 0 \qquad P(C) \neq 0$$
$$P(AB) = P(A)P(B) \qquad P(AC) = P(A)P(C) \qquad P(BC) = P(B)P(C)$$

Subject to these six conditions, determine, for each of the following entries, whether it must be true, it might be true, or it cannot be true:
a $P(ABC) = P(A)P(B)P(C)$ **b** $P(B \mid A) = P(B \mid C)$
c $P(AB \mid C) = P(A \mid C)P(B \mid C)$
d $P(A + B + C) < P(A) + P(B) + P(C)$
e If $W = AB$ and $V = AC$, $WV = \phi$

1.20 Events E, F, and G form a list of mutually exclusive and collectively exhaustive events with $P(E) \neq 0$, $P(F) \neq 0$, and $P(G) \neq 0$. Determine, for each of the following statements, whether it must be true, it might be true, or it cannot be true:
a E', F', and G' are mutually exclusive.
b E', F', and G' are collectively exhaustive.
c E' and F' are independent.
d $P(E') + P(F') > 1.0$.
e $P(E' + EF' + EFG') = 1.0$.

1.21 Bo and Ci are the only two people who will enter the Rover Dog Food jingle contest. Only one entry is allowed per contestant, and the judge (Rover) will declare the one winner as soon as he receives a suitably inane entry, which may be never.

Bo writes inane jingles rapidly but poorly. He has probability 0.7 of submitting his entry first. If Ci has not already won the contest, Bo's entry will be declared the winner with probability 0.3. Ci writes slowly, but he has a gift for this sort of thing. If Bo has not already won the contest by the time of Ci's entry, Ci will be declared the winner with probability 0.6.
a What is the probability that the prize never will be awarded?
b What is the probability that Bo will win?
c Given that Bo wins, what is the probability that Ci's entry arrived first?
d What is the probability that the first entry wins the contest?
e Suppose that the probability that Bo's entry arrived first were P instead of 0.7. Can you find a value of P for which "First entry wins" and "Second entry does not win" are independent events?

1.22

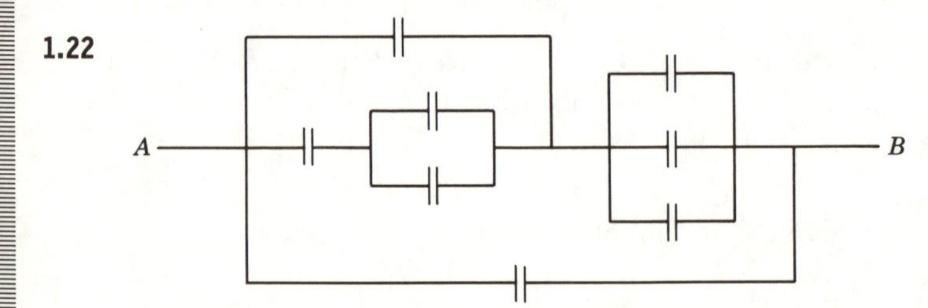

Each ⊣⊢ represents one communication link. Link failures are independent, and each link has a probability of 0.5 of being out of service. Towns A and B can communicate as long as they are connected in the communication network by at least one path which contains only in-service links. Determine, in an efficient manner, the probability that A and B can communicate.

1.23

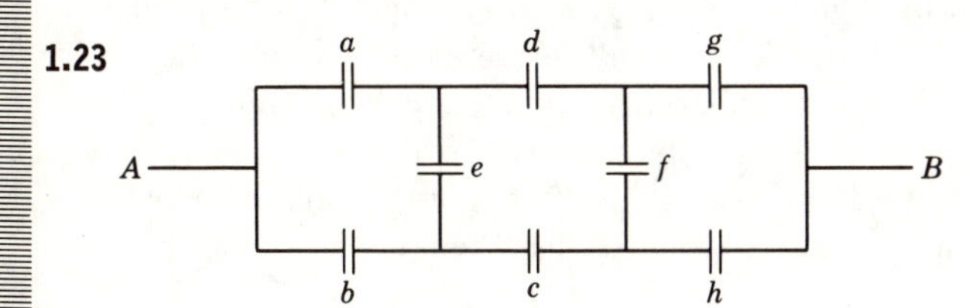

In the above communication network, link failures are independent, and each link has a probability of failure of p. Consider the physical situation before you write anything. A can communicate with B as long as they are connected by at least one path which contains only in-service links.

a Given that exactly five links have failed, determine the probability that A can still communicate with B.

b Given that exactly five links have failed, determine the probability that either g or h (*but not both*) is still operating properly.

c Given that a, d, and h have failed (but no information about the condition of the other links), determine the probability that A can communicate with B.

1.24 The events X, Y, and Z are defined on the same experiment. For each of the conditions below, state whether or not each statement following it must necessarily be true if the condition is true, and give a reason for your answer. (Remember that a simple way to demonstrate that a statement is not necessarily true may be to find a counterexample.)

a If $P(X + Y) = P(X) + P(Y)$, then:
i $P(XY) = P(X)P(Y)$ **iii** $P(Y) = 0$
ii $X + Y = X + X'Y$ **iv** $P(XY) = 0$

b If $P(XY) = P(X)P(Y)$ and $P(X) \neq 0$, $P(Y) \neq 0$, then:
i $P(X \mid Y) = P(Y \mid X)$ **iii** $P(XY) \neq 0$
ii $P(X + Y) = P(X) + P(Y) - P(XY)$ **iv** $X \neq Y$ unless $X = U$

c If $P(XY \mid Z) = 1$, then:
i $P(XY) = 1$ **iii** $P(XYZ) = P(Z)$
ii $Z = U$ **iv** $P(X' \mid Z) = 0$

d If $P[(X + Y) \mid Z] = 1$ and $X \neq \phi$, $Y \neq \phi$, $Z \neq \phi$, then:
i $P(XY \mid Z) = 0$ **iii** $P(X'Y' \mid Z) = 0$
ii $P(X + Y) < 1$ **iv** $P(X \mid Z) < 1$

e Define the events $R = X$, $S = X'Z$, $T = X'Y'Z'$. If the events R,

S, and T are collectively exhaustive, then:

i R, S, and T are mutually exclusive

ii $P(RST) = P(S)P(RT \mid S)$

iii $P(X'YZ') = 0$

iv $P[(R + S + T) \mid XYZ] = 1$

1.25 Only at midnight may a mouse go from either of two rooms to the other. Each midnight he stays in the same room with probability 0.9, or he goes to the other room with probability 0.1. We can't observe the mouse directly, so we use the reports of an unreliable observer, who each day reports correctly with probability 0.7 and incorrectly with probability 0.3. On day 0, we installed the mouse in room 1. We use the notation:

Event $R_k(n)$: Observer reports mouse in room k on day n.

Event $S_k(n)$: Mouse is in room k on day n.

a What is the a priori probability that on day 2 we shall receive the report $R_2(2)$?

b If we receive the reports $R_1(1)$ and $R_2(2)$, what is the conditional probability that the observer has told the truth on both days?

c If we receive the reports $R_1(1)$ and $R_2(2)$, what is the conditional probability of the event $S_2(2)$? Would we then believe the observer on the second day? Explain.

d Determine the conditional probability $P[R_2(2) \mid R_1(1)]$, and compare it with the conditional probability $P[S_2(2) \mid S_1(1)]$.

1.26 **a** The Pogo Thorhead rocket will function properly only if all five of its major systems operate simultaneously. The systems are labeled A, B, C, D, and E. System failures are independent, and each system has a probability of failure of 1/3. Given that the Thorhead fails, determine the probability that system A is *solely* at fault.

b The Pogo Thorhead II is an improved configuration using the same five systems, each still with a probability of failure of 1/3. The Thorhead II will fail if system A fails or if (at least) any two systems fail. Alas, the Thorhead II also fails. Determine the probability that system A is *solely* at fault.

1.27 The individual shoes from eight pairs, each pair a different style, have been put into a barrel. For \$1, a customer may randomly draw and keep two shoes from the barrel. Two successive customers do this. What is the probability that at least one customer obtains two shoes from the same pair? How much would an individual customer improve his chances of getting at least one complete pair by investing \$2 to be allowed to draw and keep four shoes? (Some combinatorial problems are quite trying. Consider, if you dare, the solution to this problem for eight customers instead of two.)

1.28 Companies A, B, C, D, and E each send three delegates to a conference. A committee of four delegates, selected by lot, is formed. Determine the probability that:

a Company A is not represented on the committee.
b Company A has exactly one representative on the committee.
c Neither company A nor company E is represented on the committee.

1.29 A box contains N items, K of which are defective. A sample of M items is drawn from the box at random. What is the probability that the sample includes at least one defective item if the sample is taken:

a With replacement
b Without replacement

1.30 The Jones family household includes Mr. and Mrs. Jones, four children, two cats, and three dogs. Every six hours there is a Jones family stroll. The rules for a Jones family stroll are:

Exactly five things (people + dogs + cats) go on each stroll.
Each stroll must include at least one parent and at least one pet.
There can never be a dog and a cat on the same stroll unless both parents go.
All acceptable stroll groupings are equally likely.

Given that exactly one parent went on the 6 P.M. stroll, what is the probability that Rover, the oldest dog, also went?

CHAPTER TWO

random variables

Often we have reasons to associate one or more numbers (in addition to probabilities) with each possible outcome of an experiment. Such numbers might correspond, for instance, to the cost to us of each experimental outcome, the amount of rainfall during a particular month, or the height and weight of the next football player we meet.

This chapter extends and specializes our earlier work to develop effective methods for the study of experiments whose outcomes may be described numerically.

2-1 Random Variables and Their Event Spaces

For the study of experiments whose outcomes may be specified numerically, we find it useful to introduce the following definition:

A *random variable* is defined by a function which assigns a value of the random variable to each sample point in the sample space of an experiment.

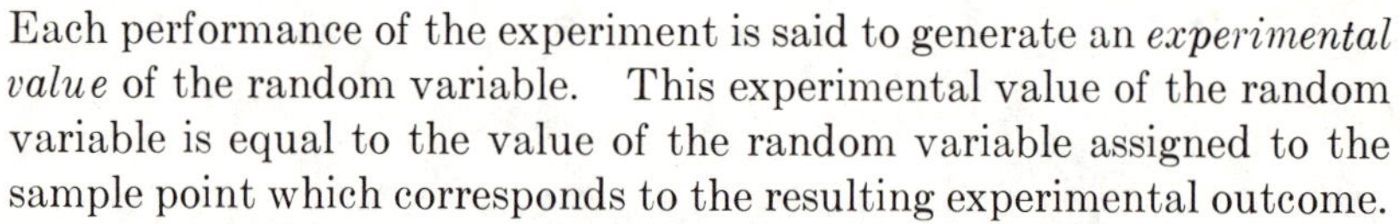

Each performance of the experiment is said to generate an *experimental value* of the random variable. This experimental value of the random variable is equal to the value of the random variable assigned to the sample point which corresponds to the resulting experimental outcome.

Consider the following example, which will be referred to in several sections of this chapter. Our experiment consists of three independent flips of a fair coin. We again use the notation

$$\text{Event} \begin{Bmatrix} H_n \\ T_n \end{Bmatrix} : \begin{Bmatrix} \text{Heads} \\ \text{Tails} \end{Bmatrix} \text{ on the } n\text{th flip}$$

We may define any number of random variables on the sample space of this experiment. We choose the following definitions for two random variables, h and r:

h = total number of heads resulting from the three flips

r = length of longest run resulting from the three flips (a run is a set of successive flips all of which have the same outcome)

We now prepare a fully labeled sequential sample space for this experiment. We include the branch traversal conditional probabilities, the probability of each experimental outcome, and the values of random variables h and r assigned to each sample point. The resulting sample space is shown at the top of the following page.

If this experiment were performed once and the experimental outcome were the event $H_1T_2T_3$, we would say that, for this performance of the experiment, the resulting experimental values of random variables h and r were 1 and 2, respectively.

Although we may require the full sample space to describe the detailed probabilistic structure of an experiment, it may be that our only practical interest in each performance of the experiment will relate to the resulting experimental values of one or more random variables.

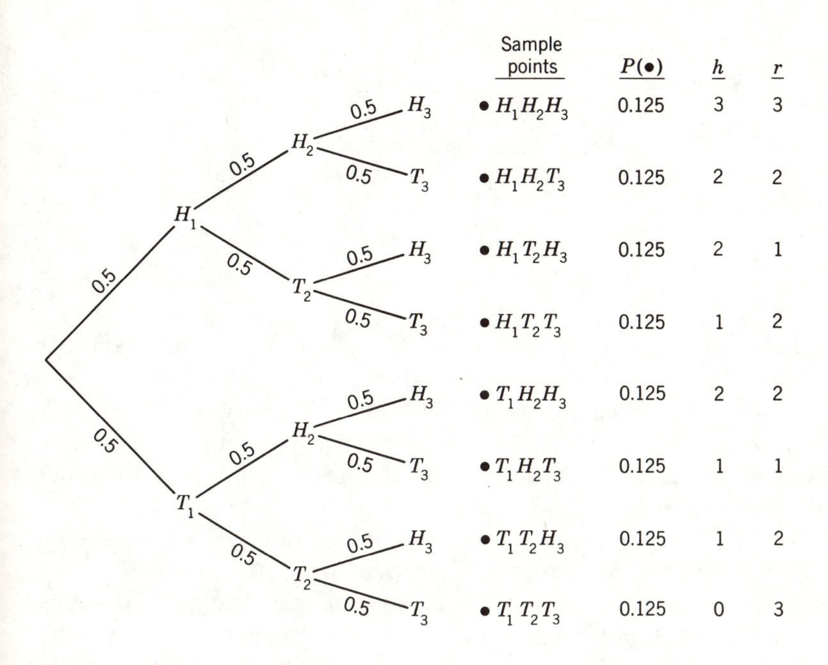

When this is the case, we may prefer to work in an *event* space which distinguishes among outcomes only in terms of the possible experimental values of the random variables of interest. Let's consider this for the above example.

Suppose that our only interest in a performance of the experiment has to do with the resulting experimental value of random variable h. We might find it desirable to work with this variable in an event space of the form

$h_0 = 0$ $h_0 = 1$ $h_0 = 2$ $h_0 = 3$ $\rightarrow h_0$

The four event points marked along the h_0 axis form a mutually exclusive collectively exhaustive listing of all possible experimental outcomes. The event point at any h_0 corresponds to the event "The experimental value of random variable h generated on a performance of the experiment is equal to h_0," or, in other words, "On a performance of the experiment, random variable h takes on experimental value h_0."

Similarly, if our concern with each performance of the experiment depended only upon the resulting experimental values of random variables h and r, a simple event space would be

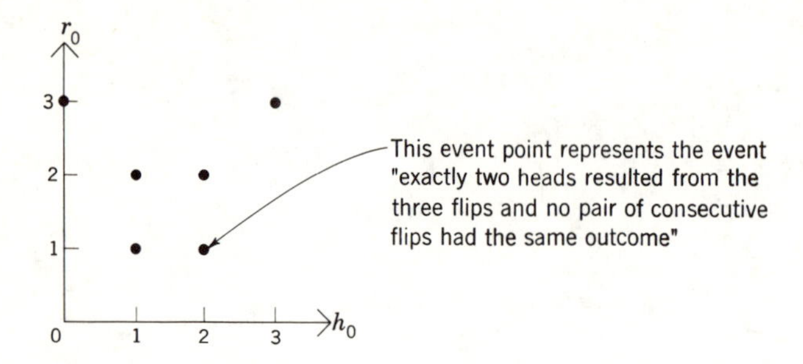

An event point in this space with coordinates h_0 and r_0 corresponds to the event "On a performance of the experiment, random variables h and r take on, respectively, experimental values h_0 and r_0." The probability assignment for each of these six event points may, of course, be obtained by collecting these events and their probabilities in the original sequential sample space.

The random variables discussed in our example could take on only experimental values selected from a set of discrete numbers. Such random variables are known as *discrete* random variables. Random variables of another type, known as *continuous* random variables, may take on experimental values anywhere within continuous ranges. Examples of continuous random variables are the exact instantaneous voltage of a noise signal and the precise reading after a spin of an infinitely finely calibrated wheel of fortune (as in the last example of Sec. 1-2).

Formally, the distinction between discrete and continuous random variables can be avoided. But the development of our topics is easier to visualize if we first become familiar with matters with regard to discrete random variables and later extend our coverage to the continuous case. Our discussions through Sec. 2-8 deal only with the discrete case, and Sec. 2-9 begins the extension to the continuous case.

2-2 The Probability Mass Function

We have learned that a random variable is defined by a function which assigns a value of that random variable to each sample point. These assigned values of the random variable are said to represent its possible experimental values. Each performance of the experiment generates an experimental value of the random variable. For many purposes, we shall find the resulting experimental value of a random variable to be an adequate characterization of the experimental outcome.

In the previous section, we indicated the form of a simple event space for dealing with a single discrete random variable. To work with a random variable x, we mark, on an x_0 axis, the points corresponding to all possible experimental values of the random variable. One such event space could be

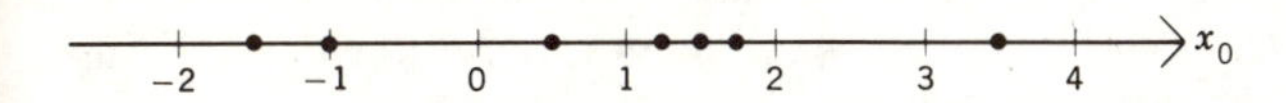

Each point in this event space corresponds to the event "On a performance of the experiment, the resulting experimental value of random variable x is equal to the indicated value of x_0."

We next define a function on this event space which assigns a probability to each event point. The function $p_x(x_0)$ is known as the *probability mass function* (PMF) for discrete random variable x, defined by

$p_x(x_0)$ = probability that the experimental value of random variable x obtained on a performance of the experiment is equal to x_0

We often present the probability mass function as a bar graph drawn over an event space for the random variable. One possible PMF is sketched below:

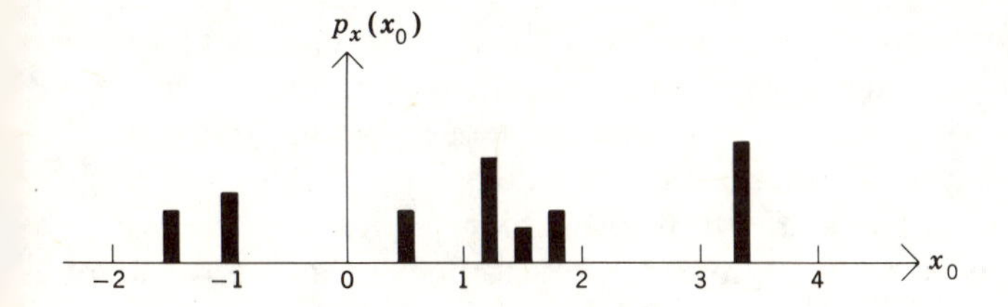

Since there must be some value of random variable x associated with every sample point, we must have

$$\sum_{x_0} p_x(x_0) = 1$$

and, of course, from the axioms of probability theory we also have

$$0 \leq p_x(x_0) \leq 1 \qquad \text{for all values of } x_0$$

Note that the argument of a PMF is a dummy variable, and the PMF for random variable x could also be written as $p_x(y)$, $p_x(\Rightarrow)$, or, as some people prefer, $p_x(\cdot)$. We shall generally use the notation $p_x(x_0)$ for a PMF. However, when another notation offers advantages of clarity or brevity for the detailed study of a particular process, we shall adopt it for that purpose.

For an example of the PMF for a random variable, let's return to the experiment of three flips of a fair coin, introduced in the previous section. We may go back to the original sample space of that experiment to collect $p_h(h_0)$, the PMF for the total number of heads resulting from the three flips. We obtain

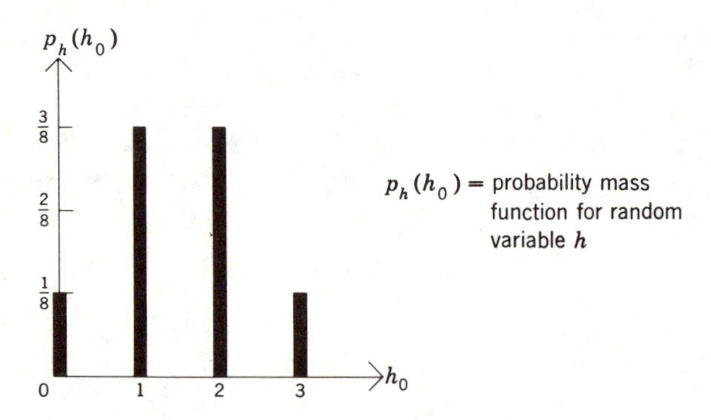

2-3 Compound Probability Mass Functions

We wish to consider situations in which values of more than one random variable are assigned to each point in the sample space of an experiment. Our discussion will be for two discrete random variables, but the extension to more than two is apparent.

For a performance of an experiment, the probability that random variable x will take on experimental value x_0 *and* random variable y will take on experimental value y_0 may be determined in sample space by summing the probabilities of each sample point which has this compound attribute. To designate the probability assignment in an x_0,y_0 event space, we extend our previous work to define the *compound* (or *joint*) PMF for two random variables x and y,

$p_{x,y}(x_0,y_0)$ = probability that the experimental values of random variables x and y obtained on a performance of the experiment are equal to x_0 and y_0, respectively

A picture of this function would have the possible event points marked on an x_0,y_0 coordinate system with each value of $p_{x,y}(x_0,y_0)$ indicated as a bar perpendicular to the x_0,y_0 plane above each event point. [We use the word *event* point here since each possible (x_0,y_0) point might represent the union of several sample points in the finest-grain description of the experiment.]

By considering an x_0,y_0 event space and recalling that an event space is a mutually exclusive, collectively exhaustive listing of all possible experimental outcomes, we see that the following relations hold:

$$\sum_{x_0} \sum_{y_0} p_{x,y}(x_0,y_0) = 1$$

$$\sum_{x_0} p_{x,y}(x_0,y_0) = p_y(y_0) \qquad \sum_{y_0} p_{x,y}(x_0,y_0) = p_x(x_0)$$

In situations where we are concerned with more than one random variable, the PMF's for single random variables, such as $p_x(x_0)$, are referred to as *marginal* PMF's. No matter how many random variables may be defined on the sample space of the experiment, this function $p_x(x_0)$ always has the same physical interpretation. For instance, $p_x(2)$ is the probability that the experimental value of discrete random variable x resulting from a performance of the experiment will be equal to 2.

Let's return to the example of three flips of a fair coin in Sec. 2-1 to obtain the compound PMF for random variables h (number of heads) and r (length of longest run). By collecting the events of interest and their probabilities from the sequential sample space in Sec. 2-1, we obtain $p_{h,r}(h_0,r_0)$. We indicate the value of $p_{h,r}(h_0,r_0)$ associated with each event by writing it beside the appropriate event point.

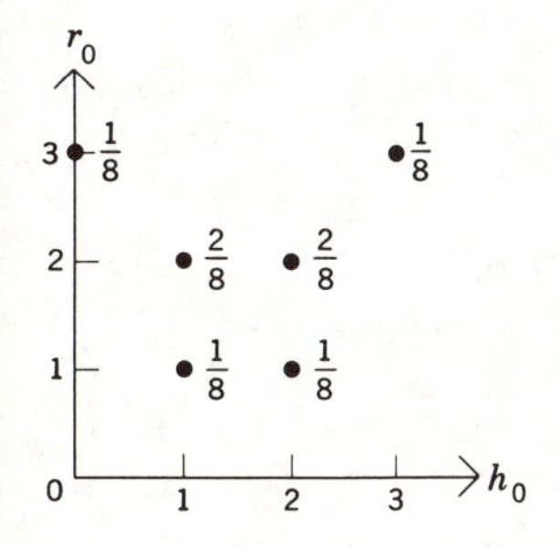

The probability of *any* event described in terms of the experimental values of random variables h and r may be found easily in this event space once $p_{h,r}(h_0,r_0)$ has been determined.

For instance, we may obtain the marginal PMF's, $p_h(h_0)$ and $p_r(r_0)$, simply by collecting the probabilities of the appropriate events in the h_0,r_0 sample space.

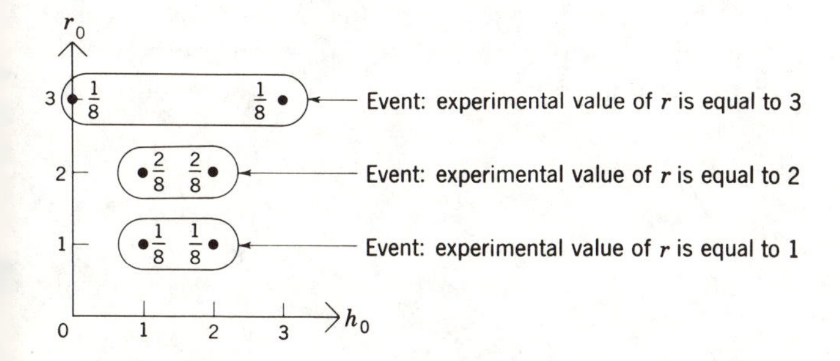

The reader can check that the above procedure and a similar operation for random variable h lead to the marginal PMF's

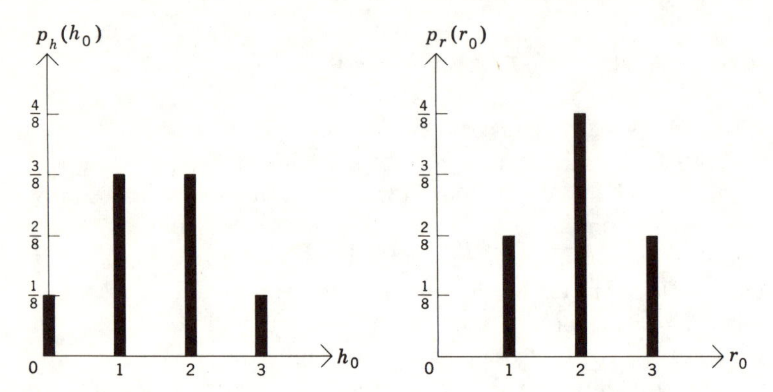

It is important to note that, in general, there is no way to go back from the marginal PMF's to determine the compound PMF.

2-4 Conditional Probability Mass Functions

Our interpretation of conditional probability in Chap. 1 noted that conditioning a sample space does one of two things to the probability of each finest-grain event. If an event does not have the attribute of the conditioning event, the conditional probability of that event is set to zero. For all finest-grain events having the attribute of the conditioning event, the conditional probability associated with each such event is equal to its original probability scaled up by a constant $[1/P(A)$, where A is the conditioning event] such that the sum of the conditional probabilities in the conditional sample space is unity. We can use the same concepts in an event space, as long as we can answer either "yes" or "no" for each event point to the question "Does this event have the attribute of the conditioning event?" Difficulty would arise only when the conditioning event was of finer grain than the event space. This matter was discussed near the end of Sec. 1-4.

When we consider a discrete random variable taking on a particular experimental value as a result of a performance of an experiment, this is simply an event like those we denoted earlier as A, B, or anything else and all our notions of conditional probability are carried over to the discussion of discrete random variables. We define the conditional PMF by

$p_{x|y}(x_0 \mid y_0)$ = conditional probability that the experimental value of random variable x is x_0, given that, on the same performance of the experiment, the experimental value of random variable y is y_0

From the definition of conditional probability, there follows

$$p_{x|y}(x_0 \mid y_0) = \frac{p_{x,y}(x_0,y_0)}{p_y(y_0)} \qquad \text{and, similarly,} \qquad p_{y|x}(y_0 \mid x_0) = \frac{p_{x,y}(x_0,y_0)}{p_x(x_0)}$$

As was the case in the previous chapter, the conditional probabilities are not defined for the case where the conditioning event is of probability zero.

Writing these definitions another way, we have

$$p_{x,y}(x_0,y_0) = p_x(x_0)p_{y|x}(y_0 \mid x_0) = p_y(y_0)p_{x|y}(x_0 \mid y_0)$$

Notice that, in general, the marginal PMF's $p_x(x_0)$ and $p_y(y_0)$ *do not* specify $p_{x,y}(x_0,y_0)$ just as, in general, $p(A)$ and $p(B)$ do not specify $p(AB)$.

Finally, we need a notation for a conditional joint probability mass function where the conditioning event is other than an observed experimental value of one of the random variables. We shall use $p_{x,y|A}(x_0,y_0 \mid A)$ to denote the conditional compound PMF for random variables x and y given event A. This is, by the definition of conditional probability,

$$p_{x,y|A}(x_0,y_0 \mid A) = \begin{cases} \dfrac{p_{x,y}(x_0,y_0)}{P(A)} & \text{if } (x_0,y_0) \text{ in } A \\ 0 & \text{if } (x_0,y_0) \text{ in } A' \end{cases}$$

We return to the h_0,r_0 event space of the previous sections and its compound PMF in order to obtain some experience with conditional probability mass functions

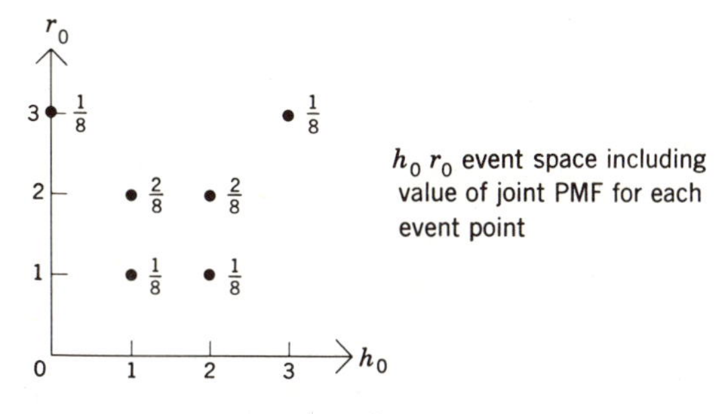

We begin by finding the conditional PMF for random variable r, the length of the longest run obtained in three flips, given that the experimental value of h, the number of heads, is equal to 2. Thus, we wish to find $p_{r|h}(r_0 \mid 2)$. Only two of the event points in the original h_0,r_0

event space have the attribute of the conditioning event ($h_0 = 2$), and the relative likelihood of these points must remain the same in the conditional event space. Either by direct use of the definition or from our reasoning in sample space, there results

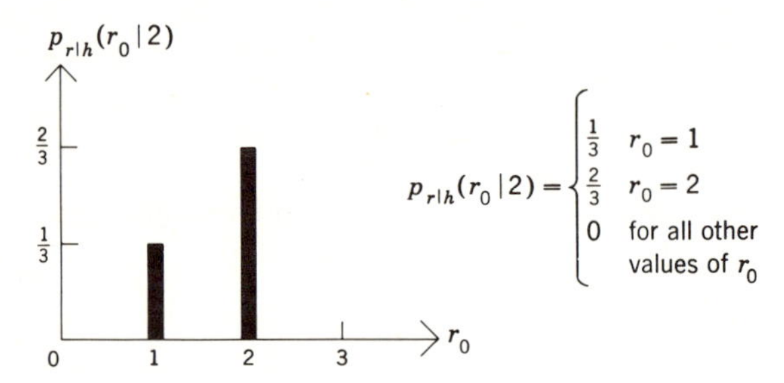

Had the conditioning event been that the experimental value of h were equal to 3, this would specify the experimental value of r for this experiment, because there is only one possible h_0,r_0 event point with $h_0 = 3$. The resulting conditional PMF would, of course, be

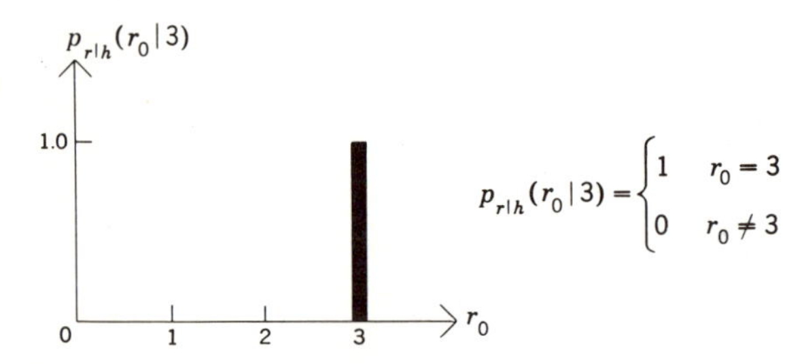

Suppose that we wish to condition the compound PMF in the h_0,r_0 event space by the event that the experimental values of h and r resulting from a performance of the experiment are not equal. In going to the appropriate conditional event space and allocation of conditional probability, we remove event points incompatible with the conditioning event and renormalize to obtain

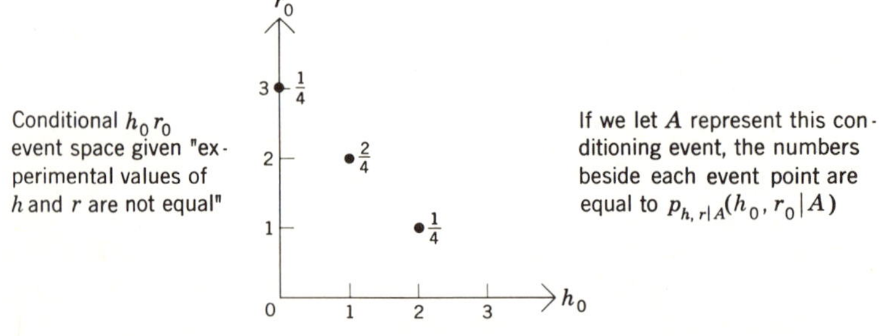

Finally, we can note one reasonable example of a phenomenon which was mentioned earlier. We have stated that we cannot always

directly condition an *event* space by an arbitrary event defined on the experiment. For our example, if we were told that the second flip resulted in heads, our simple conditioning argument cannot be applied in an h_0,r_0 event space because we can't answer uniquely "yes" or "no" as to whether or not each event point has this attribute. The conditioning requires information which appeared in the sequential *sample* space of Sec. 2-1 but which is no longer available in an h_0,r_0 *event* space.

2-5 Independence and Conditional Independence of Random Variables

In Sec. 1-6 we obtained our definition of the independence of two events by stating formally an intuitive notion of independence. For two random variables to be independent, we shall require that *no* possible experimental value of one random variable be able to give us any new information about the probability of *any* experimental value of the other random variable. A formal statement of this notion of the independence of two random variables is

Random variables x and y are defined to be independent if and only if $p_{y|x}(y_0 \mid x_0) = p_y(y_0)$ for all possible values of x_0 and y_0.

From the definition of the conditional PMF's, as long as the conditioning event is of nonzero probability, we may always write

$$p_{x,y}(x_0,y_0) = p_x(x_0)p_{y|x}(y_0 \mid x_0) = p_y(y_0)p_{x|y}(x_0 \mid y_0)$$

and, substituting the above definition of independence into this equation, we find that $p_{y|x}(y_0 \mid x_0) = p_y(y_0)$ for all x_0,y_0 requires that $p_{x|y}(x_0 \mid y_0) = p_x(x_0)$ for all x_0,y_0; thus, one equivalent definition of the independence condition would be to state that random variables x and y are independent if and only if $p_{x,y}(x_0,y_0) = p_x(x_0)p_y(y_0)$ for all x_0,y_0.

We define any number of random variables to be *mutually* independent if the compound PMF for all the random variables factors into the product of all the marginal PMF's for all arguments of the compound PMF.

It is also convenient to define the notion of conditional independence. One of several equivalent definitions is

Random variables x and y are *defined* to be conditionally independent given event A [with $P(A) \neq 0$] if and only if

$$p_{x,y|A}(x_0,y_0 \mid A) = p_{x|A}(x_0 \mid A)p_{y|A}(y_0 \mid A) \qquad \text{for all } (x_0,y_0)$$

Of course, the previous unconditional definition may be obtained by setting $A = U$ in the conditional definition. The function $p_{x|A}(x_0 \mid A)$ is referred to as the *conditional marginal* PMF for random variable x given that the experimental outcome on the performance of the experiment had attribute A.

We shall learn later that the definition of independence has implications beyond the obvious one here. In studying situations involving several random variables, it will normally be the case that, if the random variables are mutually independent, the analysis will be greatly simplified and several powerful theorems will apply.

The type of independence we have defined in this section is often referred to as *true*, or *statistical*, independence of random variables. These words are used to differentiate between this complete form of independence and a condition known as *linear* independence. The latter will be defined in Sec. 2-7.

The reader may wish to use our three-flip experiment and its h_0,r_0 event space to verify that, although h and r are clearly not independent in their original event space, they are conditionally independent given that the longest run was shorter than three flips.

2-6 Functions of Random Variables

A function of some random variables is just what the name indicates—it is a function whose experimental value is determined by the experimental values of the random variables. For instance, let h and r again be the number of heads and the length of the longest run for three flips of a fair coin. Some functions of these random variables are

$$v(h,r) = h^2 \qquad w(h,r) = |h - r| \qquad x(h,r) = e^{-h} \log (r \cos h)$$

$$y(h,r) = \max\left(\frac{h}{r}, \frac{r}{2h}\right) \qquad z(h,r) = \begin{cases} h + r & r < 2h \\ 3h - r & r \geq 2h \end{cases}$$

and, of course, h and r.

Functions of random variables are thus new random variables themselves. The experimental values of these new random variables may be displayed in the event space of the original random variables, for instance, by adding some additional markers beside the event points. Once this is done, it is a simple matter to assemble the PMF for a function of the original random variables.

For example, let random variable w be defined by $w = |h - r|$. We'll write the value of w assigned to each event point in a box beside the point in the h_0,r_0 event space,

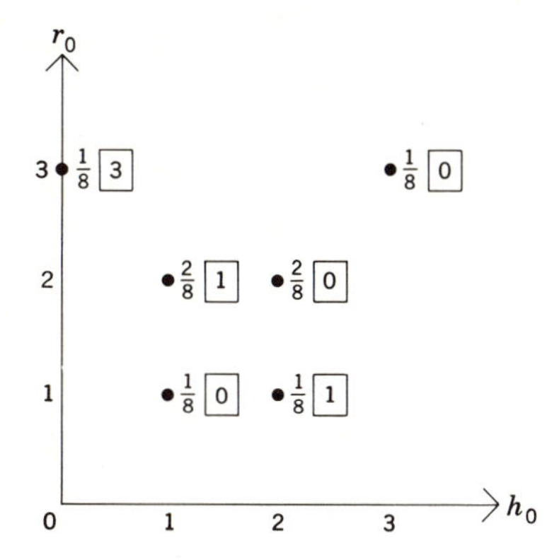

and then working in the above event space, we can rapidly collect $p_w(w_0)$, to obtain

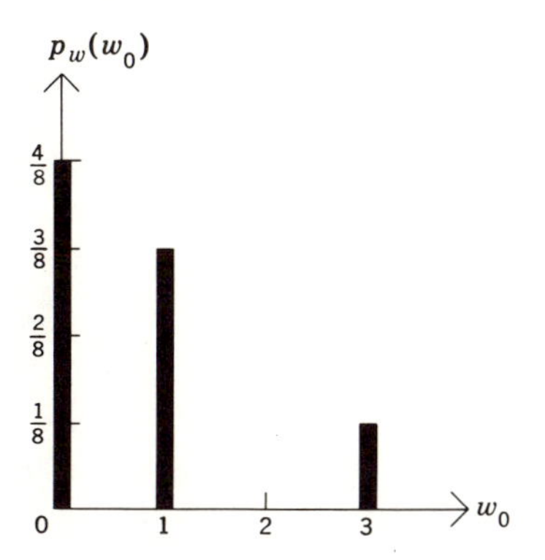

2-7 Expectation and Conditional Expectation

Let x be a random variable, and let $g(x)$ be any single-valued function of its argument. Then $g(x)$ is a function of a random variable and is itself a random variable. We define $E[g(x)]$, the *expectation*, or *expected value*, of this function of random variable x, to be

$$E[g(x)] \equiv \sum_{x_0} g(x_0)p_x(x_0) \equiv \overline{g(x)}$$

and we also define $E[g(x) \mid A]$, the conditional expectation of $g(x)$ given that the experimental outcome has attribute A, to be

$$E[g(x) \mid A] \equiv \sum_{x_0} g(x_0)p_{x|A}(x_0 \mid A) \equiv \overline{g(x \mid A)}$$

As usual, this definition for the conditional case includes the unconditional case (obtained by setting A equal to the universal event).

Consider the event-space interpretation of the definition of $E[g(x)]$ in an x_0 event space. For each event point x_0, we multiply $g(x_0)$ by the probability of the event point representing that experimental outcome, and then we sum all such products. Thus, the expected value of $g(x)$ is simply the weighted sum of all possible experimental values of $g(x)$, each weighted by its probability of occurrence on a performance of the experiment. We might anticipate a close relationship between $E[g(x)]$ and the average of the experimental values of $g(x)$ generated by many performances of the experiment. This type of relation will be studied in the last two chapters of this book.

Certain cases of $g(x)$ give rise to expectations of frequent interest and these expectations have their own names.

If $g(x) = x^n$:

$$E[g(x)] = \sum_{x_0} x_0{}^n p_x(x_0) = \overline{x^n}$$

If $g(x) = [x - E(x)]^n$:

$$E[g(x)] = \sum_{x_0} [x_0 - E(x)]^n p_x(x_0) = \overline{(x - \bar{x})^n}$$

The quantity $\overline{x^n}$ is known as the *nth moment*, and the quantity $\overline{(x - \bar{x})^n}$ is known as the *nth central moment* of random variable x.

Often we desire a few simple parameters to characterize the PMF for a particular random variable. Two choices which have both intuitive appeal and physical significance are the expected value and the second central moment of the random variable. We shall discuss the intuitive interpretation of these quantities here. Their physical significance will become apparent in our later work on limit theorems and statistics.

The *first moment* (or *expected value* or *mean value*) of random variable x is given by

$$E(x) = \sum_{x_0} x_0 p_x(x_0)$$

and if we picture a PMF bar graph for random variable x to be composed of broomsticks on a weightless axis, we may say that $E(x)$ specifies the location of the *center of mass* of the PMF.

The second central moment $E\{[x - E(x)]^2\}$ is a measure of the second power of the spread of the PMF for random variable x about its expected value. The second central moment of random variable x is known as its *variance* and is denoted by $\sigma_x{}^2$. The square root of the variance, σ_x, is known as the *standard deviation* of random variable x and may be considered to be one characterization of the spread of a PMF about $E(x)$. Here are a few PMF's for random variable x, each with the same mean but different standard deviations.

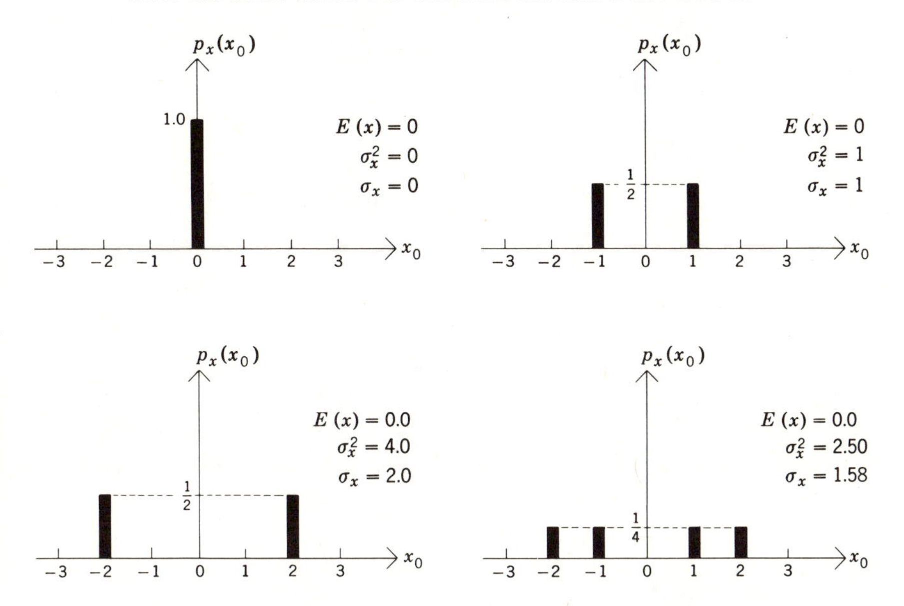

A *conditional* central moment is a measure of the nth power of the spread of a conditional PMF for a random variable about its *conditional* mean. For instance, given that the experimental outcome had attribute A, the conditional variance of random variable x, $\sigma_{x|A}^2$, is given by

$$\sigma_{x|A}^2 = \sum_{x_0} [x_0 - E(x \mid A)]^2 p_{x|A}(x_0 \mid A)$$

For functions of several random variables, we again define expectation to be the weighted sum of all possible experimental values of the function, with each such value weighted by the probability of its occurrence on a performance of the experiment. Let $g(x,y)$ be a single-valued function of random variables x and y. By now we are familiar enough with the ideas involved to realize that a definition of $E[g(x,y) \mid A]$, the conditional expectation of $g(x,y)$, will include the definition for the unconditional case.

$$E[g(x,y) \mid A] = \sum_{x_0} \sum_{y_0} g(x_0,y_0) p_{x,y|A}(x_0,y_0 \mid A)$$

We should remember that in order to determine the expectation of $g(x)$ [or of $g(x,y)$ or of any function of $g(x,y)$] it is not necessary that we first determine the PMF $p_g(g_0)$. The calculation of $E[g(x,y)]$ can always be carried out directly in the x_0,y_0 event space (see Prob. 2.10).

We wish to establish some definitions and results regarding the expected values of sums and products of random variables. From the definition of the expectation of a function of several random variables, we may write

$$E(x+y) = \sum_{x_0}\sum_{y_0}(x_0+y_0)p_{x,y}(x_0,y_0)$$

$$= \underbrace{\sum_{x_0}\sum_{y_0} x_0p_{x,y}(x_0,y_0)}_{\text{We'll sum this over } y_0 \text{ first}} + \underbrace{\sum_{x_0}\sum_{y_0} y_0p_{x,y}(x_0,y_0)}_{\text{We'll sum this over } x_0 \text{ first}}$$

$$= \sum_{x_0} x_0p_x(x_0) + \sum_{y_0} y_0p_y(y_0) = E(x) + E(y) = E(x+y)$$

The expected value of the sum of two random variables is always equal to the sum of their expected values. This holds with no restrictions on the random variables, which may, of course, be functions of other random variables. The reader should, for instance, be able to use this result directly in the definition of the variance of random variable x to show

$$\sigma_x^2 = E\{[x - E(x)]^2\} = E(x^2) - [E(x)]^2$$

Now, consider the expected value of the product xy,

$$E(xy) = \sum_{x_0}\sum_{y_0} x_0y_0p_{x,y}(x_0,y_0)$$

In general, we can carry this operation no further without some knowledge about $p_{x,y}(x_0,y_0)$. Clearly, if x and y are independent, the above expression will factor to yield the result $E(xy) = E(x)E(y)$. Even if x and y are not independent, it is still possible that the numerical result would satisfy this condition.

If $E(xy) = E(x)E(y)$, random variables x and y are said to be *linearly independent.* (Truly independent random variables will always satisfy this condition.)

An important expectation, the *covariance* of two random variables, is introduced in Prob. 2.33. Chapter 3 will deal almost exclusively with the properties of some other useful expected values, the *transforms* of PMF's. We shall also have much more to say about expected values when we consider limit theorems and statistics in Chaps. 6 and 7.

2-8 Examples Involving Discrete Random Variables

We have dealt with only one example related to our study of discrete random variables. We now work out some more detailed examples.

example 1 A biased four-sided die is rolled once. Random variable N is defined to be the down-face value and is described by the PMF,

$$p_N(N_0) = \begin{cases} \dfrac{N_0}{10} & \text{for } N_0 = 1, 2, 3, 4 \\ 0 & \text{for all other values of } N_0 \end{cases}$$

Based on the outcome of this roll, a coin is supplied for which, on any flip, the probability of an outcome of heads is $(N + 1)/2N$. The coin is flipped once, and the outcome of this flip completes the experiment.

Determine:

(a) The expected value and variance of discrete random variable N.

(b) The conditional PMF, conditional expected value, and conditional variance for random variable N given that the coin came up heads.

(c) If we define the events

Event A: Value of down face on die roll is either 1 or 4

Event H: Outcome of coin flip is heads

are the events A and H independent?

We'll begin by drawing a sequential sample space for the experiment, labeling all branches with the appropriate branch-traversal probabilities, and collecting some relevant information.

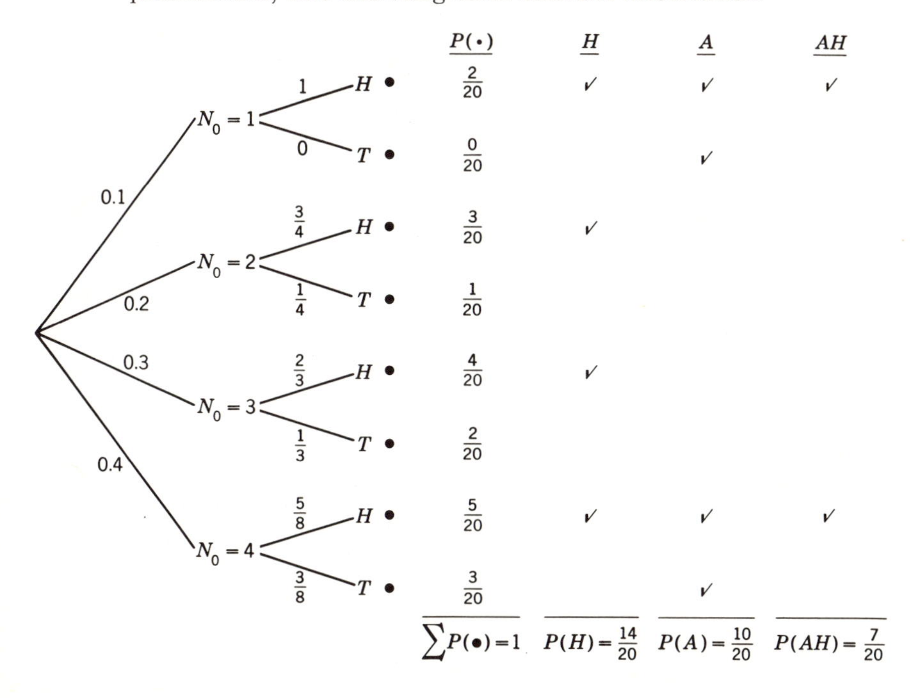

a Applying the definitions of mean and variance to random variable N, described by the PMF $p_N(N_0)$ given above, we have

$$E(N) = \sum_{N_0} N_0 p_N(N_0) = 1 \cdot \tfrac{1}{10} + 2 \cdot \tfrac{2}{10} + 3 \cdot \tfrac{3}{10} + 4 \cdot \tfrac{4}{10} = 3.0$$

$$\sigma_N{}^2 = \sum_{N_0} [N_0 - E(N)]^2 p_N(N_0)$$

$$= (-2)^2 \cdot \tfrac{1}{10} + (-1)^2 \cdot \tfrac{2}{10} + 0^2 \cdot \tfrac{3}{10} + 1^2 \cdot \tfrac{4}{10} = 1.0$$

b Given that event H did occur on this performance of the experiment, we may condition the above space by event H by removing all points with attribute T and scaling up the probabilities of all remaining events by multiplying by $1/P(H) = 10/7$. This results in the four-point conditional event space shown below. To the right of this conditional event space we present the resulting conditional PMF for random variable N, given that event H did occur.

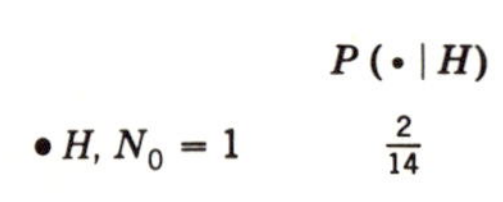

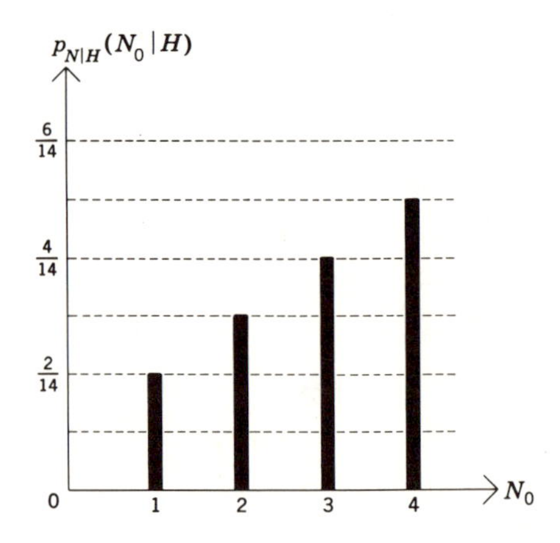

Applying the definitions of the mean and variance in this conditional event space, we have

$$E(N \mid H) = \sum_{N_0} N_0 p_{N|H}(N_0 \mid H) = 1 \cdot \tfrac{2}{14} + 2 \cdot \tfrac{3}{14} + 3 \cdot \tfrac{4}{14} + 4 \cdot \tfrac{5}{14} = \tfrac{20}{7}$$

$$\sigma_{N|A}{}^2 = \sum_{N_0} [N_0 - E(N \mid H)]^2 p_{N|H}(N_0 \mid H)$$

$$= (-\tfrac{13}{7})^2 \cdot \tfrac{2}{14} + (-\tfrac{6}{7})^2 \cdot \tfrac{3}{14} + (\tfrac{1}{7})^2 \cdot \tfrac{4}{14} + (\tfrac{8}{7})^2 \cdot \tfrac{5}{14} = \tfrac{55}{49}$$

c We wish to test $P(AH) \stackrel{?}{=} P(A)P(H)$, and we have already collected each of these three quantities in the sample space for the experiment.

$$P(A) = \tfrac{10}{20} \qquad P(H) = \tfrac{14}{20} \qquad P(AH) = \tfrac{7}{20}$$

So the events A and H are independent.

example 2 Patrolman G. R. Aft of the local constabulary starts each day by deciding how many parking tickets he will award that day. For any day, the probability that he decides to give exactly K tickets is given by the PMF

$$p_K(K_0) = \begin{cases} \dfrac{5 - K_0}{10} & \text{for } K_0 = 1, 2, 3, \text{ or } 4 \\ 0 & \text{for all other values of } K_0 \end{cases}$$

But the more tickets he gives, the less time he has to assist old ladies at street crossings. Given that he has decided to have a K-ticket day, the conditional probability that he will also help exactly L old ladies cross the street that day is given by

$$p_{L|K}(L_0 \mid K_0) = \begin{cases} \dfrac{1}{5 - K_0} & \text{if } 1 \leq L_0 \leq 5 - K_0 \\ 0 & \text{if } L_0 < 1 \text{ or if } L_0 > 5 - K_0 \end{cases}$$

His daily salary S is computed according to the formula

$$S = 2K + L \qquad \text{(dollars)}$$

Before we answer some questions about Officer Aft, we should be sure that we understand the above statements. For instance, on a day when Officer Aft has decided to give two tickets, the conditional PMF states that he is equally likely to help one, two, or three old ladies. Similarly, on a day when he has decided to give exactly four tickets, it is *certain* that he will help exactly one old lady cross the street.

(a) Determine the marginal PMF $p_L(L_0)$. This marginal PMF tells us the probability that Officer Aft will assist exactly L_0 old ladies on any day. Determine also the expected value of random variable L.

(b) Random variable S, Officer Aft's salary on any given day, is a function of random variables K and L. Determine the expected value of the quantity $S(L,K)$.

(c) Given that, on the day of interest, Officer Aft earned at least \$6, determine the conditional marginal PMF for random variable K, the number of traffic tickets he awarded on that particular day.

(d) We define
Event A: Yesterday he gave a total of one or two parking tickets.
Event B: Yesterday he assisted a total of one or two old ladies.
Determine whether or not random variables K and L are conditionally independent given event AB.

From the statement of the example we can obtain a sample space and the assignment of a priori probability measure for the experiment. We could begin with a sequential picture of the experiment such as

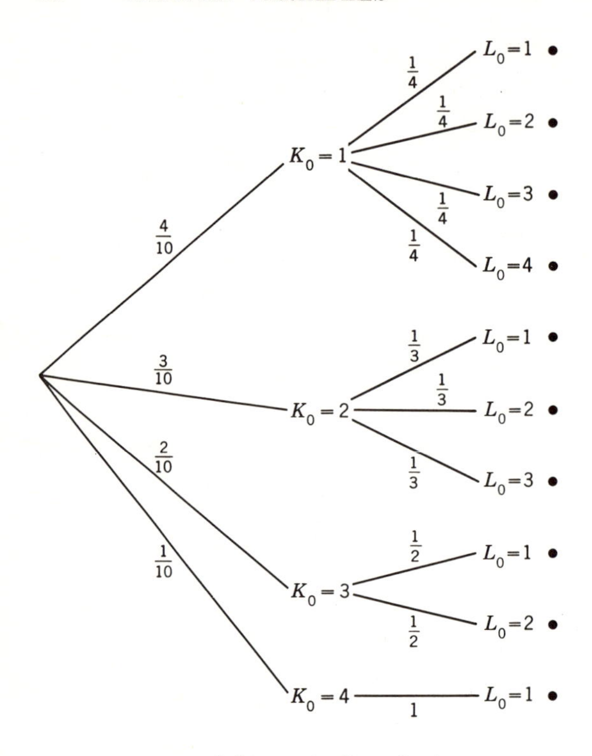

or we might work directly in a K_0,L_0 coordinate event space with the probability assignment $p_{K,L}(K_0,L_0)$ determined by

$$p_{K,L}(K_0,L_0) = p_K(K_0)p_{L|K}(L_0 \mid K_0)$$

$$= \begin{cases} \dfrac{5 - K_0}{10} \cdot \dfrac{1}{5 - K_0} & \text{if } K_0 = 1, 2, 3, 4 \text{ and } 1 \leq L_0 \leq 5 - K_0 \\ 0 & \text{otherwise} \end{cases}$$

$$= \begin{cases} 0.1 & \text{if } K_0 = 1, 2, 3, 4 \text{ and } 1 \leq L_0 \leq 5 - K_0 \\ 0.0 & \text{otherwise} \end{cases}$$

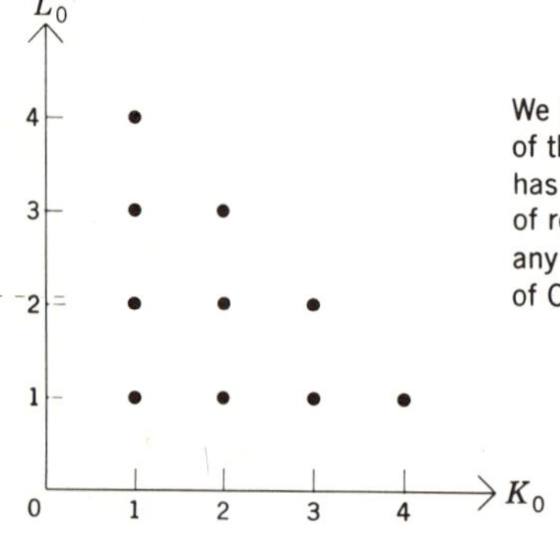

We have established that each of the ten possible event points has an a priori probability of 0.1 of representing the outcome of any particular day in the life of Officer G. R. Aft

a The calculation $p_L(L_0) = \sum_{K_0} p_{L,K}(L_0,K_0)$ is easily performed in our event space to obtain

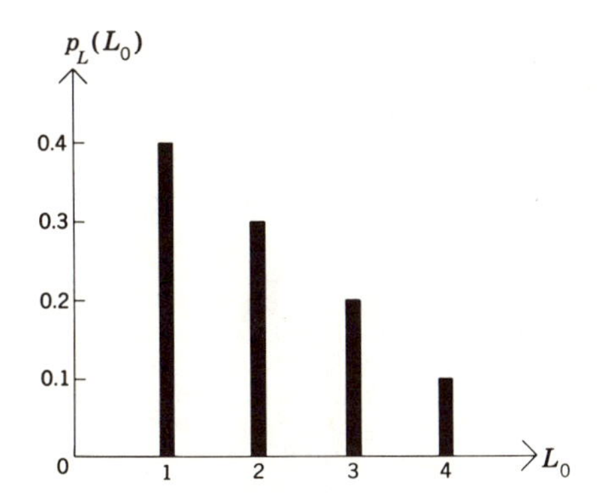

To find the expectation of L, we could multiply the experimental value of L corresponding to each sample point by the probability measure assigned to that point and sum these products. But since we already have $p_L(L_0)$, it is quicker to work in the event space for random variable L to obtain

$$E(L) = \sum_{L_0} L_0 p_L(L_0) = 1 \cdot \tfrac{4}{10} + 2 \cdot \tfrac{3}{10} + 3 \cdot \tfrac{2}{10} + 4 \cdot \tfrac{1}{10} = 2.0$$

Although it has happened in this example, we should note that there is no reason whatever why the expected value of a random variable need be equal to a possible experimental value of that random variable.

b $E(S) = \sum_{K_0} \sum_{L_0} (2K_0 + L_0) p_{K,L}(K_0,L_0)$. We can simply multiply the experimental value of S corresponding to each event point by the probability assignment of that event point. Let's write the corresponding experimental value of S beside each event point and then compute $E(S)$.

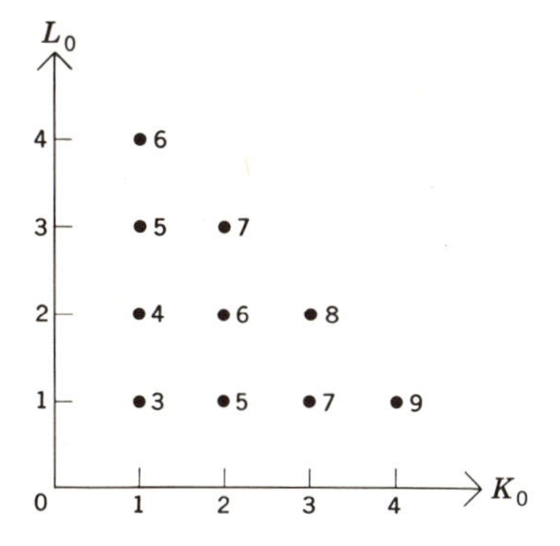

$$E(S) = \tfrac{1}{10}(3 + 4 + 5 + 6 + 5 + 6 + 7 + 7 + 8 + 9) = \$6$$

c Given that Officer Aft earned at least \$6, we can easily condition the above event space to get the conditional event space (still with the experimental values of S written beside the event points representing the remaining possible experimental outcomes)

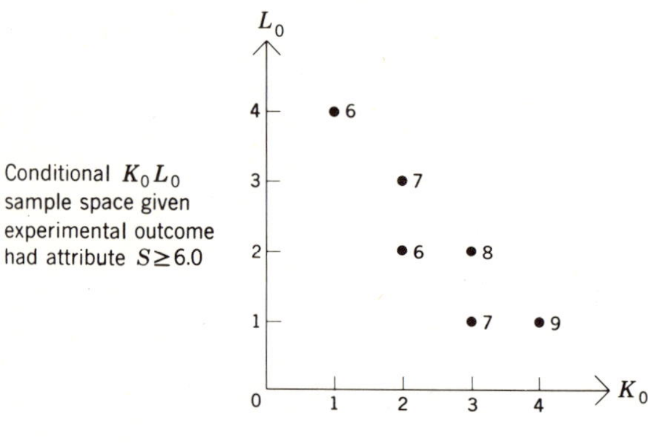

Thus, by using the notation Event C: $S \geq 6$, we have

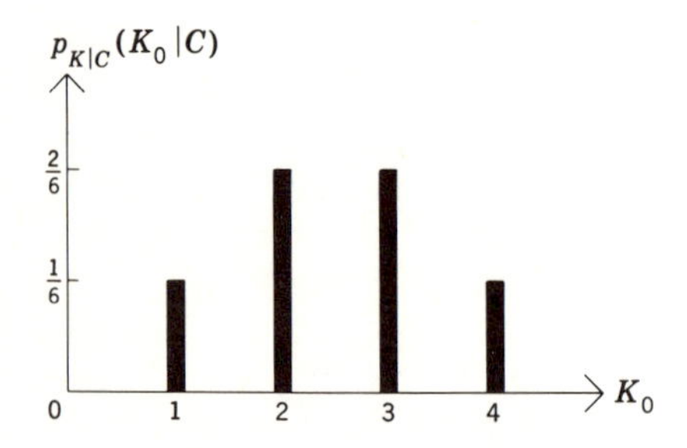

Recall that we can use this simple interpretation of conditional probability only if the event space to which we wish to apply it is of fine enough grain to allow us to classify each event point as being wholly in C or being wholly in C'.

d There are only four (equally likely) event points in the conditional K_0,L_0 event space given that the experimental outcome has attribute AB.

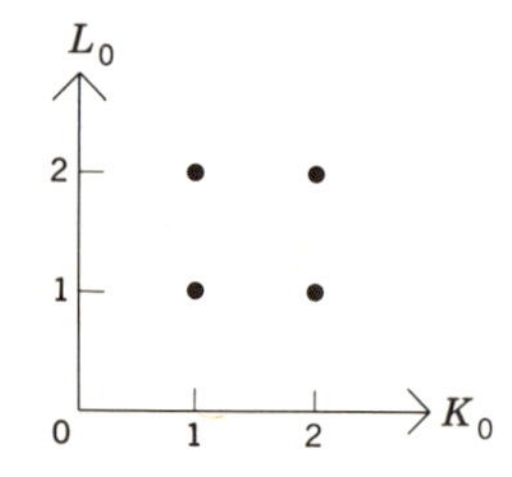

We wish to check

$$p_{K,L|AB}(K_0,L_0 \mid AB) \stackrel{?}{=} p_{K|AB}(K_0 \mid AB)p_{L|AB}(L_0 \mid AB) \qquad \text{for all } K_0,L_0$$

Each of these three PMF's is found from the conditional K_0,L_0 event space presented above.

$$p_{K,L|AB}(K_0,L_0 \mid AB) = \begin{cases} \frac{1}{4} & \text{if } K_0 = 1, 2 \text{ and } L_0 = 1, 2 \\ 0 & \text{otherwise} \end{cases}$$

$$p_{K|AB}(K_0 \mid AB) = \begin{cases} \frac{1}{2} & \text{if } K_0 = 1, 2 \\ 0 & \text{otherwise} \end{cases}$$

$$p_{L|AB}(L_0 \mid AB) = \begin{cases} \frac{1}{2} & \text{if } L_0 = 1, 2 \\ 0 & \text{otherwise} \end{cases}$$

The definition of conditional independence is found to be satisfied, and we conclude that random variables K and L, which were not independent in the original sample space, are conditionally independent given that the experimental outcome has attribute AB. Thus, for instance, given that AB has occurred, the conditional marginal PDF of variable L will be unaffected by the experimental value of random variable K.

In this text, the single word *independence* applied to random variables is always used to denote *statistical* independence.

2-9 A Brief Introduction to the Unit-impulse Function

To prepare for a general discussion of continuous random variables, it is desirable that we become familiar with some properties of the unit-impulse function. Our introduction to this function, though adequate for our purposes, will lack certain details required to deal with more advanced matters.

The unit-impulse function $\mu_0(x_0 - a)$ is a function of x_0 which is equal to infinity at $x_0 = a$ and which is equal to zero for all other values of x_0. The integral of $\mu_0(x_0 - a)$ over any interval which includes the point where the unit impulse is nonzero is equal to unity.

One way to obtain the unit impulse $\mu_0(x_0 - a)$ is to consider the *limit*, as Δ goes to zero, of

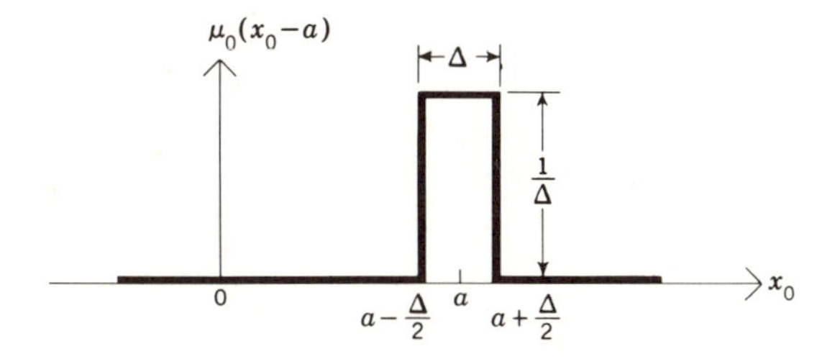

which, after the limit is taken, is normally represented as

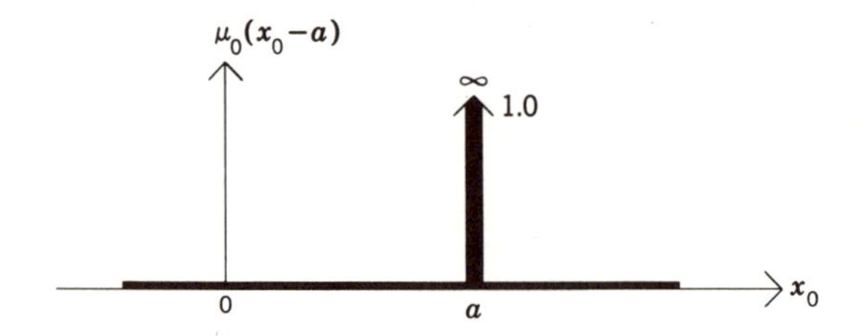

where the scale height of the impulse is irrelevant and the total area under the impulse function is written beside the arrowhead. The integral of the unit-impulse function,

$$\int_{x_0=-\infty}^{x_0} \mu_0(x_0 - a)\, dx_0$$

is a function known as the unit-step function and written as $\mu_{-1}(x_0 - a)$.

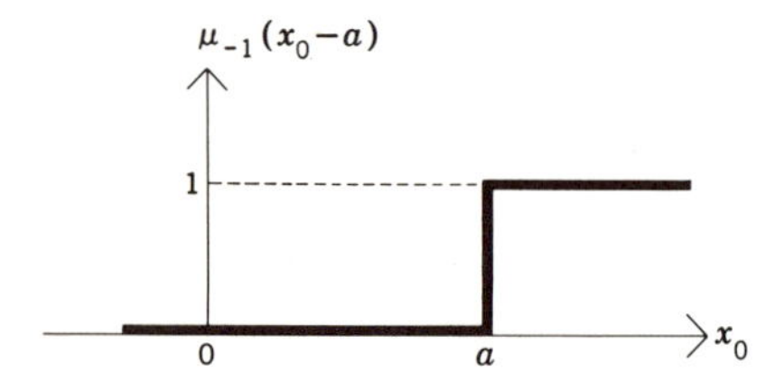

Thus an impulse may be used to represent the derivative of a function at a point where the function has a vertical discontinuity.

As long as a function $g(x_0)$ does not have a discontinuity at $x_0 = a$, the integral

$$\int g(x_0)\mu_0(x_0 - a)\, dx_0$$

over any interval which includes $x_0 = a$ is simply $g(a)$. This results from the fact that, over the very small range where the impulse is nonzero, $g(x_0)$ may be considered to be constant, equal to $g(a)$, and factored out of the integral.

2-10 The Probability Density Function for a Continuous Random Variable

We wish to extend our previous work to include the analysis of situations involving random variables whose experimental values may fall anywhere within continuous ranges. Some examples of such continuous random variables have already occurred in Secs. 1-2 and 2-1.

The assignment of probability measure to continuous sample and event spaces will be given by a probability *density* function (PDF). Let us begin by considering a single continuous random variable x whose event space is the real line from $x_0 = -\infty$ to $x_0 = \infty$. We define the PDF for random variable x, $f_x(x_0)$, by

$$\text{Prob}(a < x \leq b) = \int_a^b f_x(x_0)\, dx_0$$

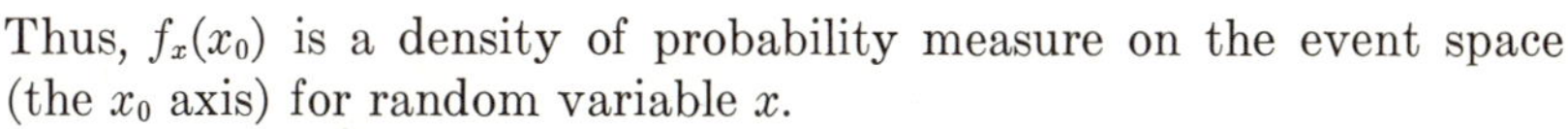

Thus, $f_x(x_0)$ is a density of probability measure on the event space (the x_0 axis) for random variable x.

Any event can be collected by selecting those parts of the x_0 axis which have the attribute of the event. For instance,

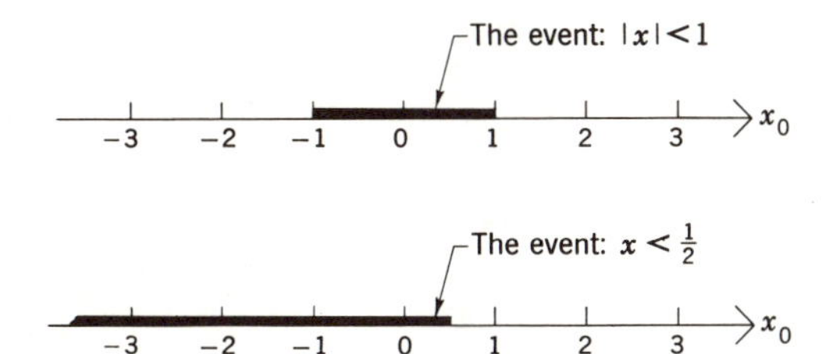

The probability of any event is found by evaluating the integral of $f_x(x_0)$ over those parts of the event space included in the event.

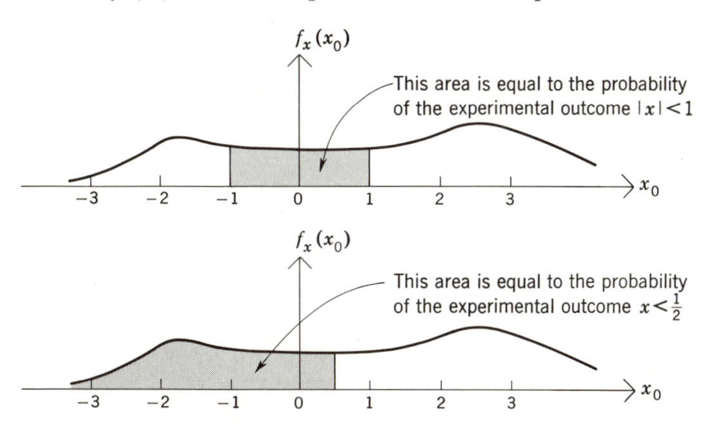

Should the PDF $f_x(x_0)$ contain impulses at either a or b, the integral $\int_a^b f_x(x_0)\, dx_0$ is defined to include the area of any impulse at the upper limit but *not* the area of any impulse at the lower limit. Note that this convention is determined by our choice of the inequality signs in the definition of $f_x(x_0)$.

Based on our understanding of event space and probability measure, we note that any PDF must have the following properties:

$$\int_{-\infty}^{\infty} f_x(x_0)\,dx_0 = 1 \qquad 0 \leq f_x(x_0) \leq \infty$$

If we wish to reason in terms of the probability of events (of nonzero probability), it is important to realize that it is *not* the PDF itself, but rather its integral over regions of the event space, which has this interpretation. As a matter of notation, we shall always use $f_x(x_0)$ for PDF's and reserve the letter p for denoting the probability of events. This is consistent with our use of $p_x(x_0)$ for a PMF.

Note that, unless the PDF happens to have an impulse at an experimental value of a random variable, the probability assignment to any single exact experimental value of a continuous random variable is zero. [The integral of a finite $f_x(x_0)$ over an interval of zero width is equal to zero.] This doesn't mean that every particular precise experimental value is an impossible outcome, but rather that such an event of probability zero is one of an infinite number of possible outcomes.

We next define the cumulative distribution function (CDF) for random variable x, $p_{x\leq}(x_0)$, by

$$p_{x\leq}(x_0) = \text{Prob}(x \leq x_0) = \int_{-\infty}^{x_0} f_x(x_0)\,dx_0$$

The function $p_{x\leq}(x_0)$ denotes the probability that, on any particular performance of the experiment, the resulting experimental value of random variable x will be less than or equal to x_0. Note the following properties of the CDF:

$$p_{x\leq}(\infty) = 1 \qquad p_{x\leq}(-\infty) = 0$$

$$p_{x\leq}(b) \geq p_{x\leq}(a) \qquad \text{for } b \geq a$$

$$\text{Prob}(a < x \leq b) = p_{x\leq}(b) - p_{x\leq}(a) \qquad \text{for } b \geq a$$

$$\frac{d}{dx_0}\,p_{x\leq}(x_0) = f_x(x_0)$$

The CDF will be especially useful for some of our work with continuous random variables. For a discrete random variable, the CDF is discontinuous and therefore, to some tastes, not differentiable. For our purposes, we have defined the derivative of the CDF at such a discontinuity to be an impulse of infinite height, zero width, and area equal to the discontinuity.

Let's consider an experiment involving a random variable x defined in the following manner: A fair coin is flipped once. If the outcome is heads, the experimental value of x is to be 0.2. If the outcome is tails, the experimental value of x is obtained by one spin of a fair, infinitely finely calibrated wheel of fortune whose range is from zero to unity. This gives rise to the PDF

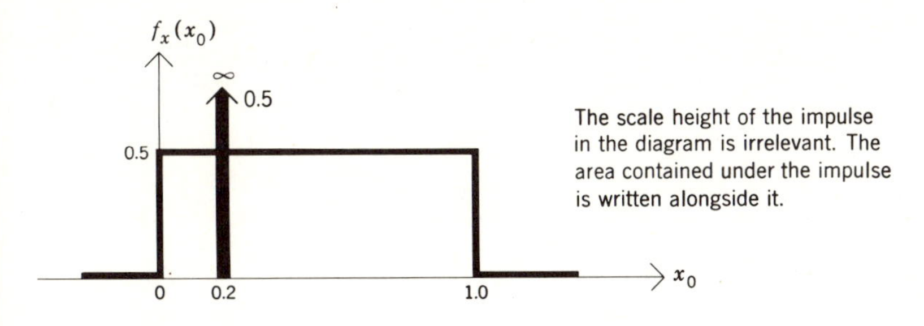

and to the CDF

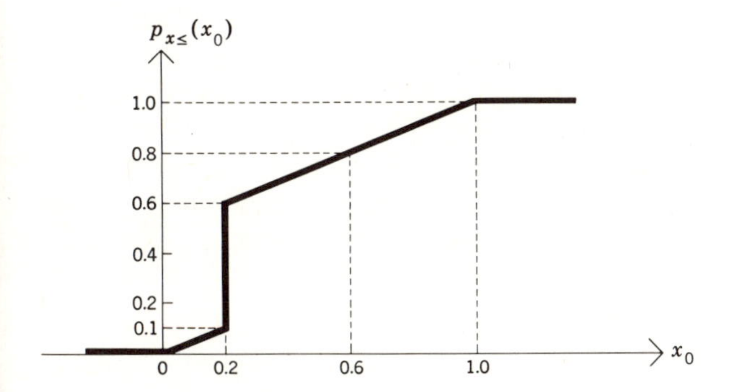

Because of the integration convention discussed earlier in this section, we can note that $p_{x\leq}(x_0)$ has (in principle) its discontinuity immediately to the left of $x_0 = 0.2$.

We consider one simple example which deals with a continuous random variable. Assume that the *lifetime* of a particular component is known to be a random variable described by the PDF

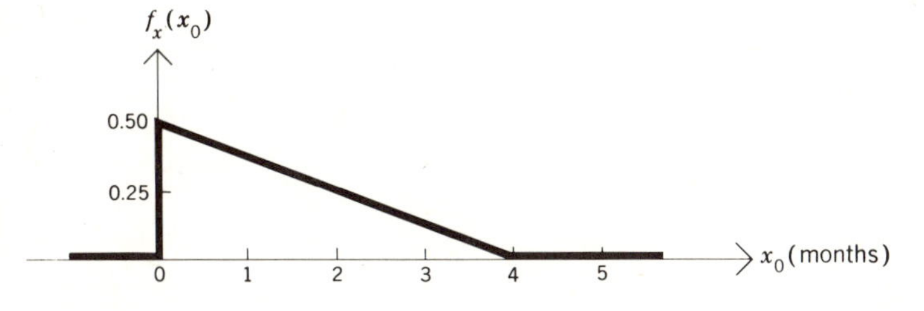

Let's begin by determining the a priori probability that the component fails during its second month of use. In an x_0 event space we can collect this event,

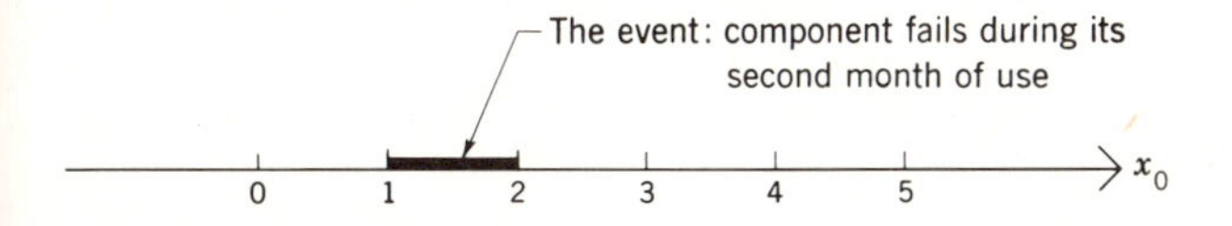

Thus, we require the quantity $\text{Prob}(1 < x \leq 2) = \int_1^2 f_x(x_0)\,dx_0 = \frac{5}{16}$, which is equal to the shaded area in the following sketch:

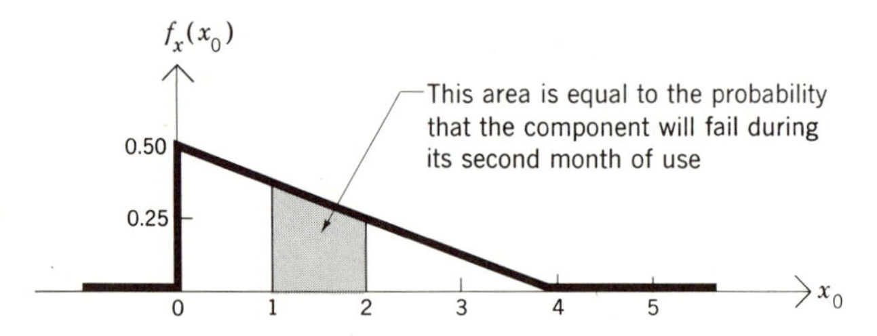

Since random variable x does not have a nonzero probability of taking on an experimental value of precisely 1.0 or 2.0, it makes no difference for this example whether we write $\text{Prob}(1 < x < 2)$ or $\text{Prob}(1 \leq x < 2)$ or $\text{Prob}(1 < x \leq 2)$ or $\text{Prob}(1 \leq x \leq 2)$.

Next, we ask for the conditional probability that the component will fail during its second month of use, given that it did not fail during the first month. We'll do this two ways. One approach is to define the events,

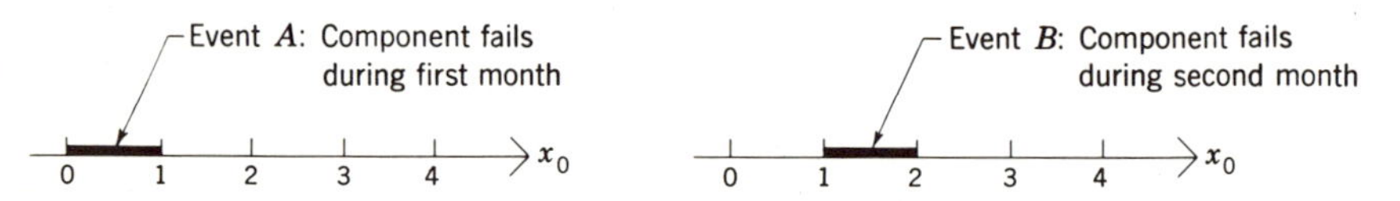

and then use the definition of conditional probability to determine the desired quantity, $P(B \mid A')$. Since it happens here that event B is included in event A', there follows

$$P(B \mid A') = \frac{P(A'B)}{P(A')} = \frac{P(B)}{P(A')}$$

$$P(B) = \tfrac{5}{16} \text{ (previous result)} \qquad\qquad P(A') = \int_1^4 f_x(x_0)\,dx_0 = \tfrac{9}{16}$$

$$P(B \mid A') = \tfrac{5}{9}$$

As we would expect from the nature of the physical situation in this problem, our result has the property $P(B|A') > P(B)$.

One other approach to this question would be to condition the event space for random variable x by noting that experimental values of x between 0.0 and 1.0 are now impossible. The relative probabilities of all events wholly contained within the conditioning event A' are to remain the same in our conditional space as they were in the original event space. To do this, the a priori PDF for the remaining event points must be multiplied by a constant so that the resulting conditional PDF for x will integrate to unity. This leads to the following conditional PDF $f_{x|A'}(x_0 \mid A')$:

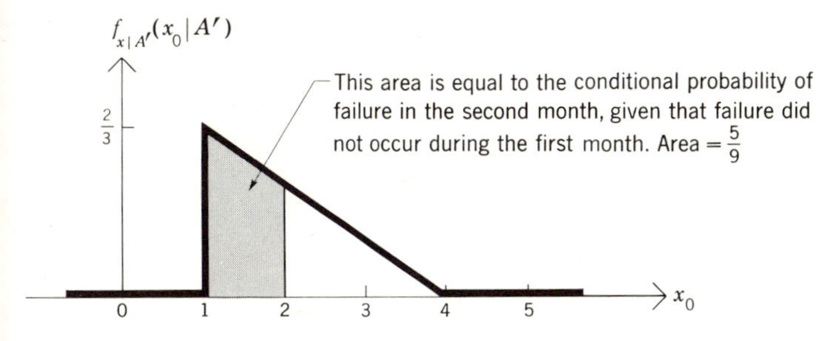

In this latter solution, we simply worked directly with the notion which was formalized in Sec. 1-4 to obtain the definition of conditional probability.

Before closing this section, let us observe that, once we are familiar with the unit-impulse function, a PMF may be considered to represent a special type of PDF. For instance, the following probability mass function and probability density function give identical descriptions of some random variable x:

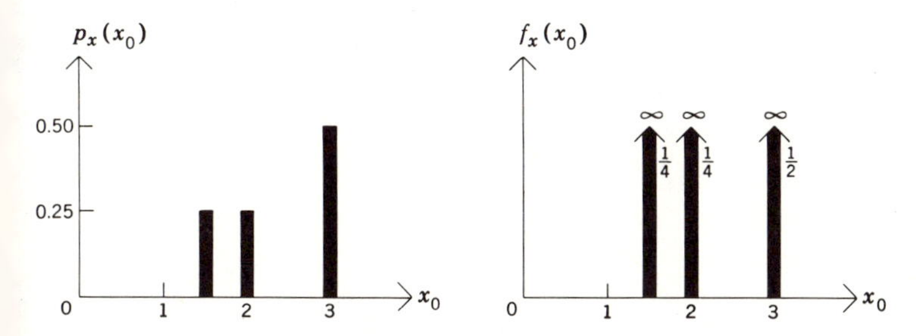

Only when there are particular advantages in doing so shall we represent the probability assignment to a purely discrete random variable by a PDF instead of a PMF.

2-11 Compound Probability Density Functions

We may now consider the case where several continuous random variables are defined on the sample space of an experiment. The assignment of probability measure is then specified by a compound probability density function in an event space whose coordinates are the experimental values of the random variables.

In the two-dimensional event space for the possible experimental values of random variables x and y, the compound PDF $f_{x,y}(x_0,y_0)$ may be pictured as a surface plotted above the x_0,y_0 plane. The volume enclosed between any area in the x_0,y_0 plane and this $f_{x,y}(x_0,y_0)$ surface is equal to the probability of the experimental outcome falling within that area. For any event A defined in the x_0,y_0 plane we have

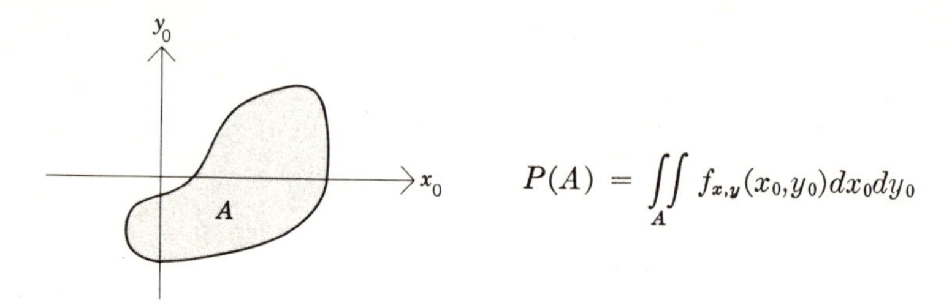

The probability that the experimental value of x will fall between x_0 and $x_0 + dx_0$, which we know must be equal to $f_x(x_0)\,dx_0$, is obtained by integrating the compound PDF over the strip in the event space which represents this event,

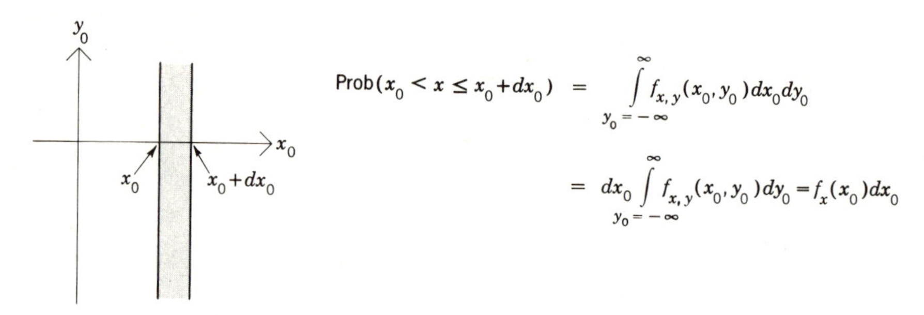

For the continuous case, we obtain the marginal PDF's by integrating over other random variables, just as we performed the same operation by summing over the other random variables in the discrete case.

$$f_x(x_0) = \int_{y_0=-\infty}^{\infty} f_{x,y}(x_0,y_0)\,dy_0 \qquad\qquad f_y(y_0) = \int_{x_0=-\infty}^{\infty} f_{x,y}(x_0,y_0)\,dx_0$$

And, in accordance with the properties of probability measure,

$$0 \leq f_{x,y}(x_0,y_0) \leq \infty \qquad\qquad \int_{y_0=-\infty}^{\infty} \int_{x_0=-\infty}^{\infty} f_{x,y}(x_0,y_0)\,dx_0\,dy_0 = 1$$

For convenience, we often use relations such as

$$\text{Prob}(x_0 < x \leq x_0 + dx_0) = f_x(x_0)\,dx_0$$

$$\text{Prob}(x_0 < x \leq x_0 + dx_0,\, y_0 < y \leq y_0 + dy_0) = f_{x,y}(x_0,y_0)\,dx_0\,dy_0$$

It should be remembered that such relations are not valid at points where a PDF contains impulses. *We shall not add this qualification each time we employ such incremental statements.* In any physical problem, as long as we are aware of the presence of impulses in the PDF (nonzero probability *mass* at a point), this situation will cause us no particular difficulty.

We close this section with a simple example. Suppose that the compound PDF for random variables x and y is specified to be

$$f_{x,y}(x_0,y_0) = \begin{cases} Ax_0 & \text{if } 0 \leq y_0 \leq x_0 \leq 1 \\ 0 & \text{otherwise} \end{cases}$$

and we wish to determine $A, f_x(x_0)$, and the probability that the product of the experimental values of x and y obtained on a performance of the experiment is less than or equal to 0.25. Three relevant sketches in the x_0, y_0 event space are

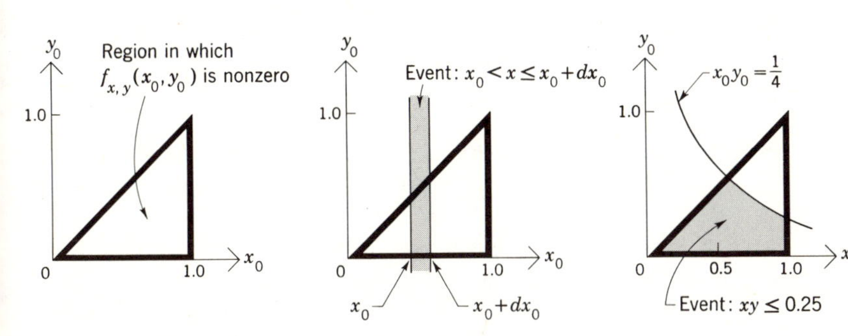

The value of A will be obtained from

$$\int_{x_0=-\infty}^{\infty} dx_0 \int_{y_0=-\infty}^{\infty} dy_0\, f_{x,y}(x_0,y_0) = 1$$

where our notation will be that the rightmost integration will always be performed first. Thus we have

$$\int_0^1 dx_0 \int_{y_0=0}^{x_0} dy_0\, Ax_0 = \int_0^1 Ax_0^2\, dx_0 = \frac{A}{3} = 1 \qquad A = 3$$

To determine the marginal PDF $f_x(x_0)$, we use

$$f_x(x_0) = \int_{y_0=-\infty}^{\infty} f_{x,y}(x_0,y_0)\, dy_0 = \begin{cases} \int_{y_0=0}^{x_0} 3x_0\, dy_0 & \text{if } 0 \leq x_0 \leq 1 \\ 0 & \text{otherwise} \end{cases}$$

Note that we must be careful to substitute the correct expression for $f_{x,y}(x_0,y_0)$ everywhere in the x_0,y_0 event space. The result simplifies to

$$f_x(x_0) = \begin{cases} 3x_0^2 & \text{if } 0 \leq x_0 \leq 1 \\ 0 & \text{otherwise} \end{cases}$$

To determine the probability of the event $xy \leq 0.25$, there are several ways to proceed. We may integrate $f_{x,y}(x_0,y_0)$ over the area representing this event in the x_0,y_0 event space. We may integrate the joint PDF over the complement of this event and subtract our result from unity. In each of these approaches we may integrate over x_0 first or over y_0 first. Note that each of these four possibilities is equivalent *but* that one of them involves far less work than the other three.

We now display this result and complete the problem by considering a more detailed sketch of the x_0,y_0 event space.

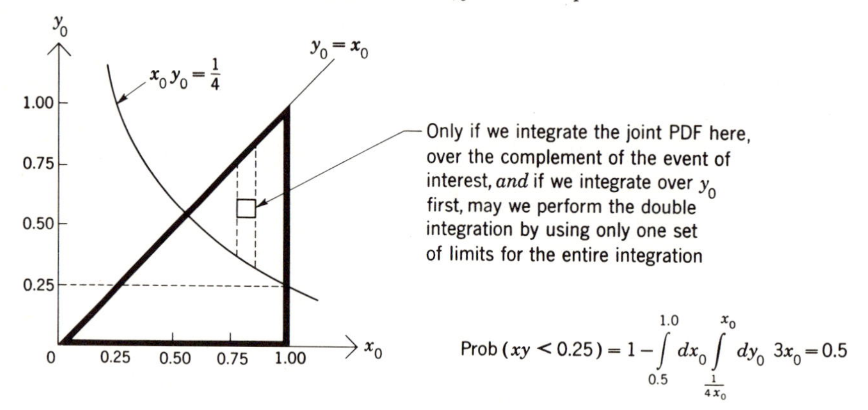

This last discussion was a matter of elementary calculus, not probability theory. However, it is important to realize how a little forethought in planning these multiple integrations can reduce the amount of calculation and improve the probability of getting a correct result.

2-12 Conditional Probability Density Functions

Consider continuous random variables x and y, defined on the sample space of an experiment. If we are given, for a particular performance of the experiment, that the experimental value of y is between y_0 and $y_0 + dy_0$, we know that the event point representing the experimental outcome must lie within the shaded strip,

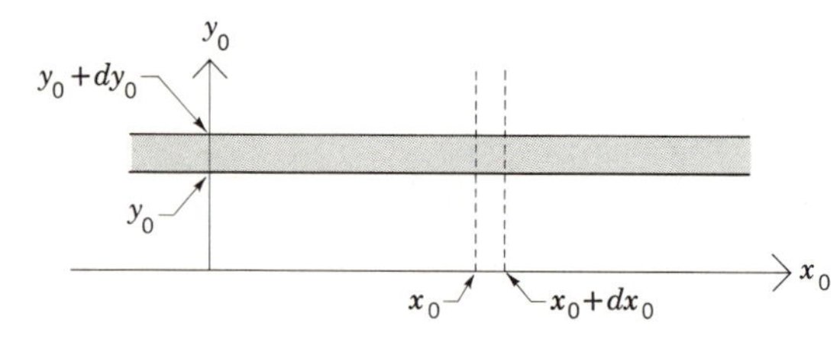

We wish to evaluate the quantity $f_{x|y}(x_0 \mid y_0)\, dx_0$, defined to be the conditional probability that the experimental value of x is between x_0 and $x_0 + dx_0$, given that the experimental value of y is between y_0 and $y_0 + dy_0$. Our procedure will be to define the incremental events of interest and substitute their probabilities into the definition of conditional probability introduced in Sec. 1-4.

Event A: $x_0 < x \leq x_0 + dx_0$ Event B: $y_0 < y \leq y_0 + dy_0$

$$P(A \mid B) = \frac{P(AB)}{P(B)} = \frac{f_{x,y}(x_0,y_0)\, dx_0\, dy_0}{f_y(y_0)\, dy_0} = \frac{f_{x,y}(x_0,y_0)}{f_y(y_0)}\, dx_0$$

Since the quantity $f_{x|y}(x_0 \mid y_0)\,dx_0$ has been defined to equal $P(A \mid B)$, we have

$$f_{x|y}(x_0 \mid y_0) = \frac{f_{x,y}(x_0,y_0)}{f_y(y_0)} \qquad \text{and, similarly,} \qquad f_{y|x}(y_0 \mid x_0) = \frac{f_{x,y}(x_0,y_0)}{f_x(x_0)}$$

The conditional PDF's are not defined when their denominators are equal to zero.

Note that the conditional PDF $f_{x|y}(x_0 \mid y_0)$, as a function of x_0 for a given y_0, is a curve of the shape obtained at the intersection of the surface $f_{x,y}(x_0,y_0)$ and a plane in three-dimensional space representing a constant value of coordinate y_0.

Using the event-space interpretation of conditional probability, we readily recognize how to condition a compound PDF given any event A of nonzero probability,

$$f_{x,y|A}(x_0,y_0 \mid A) = \begin{cases} \dfrac{f_{x,y}(x_0,y_0)}{P(A)} & \text{if } x_0,y_0 \text{ in } A \\ 0 & \text{if } x_0,y_0 \text{ in } A' \end{cases}$$

As an example of some of these concepts, let's continue with the example of the previous section, where the PDF for continuous random variables x and y is specified by

$$f_{x,y}(x_0,y_0) = \begin{cases} 3x_0 & \text{if } 0 \le y_0 \le x_0 \le 1 \\ 0 & \text{otherwise} \end{cases}$$

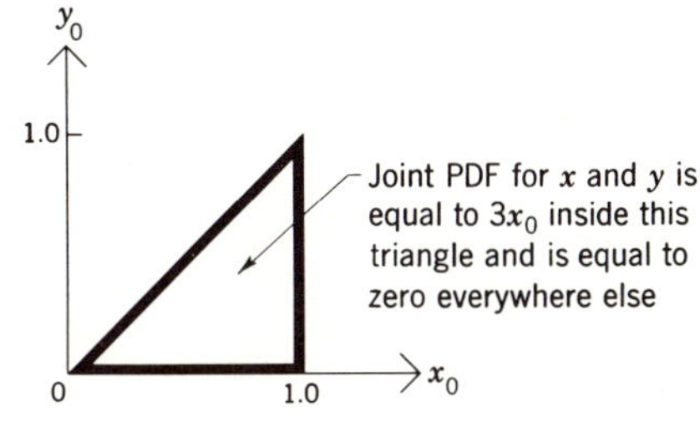

We'll calculate $f_{y|x}(y_0 \mid x_0)$, taking advantage of the fact that we have already determined $f_x(x_0)$.

$$f_{y|x}(y_0 \mid x_0) = \frac{f_{x,y}(x_0,y_0)}{f_x(x_0)} = \begin{cases} \dfrac{3x_0}{3x_0{}^2} = \dfrac{1}{x_0} & \text{if } 0 \le y_0 \le x_0 \le 1 \\ 0 & \text{otherwise} \end{cases}$$

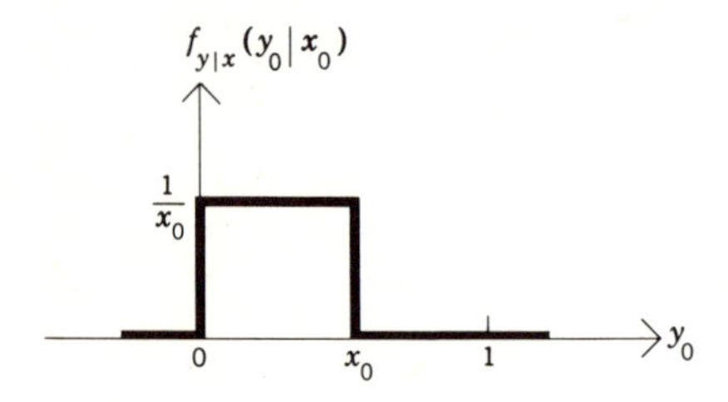

Since the a priori PDF $f_{x,y}(x_0,y_0)$ is not a function of y_0, it is reasonable that the conditional PDF $f_{y|x}(y_0 \mid x_0)$ should be uniform over the possible experimental values of random variable y. For this example, the reader might wish to establish that

$$f_{x|y}(x_0 \mid y_0) = \begin{cases} \dfrac{2x_0}{1 - y_0^2} & \text{if } 0 \leq y_0 \leq x_0 \leq 1 \\ 0 & \text{otherwise} \end{cases}$$

2-13 Independence and Expectation for Continuous Random Variables

Two continuous random variables x and y are defined to be independent (or *statistically* independent) if and only if

$$f_{x|y}(x_0 \mid y_0) = f_x(x_0) \qquad \text{for all possible } x_0,y_0$$

and since $f_{x,y}(x_0,y_0) = f_x(x_0)f_{y|x}(y_0 \mid x_0) = f_y(y_0)f_{x|y}(x_0 \mid y_0)$ is always true by the definition of the conditional PDF's, an equivalent condition for the independence of x and y is

$$f_{x,y}(x_0,y_0) = f_x(x_0)f_y(y_0) \qquad \text{for all } x_0,y_0$$

We say that any number of random variables are mutually independent if their compound PDF factors into the product of their marginal PDF's for all possible experimental values of the random variables.

The conditional expectation of $g(x,y)$, a single-valued function of continuous random variables x and y, given that event A has occurred, is defined to be

$$E[g(x,y) \mid A] = \int_{y_0=-\infty}^{\infty} \int_{x_0=-\infty}^{\infty} g(x_0,y_0)f_{x,y|A}(x_0,y_0 \mid A)\, dx_0\, dy_0$$

All the definitions and results obtained in Sec. 2-7 carry over directly to the continuous case, with summations replaced by integrations.

2-14 Derived Probability Density Functions

We have learned that $g(x,y)$, a function of random variables x and y, is itself a new random variable. From the definition of expectation, we also know that the expected value of $g(x,y)$, or any function of $g(x,y)$, may be found directly in the x_0,y_0 event space without ever determining the PDF $f_g(g_0)$.

However, if we have an interest only in the behavior of random variable g and we wish to answer several questions about it, we may desire to work in a g_0 event space with the PDF $f_g(g_0)$. A PDF obtained for a function of some random variables whose PDF is known is referred to as a *derived* PDF.

We shall introduce one simple method for obtaining a derived distribution by working in the event space of the random variables whose PDF is known. There may be more efficient techniques for particular classes of problems. Our method, however, will get us there and because we'll live in event space, we'll always know exactly what we are doing.

To derive the PDF for g, a function of some random variables, we need to perform only two simple steps in the event space of the original random variables:

FIRST STEP: Determine the probability of the event $g \leq g_0$ for all values of g_0.

SECOND STEP: Differentiate this quantity with respect to g_0 to obtain $f_g(g_0)$.

The first step requires that we calculate the cumulative probability distribution function $p_{g\leq}(g_0)$. To do so, we simply integrate the given PDF for the original random variables over the appropriate region of their event space.

Consider the following example: A fair wheel of fortune, continuously calibrated from 0.00 to 1.00, is to be spun twice. The experimental values of random variables x and y are defined to be the readings on the first and second spins, respectively. [By "fair" we mean, of course, that the wheel has no memory (different spins are independent events) and that any intervals of equal arc are equally likely to include the experimental outcome.] We wish to determine the PDF $f_g(g_0)$ for the case where random variable g is defined by $g(x,y) = \dfrac{x}{y}$.

The example states that the spins are independent. Therefore we may obtain the joint PDF

$$f_{x,y}(x_0,y_0) = f_x(x_0)f_y(y_0) = \begin{cases} 1.0 & \text{if } 0 < x_0 \leq 1, 0 < y_0 \leq 1 \\ 0.0 & \text{otherwise} \end{cases}$$

Next, in the x_0,y_0 event space, we collect the event $g \leq g_0$,

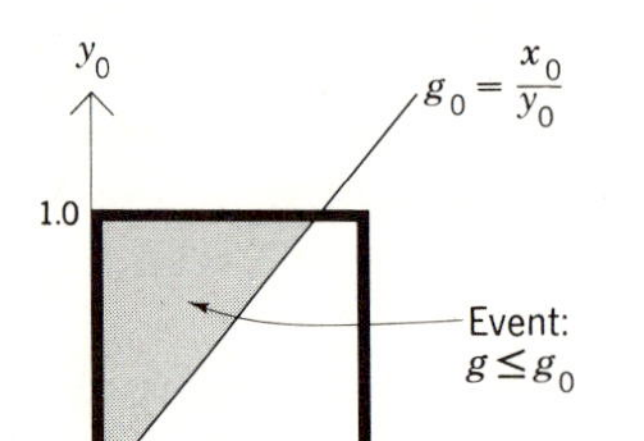

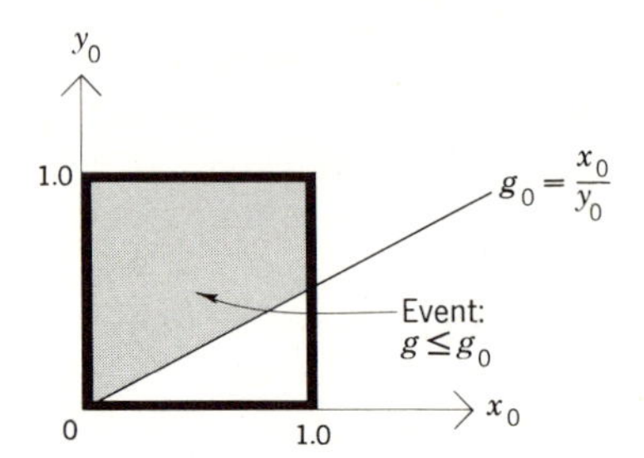

Two sketches are given to show that the boundaries of the event of interest are different for the cases $g_0 < 1$ and $g_0 > 1$. For our particular example, because of the very special fact that $f_{x,y}(x_0,y_0)$ is everywhere equal to either zero or unity, we can replace the integration

$$p_{g\le}(g_0) = \iint f_{x,y}(x_0,y_0)\,dx_0\,dy_0$$

by a simple calculation of areas to obtain, for the first step of our two-step procedure,

$$p_{g\le}(g_0) = \begin{cases} 0 & g_0 \le 0 \\ \dfrac{g_0}{2} & 0 \le g_0 \le 1 \\ 1 - \dfrac{1}{2g_0} & 1 \le g_0 \le \infty \end{cases}$$

At this point, of course, we may check that this CDF is a monotonically increasing function of x_0 and that it increases from zero at $g_0 = -\infty$ to unity at $g_0 = +\infty$. Now, for step 2, we differentiate $p_{g\le}(g_0)$ to find

$$f_g(g_0) = \begin{cases} 0 & g_0 \le 0 \\ 0.5 & 0 \le g_0 \le 1 \\ 0.5g_0^{-2} & 1 \le g_0 \le \infty \end{cases}$$

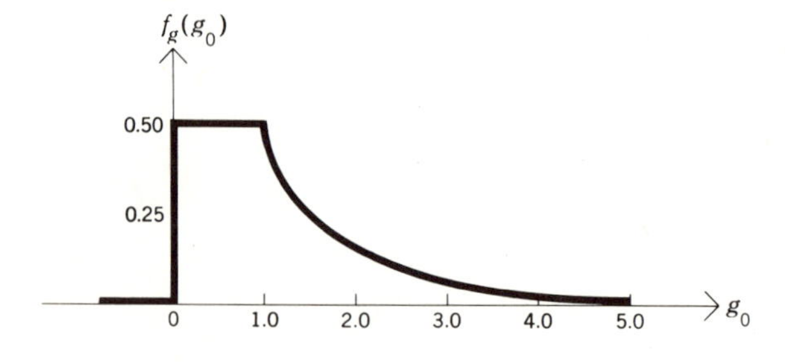

If we wish to derive a joint PDF, say, for $g(x,y)$ and $h(x,y)$, then in step 1 we collect the probability

$p_{g\leq,h\leq}(g_0,h_0)$ for all values of g_0 and h_0

which represents the joint CDF for random variables g and h. For step 2 we would perform the differentiation

$$\frac{\partial^2 p_{g\leq,h\leq}(g_0,h_0)}{\partial g_0\,\partial h_0} = f_{g,h}(g_0,h_0)$$

As the number of derived random variables increases, our method becomes unreasonably cumbersome; but more efficient techniques exist for particular types of problems.

One further detail is relevant to the mechanics of the work involved in carrying out our two-step method for derived distributions. Our method involves an integration (step 1) followed by a differentiation (step 2). Although the integrations and differentiations are generally with respect to different variables, we may wish to differentiate first before formally performing the integration. For this purpose, it is useful to remember one very useful formula and the picture from which it is obtained.

In working with a relation of the form

$$R(\alpha) = \int_{a(\alpha)}^{b(\alpha)} r(\alpha,x)\,dx$$

we have an integral over x whose lower limit, upper limit, and integrand are all functions of α. If we desire to obtain the derivative $\frac{dR(\alpha)}{d\alpha}$ it is more efficient to use the following formula directly than to first perform the integration with respect to x and then to differentiate with respect to α.

$$\frac{dR(\alpha)}{d\alpha} = -r[\alpha,a(\alpha)]\,\frac{da(\alpha)}{d\alpha} + r[\alpha,b(\alpha)]\,\frac{db(\alpha)}{d\alpha} + \int_{a(\alpha)}^{b(\alpha)} \frac{\partial r(\alpha,x)}{\partial\alpha}\,dx$$

This relation is easy to remember if we keep in mind the picture from which it may be obtained

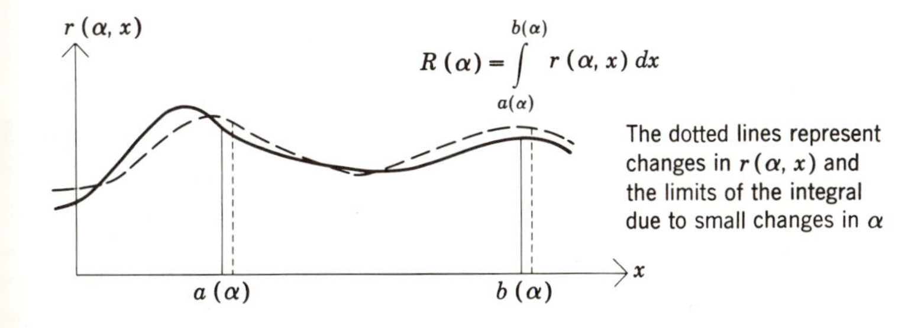

The reader will have many opportunities to benefit from the above expression for $\frac{dR(\alpha)}{d\alpha}$ in obtaining derived distributions. For instance, in Prob. 2.30 at the end of this chapter, this expression is used in order to obtain the desired PDF in a useful form.

2-15 Examples Involving Continuous Random Variables

However simple the concepts may seem, no reader should assume that he has a working knowledge of the material in this chapter until he has successfully mastered many problems like those at the end of the chapter. Two examples with solutions follow, but we must realize that these examples are necessarily a narrow representation of a very large class of problems. Our first example gives a straightforward drill on notation and procedures. The second example is more physically motivated.

example 1 Random variables x, y, and z are described by the compound probability density function,

$$f_{x,y,z}(x_0,y_0,z_0) = \begin{cases} x_0z_0 + 3y_0z_0 & \text{if } 0 \le x_0 \le 1, 0 \le y_0 \le 1, 0 \le z_0 \le 1 \\ 0 & \text{otherwise} \end{cases}$$

Determine the quantities:

(a) $p_{x\le}(3)$ (b) $f_{x,y}(x_0,y_0)$
(c) $p_{x\le,y\le,z\le}(1,2,z_0)$ (d) $f_x(x_0)$
(e) $E(xy)$ (f) $E(y \mid x)$

a From the statement of the compound PDF, note that the experimental value of random variable x can never be greater than unity. Since we are asked to determine the probability that this experimental value is less than or equal to 3.0, we can immediately answer

$$p_{x\le}(3) = 1.0$$

b When determining expressions for PDF's and CDF's we must always remember that the proper expressions must be listed for all values of the arguments.

$$f_{x,y}(x_0,y_0) = \begin{cases} \int_{z_0=0}^{1} dz_0(x_0z_0 + 3y_0z_0) & \text{if } 0 \le x_0 \le 1,\ 0 \le y_0 \le 1 \\ 0 & \text{otherwise} \end{cases}$$

which simplifies to

$$f_{x,y}(x_0,y_0) = \begin{cases} \frac{1}{2}(x_0 + 3y_0) & \text{if } 0 \le x_0 \le 1,\ 0 \le y_0 \le 1 \\ 0 & \text{otherwise} \end{cases}$$

c Because of the given ranges of possible experimental values of x and y, we note that

$$p_{x\leq,y\leq,z\leq}(1,2,z_0) = p_{z\leq}(z_0)$$

$$p_{z\leq}(z_0) = \begin{cases} 0 & z_0 \leq 0 \\ \int_{z_0=0}^{z_0} dz_0 \int_{x_0=0}^{1} dx_0 \int_{y_0=0}^{1} dy_0(x_0z_0 + 3y_0z_0) & 0 \leq z_0 \leq 1 \\ 1 & 1 \leq z_0 \end{cases}$$

which simplifies to

$$p_{z\leq}(z_0) = \begin{cases} 0 & z_0 \leq 0 \\ {z_0}^2 & 0 \leq z_0 \leq 1 \\ 1 & 1 \leq z_0 \end{cases}$$

which has all the essential properties of a CDF.

d Since we have already found $f_{x,y}(x_0,y_0)$, we can determine the marginal PDF $f_x(x_0)$ by integrating over y_0. For $0 \leq x_0 \leq 1$, we have

$$f_x(x_0) = \int_{y_0=0}^{1} dy_0 \tfrac{1}{2}(x_0 + 3y_0) = (\tfrac{1}{2}x_0 + \tfrac{3}{4})$$

and we know that $f_x(x_0)$ is zero elsewhere.

$$f_x(x_0) = \begin{cases} \tfrac{1}{2}x_0 + \tfrac{3}{4} & \text{if } 0 \leq x_0 \leq 1 \\ 0 & \text{otherwise} \end{cases}$$

Whenever possible, we perform simple checks on our answers to see whether or not they make any sense. For instance, here we'd check to see that $\int_{x_0=-\infty}^{\infty} f_x(x_0)\, dx_0$ is unity. Happily, it is.

e $E(xy) = \int_{x_0=0}^{1} dx_0 \int_{y_0=0}^{1} dy_0\, x_0y_0f_{x,y}(x_0,y_0) = \tfrac{1}{3}$

This result is at least compatible with reason, since xy is always between zero and unity.

f $E(y \mid x) = \int_{y_0=0}^{1} y_0f_{y|x}(y_0 \mid x_0)\, dy_0$

$$= \int_{y_0=0}^{1} y_0 \frac{f_{x,y}(x_0,y_0)}{f_x(x_0)}\, dy_0 \qquad \text{for all possible } x_0$$

$$= \int_{y_0=0}^{1} y_0 \frac{\tfrac{1}{2}(x_0 + 3y_0)}{\tfrac{1}{2}x_0 + \tfrac{3}{4}}\, dy_0 = \frac{\tfrac{1}{2}}{\tfrac{1}{2}x_0 + \tfrac{3}{4}} \int_{y_0=0}^{1} (x_0y_0 + 3{y_0}^2)\, dy_0$$

$$= \frac{x_0 + 2}{2x_0 + 3}$$

For any possible value of x_0, our $E(y \mid x)$ does result in a conditional expectation for y which is always between the smallest and largest possible experimental values of random variable y.

example 2 Each day as he leaves home for the local casino, Oscar spins a biased wheel of fortune to determine how much money to take with him. He takes exactly x hundred dollars with him, where x is a continuous random variable described by the probability density function

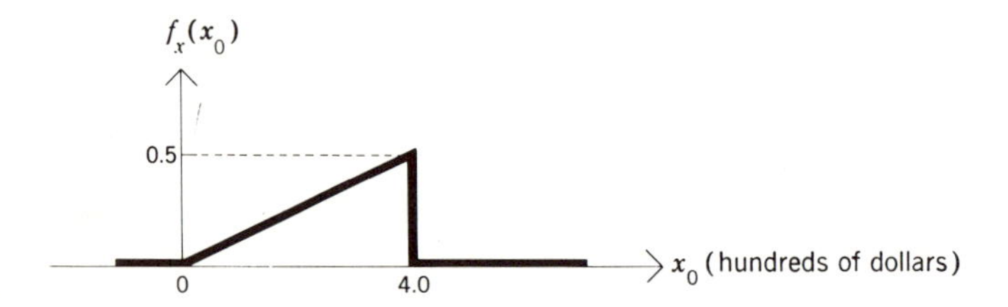

As a matter of convenience, we are assuming that the currency is infinitely divisible. (Rounding off to the nearest penny wouldn't matter much.)

Oscar has a lot of experience at this. All of it is bad. Decades of experience have shown that, over the course of an evening, Oscar never wins. In fact, the amount with which he returns home on any particular night is uniformly distributed between zero and the amount with which he started out.

Let random variable y represent the amount (in hundreds of dollars) Oscar brings home on any particular night.

(a) Determine $f_{x,y}(x_0,y_0)$, the joint PDF for his original wealth x and his terminal wealth y on any evening.

(b) Determine $f_y(y_0)$, the marginal probability density function for the amount Oscar will bring home on a randomly selected night.

(c) Determine the expected value of Oscar's loss on any particular night.

(d) On one particular night, we learn that Oscar returned home with less than \$200. For that night, determine the conditional probability of each of the following events:

(i) He started out for the casino with less than \$200.

(ii) His loss was less than \$100.

(iii) His loss that night was exactly \$75.

a From the example statement we obtain

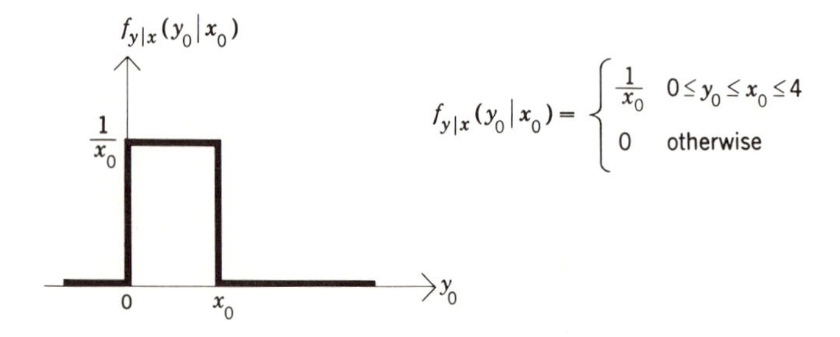

$$f_{y|x}(y_0 \mid x_0) = \begin{cases} \frac{1}{x_0} & 0 \le y_0 \le x_0 \le 4 \\ 0 & \text{otherwise} \end{cases}$$

The definition of conditional probability is used with the given $f_x(x_0)$ to determine

$$f_{x,y}(x_0,y_0) = f_x(x_0)f_{y|x}(y_0 \mid x_0) = \begin{cases} \frac{x_0}{8} \cdot \frac{1}{x_0} = \frac{1}{8} & \text{if } 0 \le y_0 \le x_0 \le 4 \\ 0 & \text{otherwise} \end{cases}$$

and this result may be displayed in an x_0,y_0 event space,

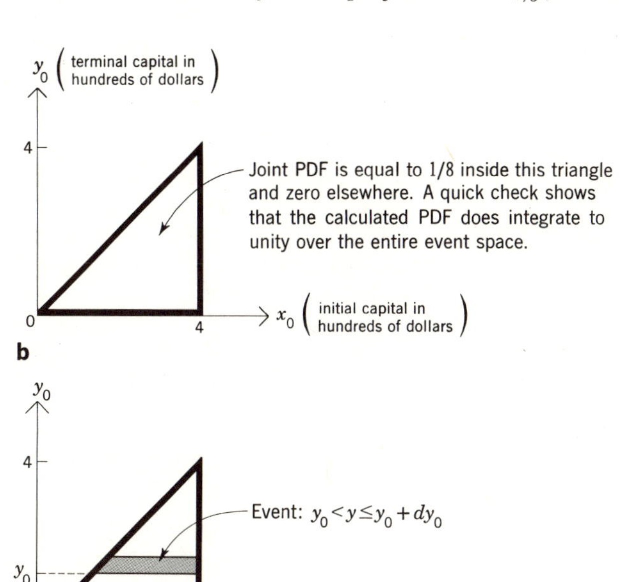

For $0 \le y_0 \le 4$,

$$f_y(y_0) = \int_{-\infty}^{\infty} dx_0\, f_{x,y}(x_0,y_0) = \int_{x_0=y_0}^{4} dx_0\, \tfrac{1}{8} = \tfrac{1}{8}(4 - y_0)$$

And we can sketch this PDF as

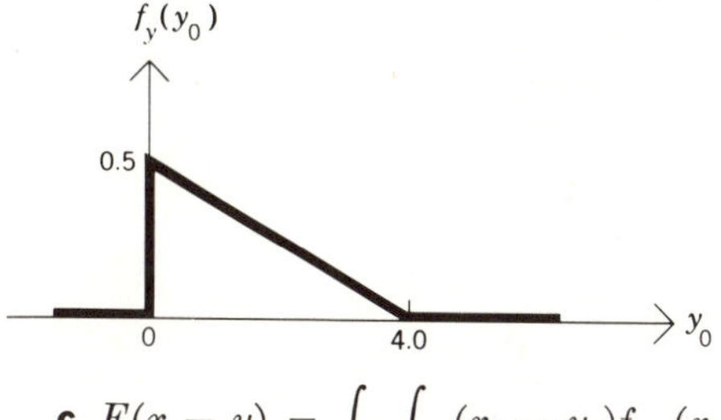

The $f_y(y_0)$ PDF does integrate to unity and, as was obvious from the above sketch (since the joint PDF was constant inside the triangle), it is linearly decreasing from a maximum at $y_0=0$ to zero at $y_0=4$

c $E(x - y) = \int_{y_0}\int_{x_0} (x_0 - y_0)f_{x,y}(x_0,y_0)\, dx_0\, dy_0$

We must always be careful of the limits on the integrals when we substitute actual expressions for the compound PDF. We'll integrate over x_0 first.

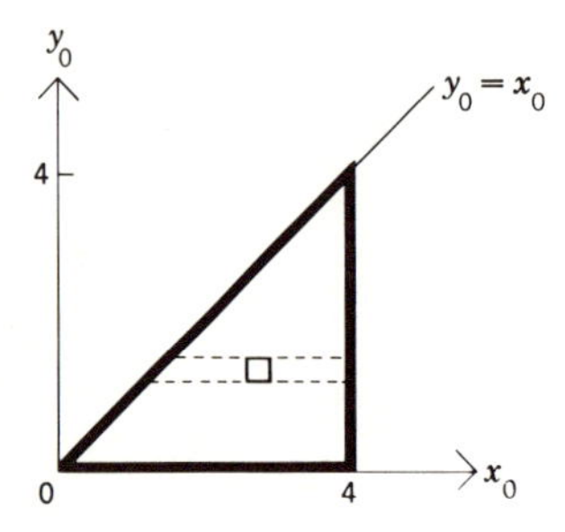

$$E(x - y) = \int_{y_0=0}^{4} dy_0 \int_{x_0=y_0}^{4} dx_0\, \tfrac{1}{8}(x_0 - y_0) = \$133.33$$

where again we are using the convention that the successive integrals are to be performed in order from right to left. By changing a sign in the proof in Sec. 2-7, we may prove the relation

$$E(x - y) = E(x) - E(y)$$

Since we already have the marginal PDF's, this relation allows us to obtain $E(x - y)$ by another route,

$$E(x) - E(y) = \int_{x_0=-\infty}^{\infty} x_0 f_x(x_0)\, dx_0 - \int_{y_0=-\infty}^{\infty} y_0 f_y(y_0)\, dy_0$$

$$= \int_0^4 x_0 \cdot \frac{x_0}{8}\, dx_0 - \int_0^4 y_0 \frac{4 - y_0}{8}\, dy_0 = \$133.33$$

d Given that Oscar returned home with less than \$200, we work in the appropriate conditional sample space. At all points consistent with the conditioning event, the conditional PDF is equal to the original joint PDF scaled up by the reciprocal of the a priori probability of the conditioning event.

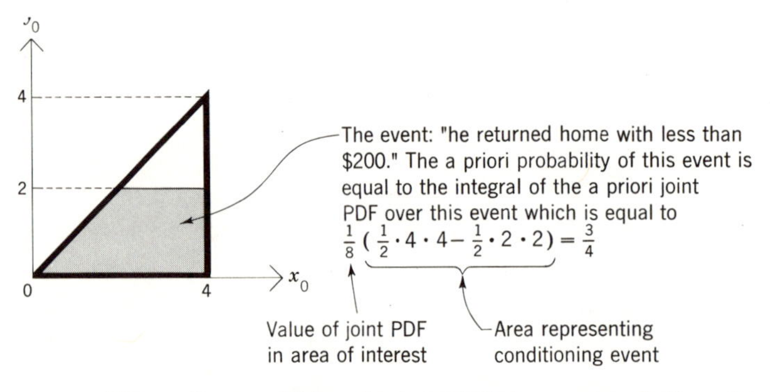

Thus, the conditional joint PDF is equal to $\frac{1}{8}/\frac{3}{4} = \frac{1}{6}$ in the region where it is nonzero. Now we may answer all questions in this conditional space.

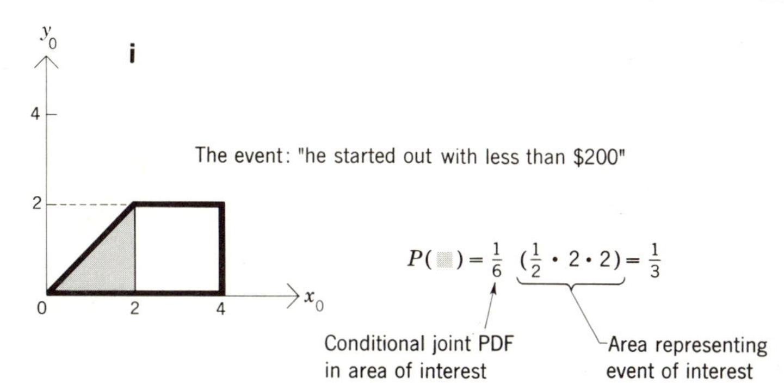

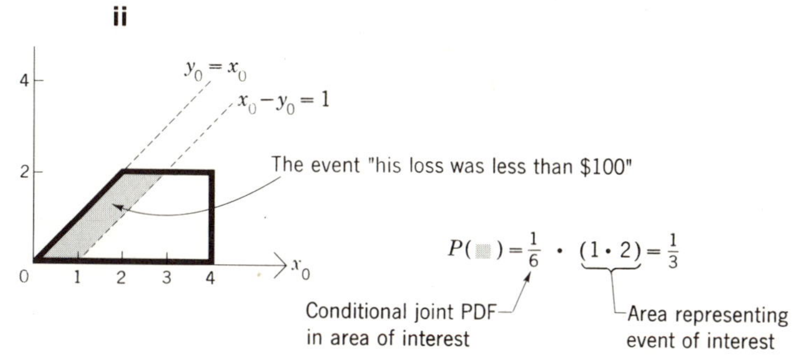

We realize that, in general, we would have to integrate the conditional PDF over the appropriate events to obtain their probabilities. Only because the conditional PDF is a constant have we been able to reduce the integrations to simple multiplications.

iii As long as we allow the currency to be infinitely divisible, the conditional probability measure associated with the event $x - y = 75$ is equal to zero. The integral of the compound PDF $f_{x,y}(x_0,y_0)$ over the *line* representing this event in the x_0,y_0 event space is equal to zero.

PROBLEMS

2.01 The *geometric* PMF for discrete random variable K is defined to be

$$p_K(K_0) = \begin{cases} C(1-P)^{K_0-1} & \text{if } K_0 = 1, 2, 3, \ldots \text{ and } 0 < P < 1 \\ 0 & \text{for all other values of } K_0 \end{cases}$$

a Determine the value of C.

b Let N be a positive integer. Determine the probability that an experimental value of K will be greater than N.

c Given that an experimental value of random variable K is greater than integer N, what is the conditional probability that it is also larger than $2N$? (We shall discuss this special result in Chap. 4.)

d What is the probability that an experimental value of K is equal to an integer multiple of 3?

2.02 The probability that any particular bulb will burn out during its Kth month of use is given by the PMF for K,

$$p_K(K_0) = \tfrac{1}{5}(\tfrac{4}{5})^{K_0-1} \qquad K_0 = 1, 2, 3, \ldots$$

Four bulbs are life-tested simultaneously. Determine the probability that

a None of the four bulbs fails during its first month of use.
b Exactly two bulbs have failed by the end of the third month.
c Exactly one bulb fails during each of the first three months.
d Exactly one bulb has failed by the end of the second month, and exactly two bulbs are still working at the start of the fifth month.

2.03 The *Poisson* PMF for random variable K is defined to be

$$p_K(K_0) = \begin{cases} \dfrac{\mu^{K_0}e^{-\mu}}{K_0!} & \text{if } K_0 = 0, 1, 2, \ldots \text{ (and } \mu > 0) \\ 0 & \text{for all other values of } K_0 \end{cases}$$

a Show that this PMF sums to unity.
b Discrete random variables R and S are defined on the sample spaces of two different, unrelated experiments, and these random variables have the PMF's

$$p_R(R_0) = \frac{\mu^{R_0}e^{-\mu}}{R_0!} \qquad R_0 = 0, 1, 2, \ldots$$

$$p_S(S_0) = \frac{\lambda^{S_0}e^{-\lambda}}{S_0!} \qquad S_0 = 0, 1, 2, \ldots$$

Use an R_0,S_0 sample space to determine the PMF $p_T(T_0)$, where discrete random variable T is defined by $T = R + S$.
c Random variable W is defined by $W = cR$, where c is a known nonzero constant. Determine the PMF $p_W(W_0)$ and the expected value of W. How will the nth central moment of W change if c is doubled?

2.04 Discrete random variable x is described by the PMF

$$p_x(x_0) = \begin{cases} K - \dfrac{x_0}{12} & \text{if } x_0 = 0, 1, 2 \\ 0 & \text{for all other values of } x_0 \end{cases}$$

Let $d_1, d_2, \ldots, d_N$ represent N successive independent experimental values of random variable x.

a Determine the numerical value of K.

b Determine the probability that $d_1 > d_2$.
c Determine the probability that $d_1 + d_2 + \cdots + d_N \leq 1.0$
d Define $r = \max(d_1,d_2)$ and $s = \min(d_1,d_2)$. Determine the following PMF's for all values of their arguments:
i $p_s(s_0)$ **ii** $p_{r|s}(r_0 \mid 0)$
iii $p_{r,s}(r_0,s_0)$ **iv** $p_t(t_0)$, with $t = (1 + d_1)/(1 + s)$
e Determine the expected value and variance of random variable s defined above.
f Given $d_1 + d_2 \leq 2.0$, determine the conditional expected value and conditional variance of random variable s defined above.

2.05 Discrete random variable x is described by the PMF $p_x(x_0)$. Before an experiment is performed, we are required to guess a value d. After an experimental value of x is obtained, we shall then be paid an amount $A - B(x - d)^2$ dollars.
a What value of d should we use to maximize the expected value of our financial gain?
b Determine the value of A such that the expected value of the gain is zero dollars.

2.06 Consider an experiment in which a fair four-sided die (with faces labeled 0, 1, 2, 3) is thrown once to determine how many times a fair coin is to be flipped. In the sample space of this experiment, random variables n and k are defined by
n = down-face value on the throw of the tetrahedral die
k = total number of heads resulting from the coin flips

Determine and sketch each of the following functions for all values of their arguments:
a $p_n(n_0)$ **b** $p_{k|n}(k_0 \mid 2)$ **c** $p_{n|k}(n_0 \mid 2)$ **d** $p_k(k_0)$
e Also determine the conditional PMF for random variable n, given that the experimental value of k is an odd number.

2.07 Joe and Helen each know that the a priori probability that her mother will be home on any given night is 0.6. However, Helen can determine her mother's plans for the night at 6 P.M., and then, at 6:15 P.M., she has only one chance each evening to shout one of two code words across the river to Joe. He will visit her with probability 1.0 if he thinks Helen's message means "Ma will be away," and he will stay home with probability 1.0 if he thinks the message means "Ma will be home."

But Helen has a meek voice, and the river is channeled for heavy barge traffic. Thus she is faced with the problem of coding for a noisy channel. She has decided to use a code containing only the code words A and B.

The channel is described by

$$P(a \mid A) = \tfrac{2}{3} \qquad P(a \mid B) = \tfrac{1}{4} \qquad P(b \mid A) = \tfrac{1}{3} \qquad P(b \mid B) = \tfrac{3}{4}$$

and these events are defined in the following sketch:

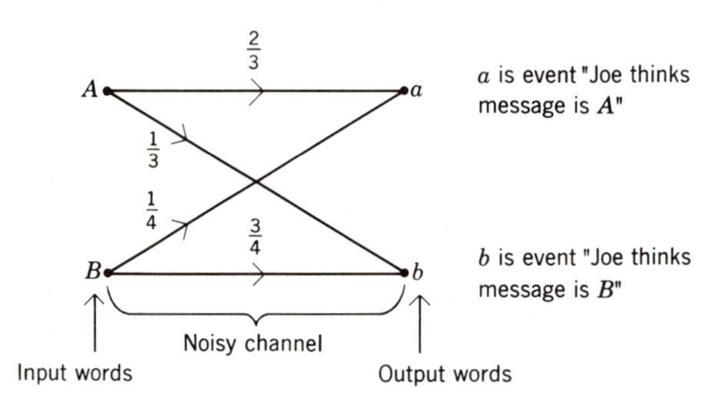

a In order to minimize the probability of error between transmitted and received messages, should Helen and Joe agree to use code I or code II?

Code I	Code II
A = Ma away	A = Ma home
B = Ma home	B = Ma away

b Helen and Joe put the following cash values (in dollars) on all possible outcomes of a day:

Ma home and Joe comes	−30
Ma home and Joe doesn't come	0
Ma away and Joe comes	+30
Ma away and Joe doesn't come	−5

Joe and Helen make all their plans with the objective of maximizing the expected value of each day of their continuing romance. Which of the above codes will maximize the expected cash value per day of this romance?

c Clara isn't quite so attractive as Helen, but at least she lives on the same side of the river. What would be the lower limit of Clara's expected value per day which would make Joe decide to give up Helen?

d What would be the maximum rate which Joe would pay the phone company for a noiseless wire to Helen's house which he could use once per day at 6:15 P.M.?

e How much is it worth to Joe and Helen to double her mother's probability of being away from home? Would this be a better or worse investment than spending the same amount of money for a telephone line (to be used once a day at 6:15 P.M.) with the following properties:

$$P(a \mid A) = P(b \mid B) = 0.9 \qquad P(b \mid A) = P(a \mid B) = 0.1$$

2.08 A frazzle is equally likely to contain zero, one, two, or three defects. No frazzle has more than three defects. The cash price of each frazzle is set at $\$(10 - K^2)$, where K is the number of defects in it. Gummed labels, each representing \$1, are placed on each frazzle to indicate its cash value (one label for a \$1 frazzle, two labels for a \$2 frazzle, etc.).

What is the probability that a randomly selected label (chosen from the pile of labels at the printing plant) will end up on a frazzle which has exactly two defects?

2.09 A pair of fair four-sided dice is thrown once. Each die has faces labeled 1, 2, 3, and 4. Discrete random variable x is defined to be the product of the down-face values. Determine the conditional variance of x^2 given that the sum of the down-face values is greater than the product of the down-face values.

2.10 Discrete random variables x and y are defined on the sample space of an experiment, and $g(x,y)$ is a single valued function of its argument. Use an event-space argument to establish that

$$E[g(x,y)] = \sum_{x_0} \sum_{y_0} g(x_0,y_0) p_{x,y}(x_0,y_0)$$

and

$$E(g) = \sum_{g_0} g_0 p_g(g_0)$$

are equivalent expressions for the expected value of $g(x,y)$.

2.11 At a particular point on a busy one-way single-lane road, a study is made of the distribution of the interarrival period T between successive car arrivals. A reasonable quantization of the data for a chain of 10,001 cars results in the following tabulation:

T, seconds	2	4	6	8	12
Number of occurrences	1,000	2,000	4,000	2,000	1,000

(Consider the cars to be as wide as the road, but very short.)

a A young wombat, who never turns back, requires five seconds to cross the street. Determine the probability that he survives if:

i He starts immediately after a car has passed.

ii He starts at a random time, selected without any dependence on the state of the traffic.

b Is the safer method of the above problem always safer, no matter what data are given?

2.12 The life span of a particular mechanical part is a random variable described by the following PDF:

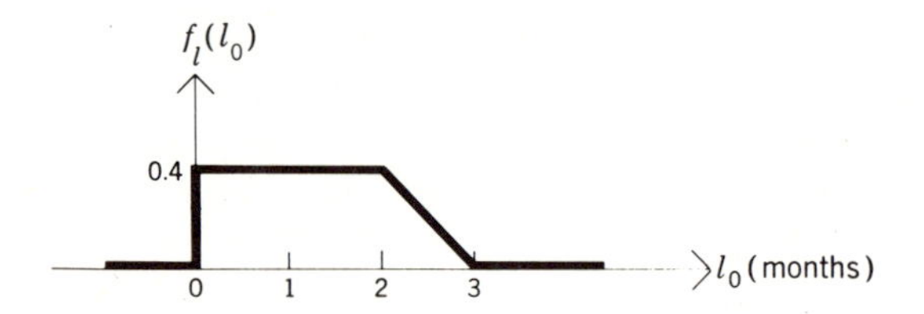

If three such parts are put into service independently at $t = 0$, determine

a The probability that the first failure will not have occurred before time t_0 $(0 \leq t_0 \leq \infty)$

b $E(l)$

c A simple expression for the expected value of the time until the majority of the parts will have failed.

2.13 Continuous random variables w, x, y, and z are described by the compound PDF $f_{w,x,y,z}(w_0,x_0,y_0,z_0)$. Determine a simple expression for the probability of the event $x = \max(w,x,y,z)$.

2.14 Random variables x and y described by the PDF

$$f_{x,y}(x_0,y_0) = \begin{cases} K & \text{if } x_0 + y_0 \leq 1 \text{ and } x_0 > 0 \text{ and } y_0 > 0 \\ 0 & \text{otherwise} \end{cases}$$

a Are x and y independent random variables?

b Are they conditionally independent given $\max(x,y) \leq 0.5$?

c Determine the expected value of random variable r, defined by $r = xy$.

d If we define events A and B by

Event A: $2(y - x) \geq y + x$ Event B: $y > \frac{3}{4}$

obtain the numerical values of $P(A)$, $P(B)$, $P(A'B)$, $P[(A'B')']$, and determine and plot the conditional probability density function $f_{x|A'B'}(x_0 \mid A'B')$.

2.15 One of two wheels of fortune, A and B, is selected by the flip of a fair coin, and the wheel chosen is spun once to determine an experimental value of random variable x. Random variable y, the reading obtained with wheel A, and random variable w, the reading obtained with wheel B, are described by the PDF's

$$f_y(y_0) = \begin{cases} 1 & \text{if } 0 < y_0 \leq 1 \\ 0 & \text{otherwise} \end{cases} \qquad f_w(w_0) = \begin{cases} 3 & \text{if } 0 < w_0 \leq \frac{1}{3} \\ 0 & \text{otherwise} \end{cases}$$

If we are told the experimental value of x was less than $\frac{1}{4}$, what is the conditional probability that wheel A was the one selected?

2.16 Four random variables are described by the probability density function

$$f_{w,x,y,z}(w_0,x_0,y_0,z_0) = \begin{cases} A\left(\dfrac{w_0x_0}{y_0z_0}\right)^2 \log \dfrac{w_0}{y_0} & \begin{aligned} &\text{if } 1 \leq w_0 \leq 2 \text{ and} \\ &\quad 1 \leq x_0 \leq 2 \text{ and} \\ &\quad 1 \leq y_0 \leq w_0 \leq 2 \text{ and} \\ &\quad 1 \leq z_0 \leq 2 \end{aligned} \\ 0 & \text{otherwise} \end{cases}$$

Determine and discuss the conditional probability density function $f_{x|w,y,z}(x_0 \mid w_0,y_0,z_0)$.

2.17 Random variables x and y are independent and are described by the probability density functions $f_x(x_0)$ and $f_y(y_0)$,

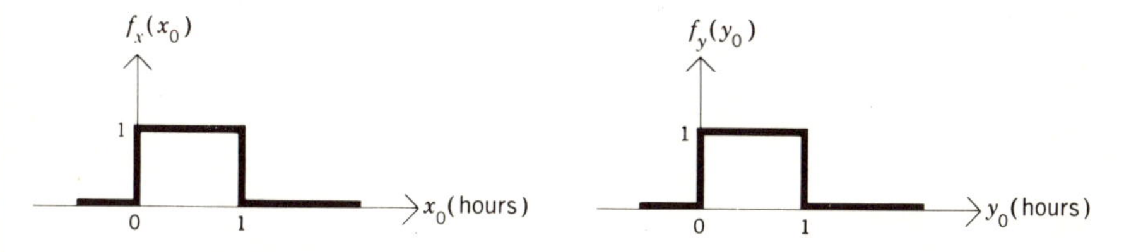

Stations A and B are connected by two *parallel* message channels. A message from A to B is sent over both channels at the same time. Random variables x and y represent the message delays over parallel channels I and II, respectively.

A message is considered "received" as soon as it arrives on any one channel, and it is considered "verified" as soon as it has arrived over both channels.

a Determine the probability that a message is received within 15 minutes after it is sent.

b Determine the probability that the message is received but not verified within 15 minutes after it is sent.

c Let μ represent the time (in hours) between transmission at A and verification at B. Determine the cumulative distribution function $p_{\mu\leq}(\mu_0)$, and then differentiate it to obtain the PDF $f_\mu(\mu_0)$.

d If the attendant at B goes home 15 minutes after the message is received, what is the probability that he is present when the message should be verified?

e If the attendant at B leaves for a 15-minute coffee break right after the message is received, what is the probability that he is present at the proper time for verification?

f The management wishes to have the maximum probability of having the attendant present for *both* reception and verification. Would they do better to let him take his coffee break as described above or simply allow him to go home 45 minutes after transmission?

2.18 The data for an experiment that could be performed only once consist of experimental values of random variables x and y, which are described by the a priori probability density function:

$$f_{x,y}(x_0,y_0) = \begin{cases} Ax_0{}^2y_0 & \text{if } 0 \le (x_0 + y_0) \le 1, \quad x_0 \ge 0, \quad y_0 \ge 0 \\ 0 & \text{otherwise} \end{cases}$$

But the experimental data were lost. All the experimenter remembers is that the experimental value of x was either 0.4 or 0.6, and he decides that he observed one or the other of these values with equal probability.

Based on the above information, determine the experimenter's probability density function for his experimental value of random variable y.

2.19 The *exponential* PDF for random variable x is given by

$f_x(x_0) = \lambda e^{-\lambda x_0}$ for $x_0 \ge 0$.

a Determine the probability that an experimental value of x will be greater than $E(x)$.

b Suppose the lifetime of a bulb is given by the above PDF and we are told that the bulb has already been on for T units of time; determine the PDF for the remaining lifetime $y = x - T$ of the bulb. (This special result will be discussed in Chap. 4.)

c Assume that each bulb is replaced as soon as it burns out. Over a very long period, determine the fraction of this interval for which illumination is supplied by those bulbs whose lifetimes are longer than $E(x)$.

2.20 **a** For three spins of a fair wheel of fortune, what is the probability that none of the resulting experimental values is within $\pm 30°$ of any other experimental value?

b What is the smallest number of spins for which the probability that at least one other reading is within $\pm 30°$ of the first reading is at least 0.9?

2.21 Random variable x is described by a PDF which is constant between $x_0 = 0$ and $x_0 = 1$ and which is zero elsewhere. K independent successive experimental values of this random variable are labeled

$d_1, d_2, \ldots, d_K$. Define the random variables

r = second largest of $d_1, d_2 \ldots, d_K$
s = second smallest of $d_1, d_2, \ldots, d_K$

and determine the joint probability density function $f_{r,s}(r_0,s_0)$ for all values of r_0 and s_0.

2.22 The probability density function for continuous random variable x is a constant in the range $a < x \leq b$ and zero elsewhere.

a Determine σ_x, the standard deviation of random variable x.
b Determine the conditional standard deviation of x, given that $|x - E(x)| > \sigma_x$.
c If $y = cx + d$, determine $E(y)$ and σ_y in terms of $E(x)$ and σ_x. Do your results depend on the form of the PDF for random variable x?

2.23 Random variable x is described by the PDF

$$f_x(x_0) = \begin{cases} 0.1 & \text{if } 0 \leq x_0 \leq 10.0 \\ 0 & \text{otherwise} \end{cases}$$

Another random variable, y, is defined by $y = -\ln x$. Determine the PDF $f_y(y_0)$.

2.24 Random variables x and y are distributed according to the joint probability density function

$$f_{x,y}(x_0,y_0) = \begin{cases} Ax_0 & \text{if } 1 \leq x_0 \leq y_0 \leq 2 \\ 0.0 & \text{otherwise} \end{cases}$$

a Evaluate the constant A.
b Determine the marginal probability density function $f_y(y_0)$.
c Determine the expected value of $1/x$, given that $y = \frac{3}{2}$.
d Random variable z is defined by $z = y - x$. Determine the probability density function $f_z(z_0)$.

2.25 Random variables x and y are described by the joint density function

$$f_{x,y}(x_0,y_0) = \begin{cases} K & \text{if } 0 \leq y_0 \leq x_0 \leq 2 \\ 0 & \text{otherwise} \end{cases}$$

Random variable z is defined by

$$z = \max(x,2y)$$

Determine and sketch $f_z(z_0)$, the probability density function for random variable z.

2.26 Melvin Fooch, a student of probability theory, has found that the hours he spends working (w) and sleeping (s) in preparation for a final exam are random variables described by

$$f_{w,s}(w_0,s_0) = \begin{cases} K & \text{if } 10 \le w_0 + s_0 \le 20, \quad 0 \le w_0, \quad 0 \le s_0 \\ 0 & \text{otherwise} \end{cases}$$

What poor Melvin doesn't know, and even his best friends won't tell him, is that working only furthers his confusion and that his grade, g, can be described by

$$g = 2.50(s - w) + 50.0$$

a Evaluate constant K.

b The instructor has decided to pass Melvin if, on the exam, he achieves $g \ge 75.0$. What is the probability that this will occur?

c Make a neat and fully labeled sketch of the probability density function $f_g(g_0)$.

d Melvin, true to form, got a grade of exactly 75.0 on the exam. Determine the conditional probability that he spent less than one hour working in preparation for this exam.

2.27 Each day Wyatt Uyrp shoots one "game" by firing at a target with the following dimensions and scores for each shot:

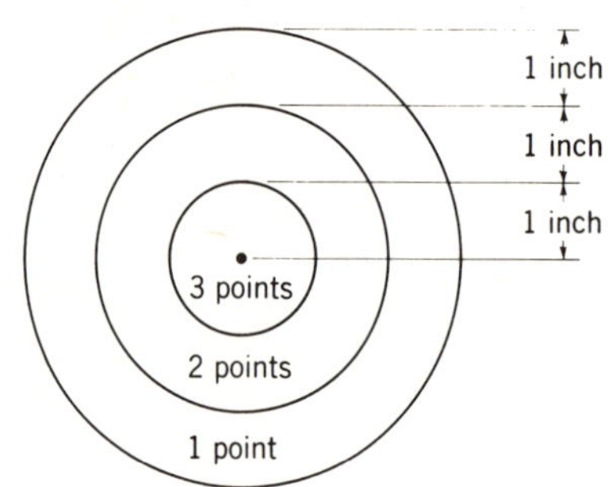

The score on any shot depends only on its distance from the center of the target

His pellet supply isn't too predictable, and the number of shots for any day's game is equally likely to be one, two, or three. Furthermore, Wyatt tires rapidly with each shot. Given that it is the kth pellet in a particular game, the value of r (distance from target center to point of impact) for a pellet is a random variable with probability density function

$$f_{r|k}(r_0 \mid k_0) = \begin{cases} \dfrac{1}{k_0} & \text{if } 0 \le r_0 \le k_0 \\ 0 & \text{otherwise} \end{cases}$$

a Determine and plot the probability mass function for random variable s_3, where s_3 is Mr. Uyrp's score on a three-shot game.

b Given only that a particular pellet was used during a three-shot game, determine and sketch the probability density function for r, the distance from the target center to where it hit.

c Given that Wyatt scored a total of exactly six points on a game, determine the probability that this was a two-shot game.

d We learn that, in a randomly selected game, there was at least one shot which scored exactly two points. Determine the conditional expected value of Wyatt's total score for that game.

e A particular pellet, marked at the factory, was used eventually by Wyatt. Determine the PMF for the number of points he scored on the shot which consumed this pellet.

2.28 Random variables x and y are described by the joint PDF

$$f_{x,y}(x_0,y_0) = \begin{cases} 1 & \text{if } 0 \le x_0 \le 1, \quad 0 \le y_0 \le 1 \\ 0 & \text{otherwise} \end{cases}$$

and random variable z is defined by $z = xy$.

Determine the conditional second moment of z, given that the equation $r^2 + xr + y = 0$ has real roots for r.

2.29 A target is located at the origin of an x,y cartesian coordinate system. One missile is fired at the target, and we assume that x and y, the coordinates of the missile impact point, are independent random variables each described by the *unit normal* PDF,

$$f_x(x_0) = f_y(x_0) = \frac{1}{\sqrt{2\pi}} e^{-x_0^2/2} \qquad -\infty \le x_0 \le \infty$$

Determine the PDF for random variable r, the distance from the target to the point of impact. Your answer should be an example of the *Rayleigh* PDF,

$$f_r(r_0) = \frac{r_0}{a^2} e^{-r_0^2/2a^2} \qquad r_0 \ge 0$$

2.30 Consider independent random variables x and y with the marginal PDF's

$$f_x(x_0) = f_y(x_0) = \frac{1}{\sqrt{2\pi}} e^{-x_0^2/2} \qquad -\infty \le x_0 \le \infty$$

Determine the PDF for random variable q, defined by $q = \frac{y}{x}\cdot$ Your answer should be a simple case of the *Cauchy* PDF,

$$f_q(q_0) = \frac{a}{\pi[a^2 + (q_0 - b)^2]} \qquad -\infty \le q_0 \le \infty$$

2.31 **a** Let x be a random variable with PDF $f_x(x_0)$. Determine the transformation $y = g(x)$ such that y will have the uniform PDF

$$f_y(y_0) = \begin{cases} 1 & \text{if } 0 \le y_0 \le 1 \\ 0 & \text{otherwise} \end{cases}$$

b How could you use a set of experimental values of a uniformly distributed random variable to obtain experimental values described by an arbitrary PDF $f_x(x_0)$?

2.32 Is it generally true that $E[g(x)]$ is the same as $g[E(x)]$? For instance, is $E\left(\frac{1}{x}\right)$ the same as $\frac{1}{E(x)}$? Please remember your result and avoid one of the most common errors in probabilistic reasoning.

2.33 **a** Variable x^*, the *standardized* random variable for random variable x, is given by $x^* = [x - E(x)]/\sigma_x$. Determine the expected value and variance of x^*.

b The *correlation coefficient* ρ, or *normalized covariance*, for two random variables x and y is defined to be

$$\rho_{xy} = E(x^*y^*) = E\left[\left(\frac{x - E(x)}{\sigma_x}\right)\left(\frac{y - E(y)}{\sigma_y}\right)\right]$$

Determine the numerical value of ρ_{xy} if:

i $x = ay$.

ii $x = -ay$.

iii x and y are linearly independent.

iv x and y are statistically independent.

v $x = ay + b$.

c For each performance of the experiment, the experimental value of random variable y^* is to be approximated by cx^*. Prove that the value of constant c which minimizes the expected *mean square error*, $E[(y^* - cx^*)^2]$, for this approximation is given by $c = \rho_{xy}$.

2.34 Al and Bo are the only participants in a race, and their elapsed times may be considered to be the random variables x and y, respectively.

$$f_x(x_0) = \begin{cases} 0.0 & x_0 < 1 \\ 1.0 & 1 \le x_0 \le 2 \\ 0.0 & 2 < x_0 \end{cases} \qquad f_y(y_0) = \begin{cases} 0.0 & y_0 < 1 \\ 0.5 & 1 \le y_0 \le 3 \\ 0.0 & 3 < y_0 \end{cases}$$

Let A be the event "Al won the race."

a Determine the conditional probability density function $f_{x|A}(x_0 \mid A)$.

b Let $w = y - x$. Determine $E(w)$ and $E(w \mid A)$.

c What is the minimum number of races which Bo must agree to enter such that the a priori probability that he will win at least one of the races is at least 0.99?

2.35 By observing the histogram of a particular random variable x, it is noted that, if we define $y = \ln(x - a)$, the behavior of y may be approximated by a Gaussian density function,

$$f_y(y_0) = \frac{1}{\sqrt{2\pi}\,\sigma_y} e^{-(y_0-m)^2/2\sigma_y^2}$$

Determine the probability density function $f_x(x_0)$.

2.36 Oscar has lost his dog in either forest A (with a priori probability 1/3) or forest B (with a priori probability 2/3). The probability that the dog will survive any particular night in forest A is 4/5 and in forest B is 3/5.

If the dog is in A (either dead or alive) and Oscar spends a day searching for him in A, the probability that he will find the dog that day is 1/2. The similar detection probability for a day of search in forest B is 2/5.

The dog cannot go from one forest to the other. Oscar can search only in the daytime and can travel from one forest to the other only at night.

Coolheaded Oscar has established the following values (in dollars):

Finding dog alive	+60
Each day (or part thereof) of search	−3
Finding dog dead	0
Not finding dog	−10
Additional cost if Oscar must actually search in both forests	−3

Oscar is incapable of figuring it all out; so he decides that he will search for just two days—looking in B on the first day and, if necessary, looking in A on the second day.

a Determine the expected value of this policy.

b Given that Oscar fails to find the dog on the first day, is the second day of search a worthwhile investment? Explain.

c If only Oscar were a thinker, he would at least have considered the following list of policies for possible two-day search efforts:

S_1: Search in A on 1st day; in B on 2d day if necessary.

S_2: Search in B on 1st day; in A on 2d day if necessary.
S_3: Search in A on 1st day; in A on 2d day if necessary.
S_4: Search in B on 1st day; in B on 2d day if necessary.
S_5: Search in A on 1st day; don't search on 2d day.
S_6: Search in B on 1st day; don't search on 2d day.
S_7: Don't search at all!

Oscar would still have to choose his own decision criteria—in fact he has already decided no dog is worth more than two days of searching in mosquito-infested forests. However, to help Oscar quantify his thinking, determine which of the above policies would:

i Maximize his expected gain (graduate student)
ii Minimize his maximum possible loss (coward)
iii Maximize his maximum possible gain (hero)
iv Maximize the probability that he will find his dog alive (idealist)

transforms and some applications to sums of independent random variables

An understanding of the concepts and applications of transform theory will contribute in several ways to our later work. Transforms are useful for the establishment of valuable general theorems, the determination of moments of random variables, the study of certain probabilistic processes, and the analysis of sums of independent random variables.

Some important applications are introduced in this chapter. However, an appreciation of the power of transform techniques will, for the most part, be developed as we study more advanced topics in later chapters.

3-1 The s Transform

Let $f_x(x_0)$ be any PDF. The *exponential transform* (or *s transform*) for this PDF, $f_x{}^T(s)$, is defined by

$$f_x{}^T(s) \equiv E(e^{-sx}) = \int_{-\infty}^{\infty} e^{-sx_0} f_x(x_0)\, dx_0$$

We are interested only in the s transforms of PDF's and not of arbitrary functions. Thus, we need note only those aspects of transform theory which are relevant to this special case.

As long as $f_x(x_0)$ is a PDF, the above integral must be finite at least for the case where s is a pure imaginary quantity (see Prob. 3.01). Furthermore, it can be proved that the s transform of a PDF is unique to that PDF.

Three examples of the calculation of s transforms follow: First, consider the PDF

$$f_x(x_0) = \begin{cases} \lambda e^{-\lambda x_0} & x_0 \geq 0 \\ 0 & x_0 < 0 \end{cases} = \mu_{-1}(x_0 - 0)\lambda e^{-\lambda x_0} \qquad -\infty \leq x_0 \leq \infty$$

$$f_x{}^T(s) = \int_{-\infty}^{\infty} e^{-sx_0} f_x(x_0)\, dx_0 = \int_0^{\infty} \lambda e^{-sx_0} e^{-\lambda x_0}\, dx_0 = \frac{\lambda}{s + \lambda}$$

[The unit step function $\mu_{-1}(x_0 - a)$ is defined in Sec. 2-9.] For a second example, we consider the *uniform* PDF

$$f_x(x_0) = \begin{cases} 1 & \text{if } 0 \leq x_0 \leq 1 \\ 0 & \text{otherwise} \end{cases} = \mu_{-1}(x_0 - 0) - \mu_{-1}(x_0 - 1)$$
$$-\infty \leq x_0 \leq \infty$$

$$f_x{}^T(s) = \int_{-\infty}^{\infty} e^{-sx_0} f_x(x_0)\, dx_0 = \int_0^1 e^{-sx_0}\, dx_0 = \frac{1 - e^{-s}}{s}$$

Our third example establishes a result to be used later. Consider the PDF for a degenerate (deterministic) random variable x which always takes on the experimental value a,

$$f_x(x_0) = \mu_0(x_0 - a) \qquad -\infty \leq x_0 \leq \infty$$

$$f_x{}^T(s) = \int_{-\infty}^{\infty} e^{-sx_0} \mu_0(x_0 - a)\, dx_0 = e^{-sa}$$

The PDF corresponding to a given s transform, $f_x{}^T(s)$, is known as the *inverse transform* of $f_x{}^T(s)$. The formal technique for obtaining inverse transforms is beyond the scope of the mathematical prerequisites assumed for this text. For our purposes, we shall often be able to obtain inverse transforms by recognition and by exploiting a few simple properties of transforms. A discussion of one simple procedure for attempting to evaluate inverse transforms will appear in our solution to the example of Sec. 3-8.

3-2 The z Transform

Once we are familiar with the impulse function, any PMF may be expressed as a PDF. To relate the PMF $p_x(x_0)$ to its corresponding PDF $f_x(x_0)$, we use the relation

$$f_x(x_0) = \sum_a p_x(a)\mu_0(x_0 - a)$$

As an example, the PMF $p_x(x_0)$ shown below

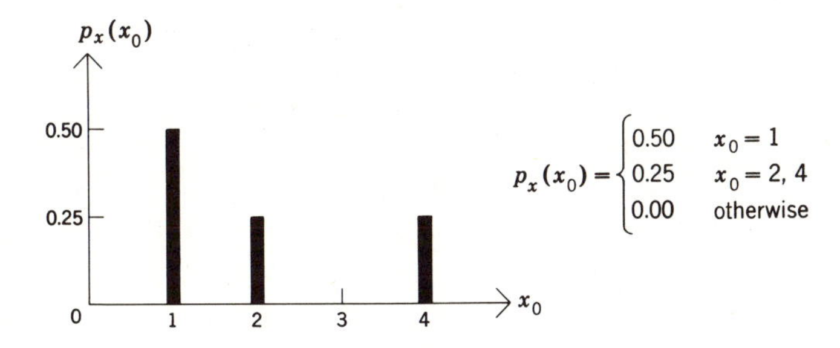

may be written as the PDF $f_x(x_0)$,

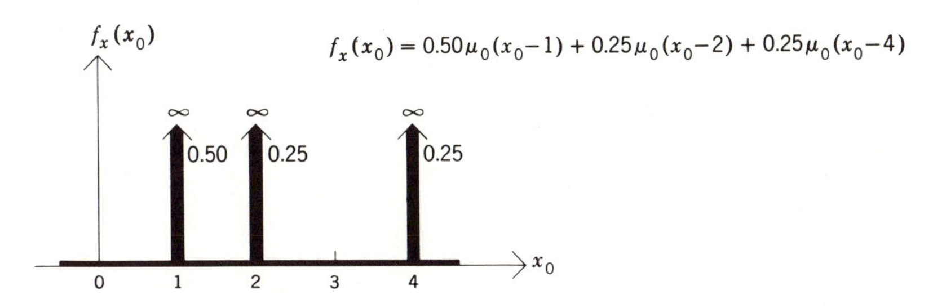

The s transform of this PDF is obtained from

$$f_x{}^T(s) = \int_{-\infty}^{\infty} e^{-sx_0} f_x(x_0)\, dx_0 = 0.50e^{-s} + 0.25e^{-2s} + 0.25e^{-4s}$$

where we have made use of the following relation from Sec. 2-9

$$\int_{-\infty}^{\infty} \mu_0(x_0 - a)g(x_0)\, dx_0 = g(a)$$

The above s transform could also have been obtained directly from the equivalent (expected value) definition of $f_x{}^T(s)$,

$$f_x{}^T(s) = E(e^{-sx}) = 0.50e^{-s} + 0.25e^{-2s} + 0.25e^{-4s}$$

Although the s transform is defined for the PDF of *any* random variable, it is convenient to define one additional type of transform for a certain type of PMF. If $p_x(x_0)$ is the PMF for a discrete random variable *which can take on only nonnegative integer experimental values* ($x_0 = 0, 1, 2, \ldots$), we define the *discrete transform* (or *z transform*) of $p_x(x_0)$ to be $p_x{}^T(z)$, given by

$$p_x{}^T(z) \equiv E(z^x) = \sum_{x_0=0}^{\infty} z^{x_0} p_x(x_0)$$

We do not find it particularly useful to define a z transform for PMF's which allow noninteger or negative experimental values. In practice, a large number of discrete random variables arise from a count of integer units and from the quantization of a positive quantity, and it is for cases like these that our nonnegative integer constraint holds.

The PMF at the start of this section allows only nonnegative integer values of its random variable. As an example, we obtain the z transform of this PMF,

$$p_x{}^T(z) = \sum_{x_0=0}^{\infty} z^{x_0} p_x(x_0) = 0.50z + 0.25z^2 + 0.25z^4$$

Note that the z transform for a PMF may be obtained from the s transform of the equivalent PDF by substituting $z = e^{-s}$.

The z transform can be shown to be finite for at least $|z| \leq 1$ and to be unique to its PMF. We shall normally go back to a PMF from its transform by recognition of a few familiar transforms. However, we can note from the definition of $p_x{}^T(z)$,

$$p_x{}^T(z) = p_x(0) + zp_x(1) + z^2p_x(2) + z^3p_x(3) + \cdots$$

that it is possible to determine the individual terms of the PMF from $p_x{}^T(z)$ by

$$p_x(x_0) = \frac{1}{x_0!}\left[\frac{d^{x_0}}{dz^{x_0}}\, p_x{}^T(z)\right]_{z=0} \qquad x_0 = 0, 1, 2, \ldots$$

3-3 Moment-generating Properties of the Transforms

Consider the nth derivative with respect to s of $f_x{}^T(s)$,

$$f_x{}^T(s) = \int_{x_0=-\infty}^{\infty} e^{-sx_0} f_x(x_0)\, dx_0 \qquad \frac{d^n f_x{}^T(s)}{ds^n} = \int_{x_0=-\infty}^{\infty} (-x_0)^n e^{-sx_0} f_x(x_0)\, dx_0$$

The right-hand side of the last equation, when evaluated at $s = 0$, may be recognized to be equal to $(-1)^n E(x^n)$. Thus, once we obtain the s transform for a PDF, we can find all the moments by repeated differentiation rather than by performing other integrations.

From the above expression for the nth derivative of $f_x^T(s)$, we may establish the following useful results:

$$[f_x^T(s)]_{s=0} = 1 \qquad E(x) = -\left[\frac{df_x^T(s)}{ds}\right]_{s=0} \qquad E(x^2) = \left[\frac{d^2f_x^T(s)}{ds^2}\right]_{s=0}$$

$$\sigma_x^2 = E\{[x - E(x)]^2\} = E(x^2) - [E(x)]^2 = \left\{\frac{d^2f_x^T(s)}{ds^2} - \left[\frac{df_x^T(s)}{ds}\right]^2\right\}_{s=0}$$

Of course, when certain moments of a PDF do not exist, the corresponding derivatives of $f_x^T(s)$ will be infinite when evaluated at $s = 0$.

As one example of the use of these relations, consider the PDF $f_x(x_0) = \mu_{-1}(x_0 - 0)\lambda e^{-\lambda x_0}$, for which we obtained $f_x^T(s) = \lambda/(s + \lambda)$ in Sec. 3-1. We may obtain $E(x)$ and σ_x^2 by use of the relations

$$E(x) = -\left[\frac{df_x^T(s)}{ds}\right]_{s=0} = -\left[\frac{-\lambda}{(s+\lambda)^2}\right]_{s=0} = \frac{1}{\lambda}$$

$$E(x^2) = (-1)^2\left[\frac{d^2f_x^T(s)}{ds^2}\right]_{s=0} = \left[\frac{2\lambda}{(s+\lambda)^3}\right]_{s=0} = \frac{2}{\lambda^2}$$

$$\sigma_x^2 = E(x^2) - [E(x)]^2 = \frac{1}{\lambda^2}$$

The moments for a PMF may also be obtained by differentiation of its z transform, although the resulting equations are somewhat different from those obtained above. Beginning with the definition of the z transform, we have

$$p_x^T(z) = \sum_{x_0=0}^{\infty} z^{x_0} p_x(x_0)$$

$$\left[\frac{dp_x^T(z)}{dz}\right]_{z=1} = \left[\sum_{x_0=0}^{\infty} x_0 z^{x_0-1} p_x(x_0)\right]_{z=1} = E(x)$$

$$\left[\frac{d^2p_x^T(z)}{dz^2}\right]_{z=1} = \left[\sum_{x_0=0}^{\infty} x_0(x_0-1) z^{x_0-2} p_x(x_0)\right]_{z=1} = E(x^2) - E(x)$$

In general, for $n = 1, 2, \ldots$, we have

$$\frac{d^np_x^T(z)}{dz^n} = \sum_{x_0=0}^{\infty} x_0(x_0-1)(x_0-2)\cdots(x_0-n+1)z^{x_0-n}p_x(x_0)$$

The right-hand side of this last equation, when evaluated at $z = 1$, is equal to some linear combination of $E(x^n)$, $E(x^{n-1})$, . . . , $E(x^2)$, and $E(x)$. What we are accomplishing here is the determination of all moments of a PMF from a single summation (the calculation of the transform itself) rather than attempting to perform a separate summation directly for each moment. This saves quite a bit of work for those PMF's whose z transforms may be obtained in closed form. We shall frequently use the following relations which are easily obtained from the above equations:

$$[p_x^T(z)]_{z=1} = 1 \qquad E(x) = \left[\frac{dp_x^T(z)}{dz}\right]_{z=1} \qquad E(x^2) = \left[\frac{d^2p_x^T(z)}{dz^2} + \frac{dp_x^T(z)}{dz}\right]_{z=1}$$

$$\sigma_x^2 = \left\{\frac{d^2}{dz^2}\,p_x^T(z) + \frac{d}{dz}\,p_x^T(z) - \left[\frac{d}{dz}\,p_x^T(z)\right]^2\right\}_{z=1}$$

We often recognize sums which arise in our work to be similar to expressions for moments of PMF's, and then we may use z transforms to carry out the summations. (Examples of this procedure arise, for instance, in the solutions to Probs. 3.10 and 3.12.)

As an example of the moment-generating properties of the z transform, consider the *geometric* PMF defined by

$$p_x(x_0) = \begin{cases} P(1-P)^{x_0-1} & \text{if } x_0 = 1, 2, 3, \ldots \\ 0 & \text{otherwise} \end{cases} \qquad 0 < P < 1$$

We shall use the z transform to obtain $E(x)$, $E(x^2)$, and σ_x^2.

$$p_x^T(z) = E(z^x) = \sum_{x_0=0}^{\infty} z^{x_0} p_x(x_0) = \sum_{x_0=1}^{\infty} P z^{x_0}(1-P)^{x_0-1} = \frac{zP}{1-z(1-P)}$$

After a calculation such as the above we may check $[p_x^T(z)]_{z=1} \stackrel{\checkmark}{=} 1$.

$$E(x) = \left[\frac{d}{dz}\,p_x^T(z)\right]_{z=1} = \frac{1}{P}$$

$$E(x^2) = \left[\frac{d^2}{dz^2}\,p_x^T(z) + \frac{d}{dz}\,p_x^T(z)\right]_{z=1} = \frac{2-P}{P^2}$$

[Try to evaluate $E(x^2)$ directly from the definition of expectation!]

$$\sigma_x^2 = E(x^2) - [E(x)]^2 = \frac{1-P}{P^2}$$

In obtaining $p_x^T(z)$, we used the relation

$$1 + a + a^2 + \cdots + a^k = \frac{1-a^{k+1}}{1-a} \qquad |a| < 1$$

A similar relation which will be used frequently in our work with z transforms is

$$1 + a + \frac{a^2}{2!} + \frac{a^3}{3!} + \frac{a^4}{4!} + \cdots = e^a \qquad -\infty < a < \infty$$

3-4 Sums of Independent Random Variables; Convolution

The properties of sums of independent random variables is an important topic in the study of probability theory. In this section we approach this topic from a sample-space point of view. A transform approach will be considered in Sec. 3-5, and, in Sec. 3-7, we extend our work to a matter involving the sum of a random number of random variables. Sums of independent random variables also will be our main concern when we discuss limit theorems in Chap. 6.

To begin, we wish to work in an x_0,y_0 event space, using the method of Sec. 2-14, to derive the PDF for w, the sum of two random variables x and y. After a brief look at the general case, we shall specialize our results to the case where x and y are independent.

We are given $f_{x,y}(x_0,y_0)$, the PDF for random variables x and y. With $w = x + y$, we go to the x_0,y_0 event space to determine $p_{w\le}(w_0)$. The derivative of this CDF is the desired PDF, $f_w(w_0)$.

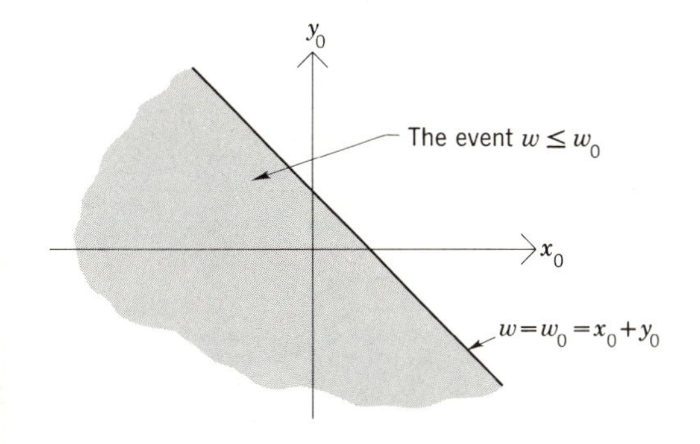

$$p_{w\le}(w_0) = \int_{x_0=-\infty}^{\infty} dx_0 \int_{y_0=-\infty}^{w_0-x_0} dy_0\, f_{x,y}(x_0,y_0)$$

$$f_w(w_0) = \frac{d}{dw_0}\, p_{w\le}(w_0) = \int_{x_0=-\infty}^{\infty} dx_0\, \frac{d}{dw_0}\left[\int_{y_0=-\infty}^{w_0-x_0} dy_0\, f_{x,y}(x_0,y_0)\right]$$

We may use the formula given in Sec. 2-14 to differentiate the quantity in the brackets to obtain

$$f_w(w_0) = \int_{x_0=-\infty}^{\infty} dx_0\, f_{x,y}(x_0,\, w_0 - x_0)$$

In general, we can proceed no further without specific knowledge of the form of $f_{x,y}(x_0,y_0)$. For the special case where x and y are independent random variables, we may write

$$f_w(w_0) = \int_{x_0=-\infty}^{\infty} dx_0\, f_x(x_0) f_y(w_0 - x_0) \qquad \text{for } w = x + y \text{ and } x,y \text{ independent}$$

This operation is known as the *convolution* of $f_x(x_0)$ and $f_y(y_0)$. Had we integrated over x_0 first instead of y_0 in obtaining $p_{w\leq}(w_0)$, we would have found the equivalent expression with x_0 and y_0 interchanged,

$$f_w(w_0) = \int_{y_0=-\infty}^{\infty} dy_0\, f_y(y_0) f_x(w_0 - y_0)$$

The convolution of two functions has a simple, and often useful, graphical interpretation. If, for instance, we wish to convolve $f_x(x_0)$ and $f_y(y_0)$ using the form

$$f_w(w_0) = \int_{x_0=-\infty}^{\infty} dx_0\, f_x(x_0) f_y(w_0 - x_0)$$

we would require plots of $f_x(x_0)$ and $f_y(w_0 - x_0)$, each plotted as a function of x_0. Then, for all possible values of w_0, these two curves may be multiplied point by point. The resulting product curve is integrated over x_0 to obtain $f_w(w_0)$. Since convolution is often easier to perform than to describe, let's try an example which requires the convolution of the following PDF's:

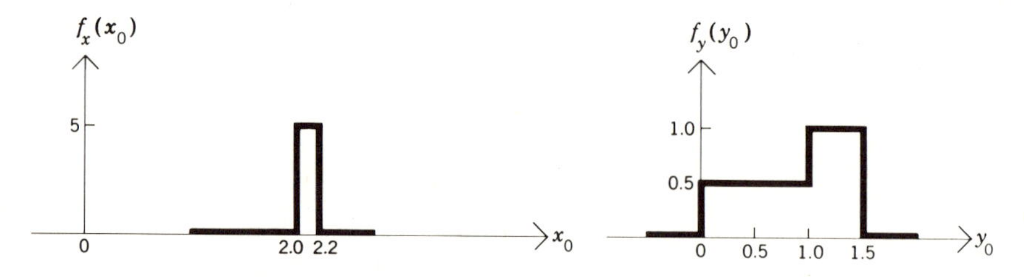

We are asked to find the PDF for $w = x + y$, given that x and y are independent random variables. To obtain the desired plot of $f_y(w_0 - x_0)$ as a function of x_0, we first "flip" $f_y(y_0)$ about the line $y_0 = 0$ to get

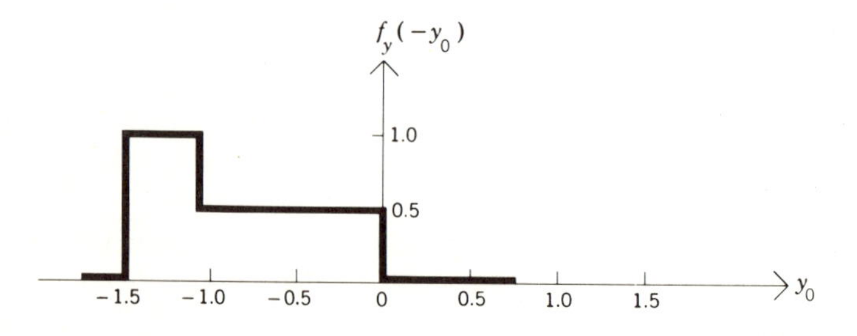

We next replace variable y_0 by x_0 and then plot, along an x_0 axis, the flipped function shifted to the right by w_0.

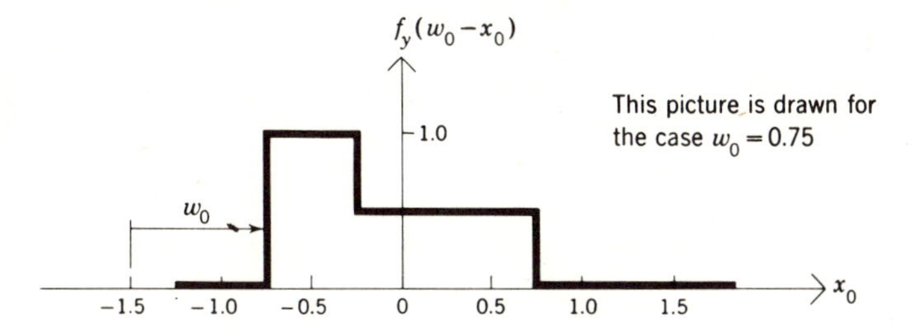

We can now present $f_y(w_0 - x_0)$ on the same plot as $f_x(x_0)$ and perform the integration of the product of the curves as a function of w_0.

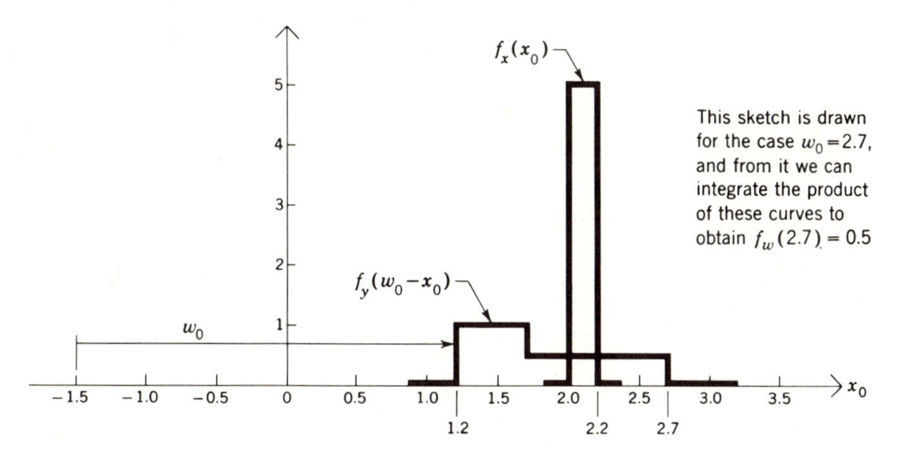

Our final step is to plot the integral of the product of these two functions for all values of w_0. We obtain

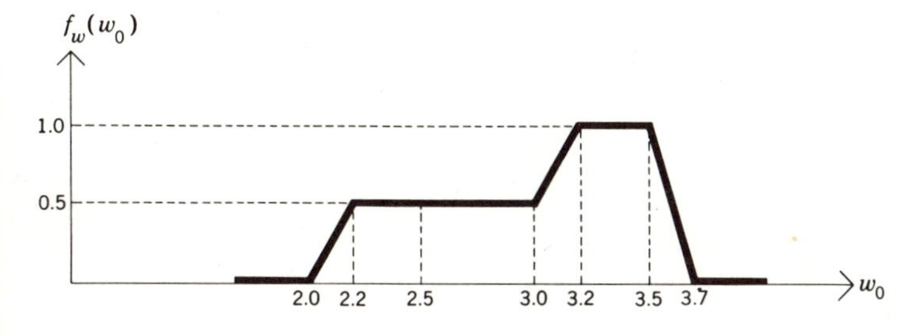

Thus, by graphical convolution, we have determined the PDF for random variable w, which, you may recall, was defined to be the sum of independent random variables x and y. Of course, we check to see that $f_w(w_0)$ is nonzero only in the range of possible values for sums of x and y $(2.0 \leq w_0 \leq 3.7)$ and that this derived PDF integrates to unity.

We are now familiar with two equivalent methods to obtain the PDF for the sum of the *independent* random variables x and y. One method would be to work directly in the x_0,y_0 event space; an alternative is to perform the convolution of their marginal PDF's. In the next section, a transform technique for this problem will be introduced.

For the special case where we are to convolve two PDF's, *each* of which contains one or more impulses, a further note is required. Our simplified definition of the impulse does not allow us to argue the following result from that definition; so we shall simply define the integral of the product of two impulses to be

$$\int_{x_0=-\infty}^{\infty} \mu_0(x_0 - a)\mu_0(x_0 - b)\, dx_0 = \mu_0(a - b)$$

Thus, the convolution of two impulses would be another impulse, with an area equal to the product of the areas of the two impulses.

A special case of convolution, the *discrete* convolution, is introduced in Prob. 3.17. The discrete convolution allows one to convolve PMF's directly without first replacing them by their equivalent PDF's.

3-5 The Transform of the PDF for the Sum of Independent Random Variables

Let $w = x + y$, where x and y are independent random variables with marginal PDF's $f_x(x_0)$ and $f_y(y_0)$. We shall obtain $f_w{}^T(s)$, the s transform of $f_w(w_0)$, from the transforms $f_x{}^T(s)$ and $f_y{}^T(s)$.

$$f_w{}^T(s) = E(e^{-sw}) = E(e^{-s(x+y)})$$
$$= \int_{x_0=-\infty}^{\infty} \int_{y_0=-\infty}^{\infty} e^{-sx_0}e^{-sy_0}f_{x,y}(x_0,y_0)\, dx_0\, dy_0$$

The compound PDF factors into $f_x(x_0)f_y(y_0)$ because of the independence of x and y, to yield

$$f_w{}^T(s) = \int_{x_0=-\infty}^{\infty} e^{-sx_0}f_x(x_0)\, dx_0 \int_{y_0=-\infty}^{\infty} e^{-sy_0}f_y(y_0)\, dy_0$$

$$f_w{}^T(s) = f_x{}^T(s)f_y{}^T(s) \qquad \text{for } w = x + y \text{ and } x,y \text{ statistically independent}$$

We have proved that the transform of the PDF of a random variable which is the sum of two *independent* random variables is the product of the transforms of their PDF's.

The proof of the equivalent result for discrete random variables which have z transforms,

$$p_w{}^T(z) = p_x{}^T(z)p_y{}^T(z) \qquad \text{for } w = x + y \text{ and } x,y \text{ statistically independent}$$

is entirely similar to the above.

Let's do one example for the discrete case, using z transforms. Independent variables x and y are described by the PMF's

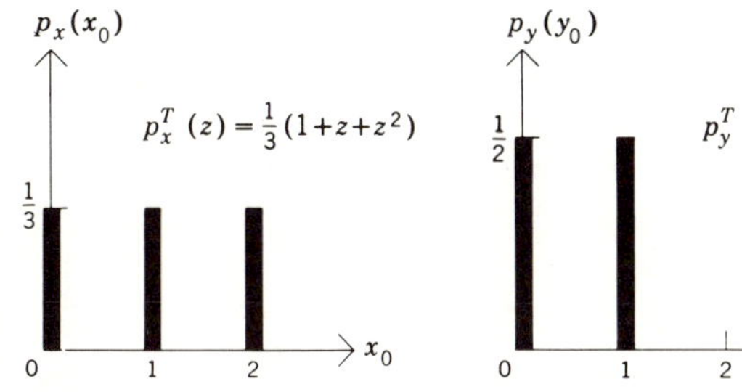

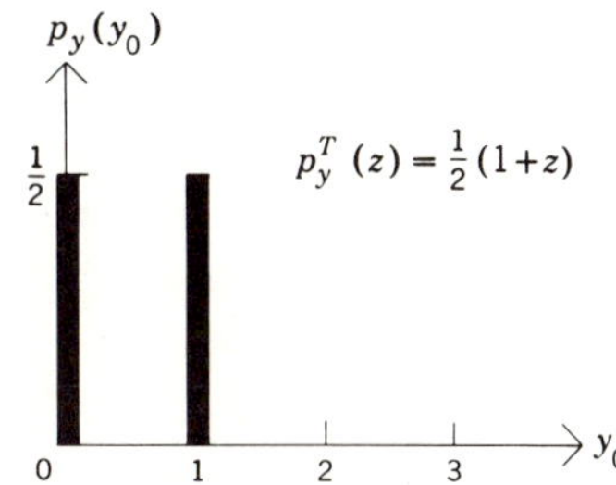

The PMF for random variable w, given that $w = x + y$, has the z transform

$$p_w{}^T(z) = p_x{}^T(z)p_y{}^T(z) = \tfrac{1}{6}(1 + 2z + 2z^2 + z^3)$$

and since we know $p_w{}^T(z) = \sum_{w_0=0}^{\infty} p_w(w_0)z^{w_0}$, we can note that the coefficient of z^{w_0} in $p_w{}^T(z)$ is equal to $p_w(w_0)$. Thus, we may take the inverse transform of $p_w{}^T(z)$ to obtain

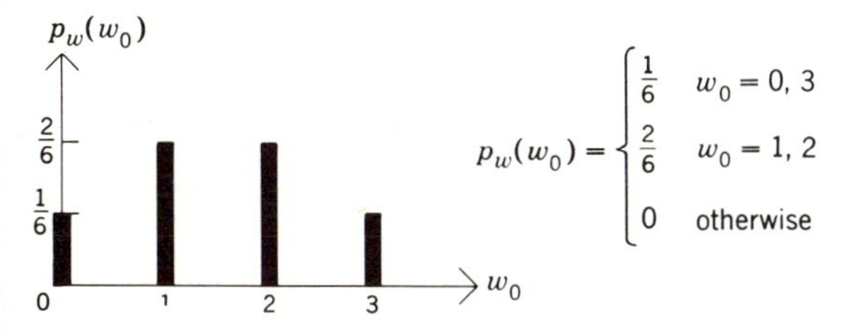

The reader is encouraged to either convolve the PMF's or work the problem in an x_0,y_0 sample space to verify the above result.

3-6 A Further Note on Sums of Independent Random Variables

For *any* random variables x and y, we proved

$$E(x + y) = E(x) + E(y)$$

in Sec. 2-7. Thus, the expected value of a sum is always equal to the sum of the expected values of the individual terms.

We next wish to note how variances combine when we add *independent* random variables to obtain new random variables. Let $w = x + y$; we then have, using the above relation for the expected value of a sum,

$$\sigma_w^2 = E\{[w - E(w)]^2\} = E\{[x + y - E(x) - E(y)]^2\} = E\{[x - E(x) + y - E(y)]^2\}$$
$$= E\{[x - E(x)]^2\} + E\{[y - E(y)]^2\} + 2E\{[x - E(x)][y - E(y)]\}$$
$$= \sigma_x^2 + \sigma_y^2 + 2E[xy - xE(y) - yE(x) + E(x)E(y)]$$

For x and y independent, the expected values of all products in the last brackets are equal. In fact, only *linear* independence is required for this to be true and we obtain the following important expression for the variance of the sum of linearly independent random variables:

$$\sigma_w^2 = \sigma_x^2 + \sigma_y^2 \qquad \text{for } w = x + y \text{ and } \quad x,y \text{ linearly independent}$$

An alternative derivation of this relation for statistically independent random variables (using transforms) is indicated in Prob. 3.14.

We now specialize our work to sums of independent random variables for the case where each member of the sum has the same PDF. When we are concerned with this case, which may be considered as a sum of independent experimental values from an experiment whose outcome is described by a particular PDF, we speak of *independent identically distributed* random variables.

Let r be the sum of n independent identically distributed random variables, each with expected value $E(x)$ and variance σ_x^2. We already know that

$$E(r) = nE(x) \qquad \sigma_r^2 = n\sigma_x^2 \qquad \sigma_r = \sqrt{n}\,\sigma_x$$

For the rest of this section *we consider only the case* $E(x) > 0$ *and* $\infty > \sigma_x^2 > 0$. The $E(x) > 0$ condition will simplify our statements and expressions. Our reasoning need not hold for $\sigma_x^2 = \infty$, and any PDF which has $\sigma_x^2 = 0$ represents an uninteresting deterministic quantity.

If we are willing to accept the standard deviation of r, σ_r, as a type of linear measure of the spread of a PDF about its mean, some interesting *speculations* follow.

As n increases, the PDF for r gets "wider" (as $\sqrt{n}$) and its expected value increases (as n). The mean and the standard deviation both grow, but the mean increases more rapidly.

This would lead us to expect that, for instance, as n goes to infinity, the probability that an experimental value of r falls within a certain absolute distance d of $E(r)$ decreases to zero. That is,

$$\lim_{n\to\infty} \text{Prob}[|r - E(r)| < d] = 0 \qquad \text{(we think)}$$

We reach this speculation by reasoning that, as the width of $f_r(r_0)$ grows as $\sqrt{n}$, the height of most of the curve should fall as $1/\sqrt{n}$ to keep its area equal to unity. If so, the area of $f_r(r_0)$ over a slit of fixed width $2d$ should go to zero.

Our second speculation is based on the fact that $E(r)$ grows faster than σ_r. We might then expect that the probability that an experimental value of r falls within $\pm A\%$ of $E(r)$ grows to unity as n goes to infinity (for $A \neq 0$). That is,

$$\lim_{n\to\infty} \text{Prob}\left[\frac{|r - E(r)|}{E(r)} < \frac{A}{100}\right] = 1 \qquad \text{for } A > 0 \qquad \text{(we think)}$$

We might reason that, while the height of the PDF in most places near $E(r)$ is probably falling as $1/\sqrt{n}$, the interval of interest, defined by

$$|r - E(r)| < \frac{A}{100} E(r)$$

grows as n. Thus the area over this interval should, as $n \to \infty$, come to include all the area of the PDF $f_r(r_0)$.

For the given conditions, we shall learn in Chap. 6 that these speculations happen to be correct. Although proofs of such theorems could be stated here, we would not have as satisfactory a physical interpretation of such *limit theorems* as is possible after we become familiar with the properties of several important PMF's and PDF's.

3-7 Sum of a Random Number of Independent Identically Distributed Random Variables

Let x be a random variable with PDF $f_x(x_0)$ and s transform $f_x^T(s)$. If r is defined to be the sum of n independent experimental values of random variable x, we know from the results of Sec. 3-5 that the transform for the PDF $f_r(r_0)$ is

$f_r^T(s) = [f_x^T(s)]^n$ where r is the sum of n (statistically) independent experimental values of x

We now wish to consider the situation when n is also a random variable [with PMF $p_n(n_0)$]. We are interested in the sum of a random (but integer) number of independent identically distributed random variables.

For instance, if $f_x(x_0)$ were the PDF for the weight of any individual in an elevator, if the weights of people in the elevator could be considered to be independent random variables, *and* if $p_n(n_0)$ were the PMF for the number of people in the elevator, random variable r would represent the total weight of the people in the elevator. *Our work will also require that x and n be independent.* In our example, the PDF for the individual weights may not depend on the number of people in the elevator.

If n can take on the experimental values $0, 1, 2, \ldots, N$, the *sample* space for each performance of our experiment is of $N + 1$ dimensions, since each performance generates one experimental value of n and up to N experimental values of random variable x. It is usually difficult to obtain the desired PDF $f_r(r_0)$ directly, but its s transform is derived quite easily. Although it may be difficult to get back to $f_r(r_0)$ in a useful form from $f_r^T(s)$, it is a simple matter to evaluate the moments and variance of random variable r.

We may determine the s transform for $f_r(r_0)$ by working in an event space for random variable n (the number of independent experimental values of x in the sum) and r (the value of the sum). This event space, perhaps a strange choice at first sight, consists of a set of parallel lines in one quadrant and one point at the origin of the r_0,n_0 plane.

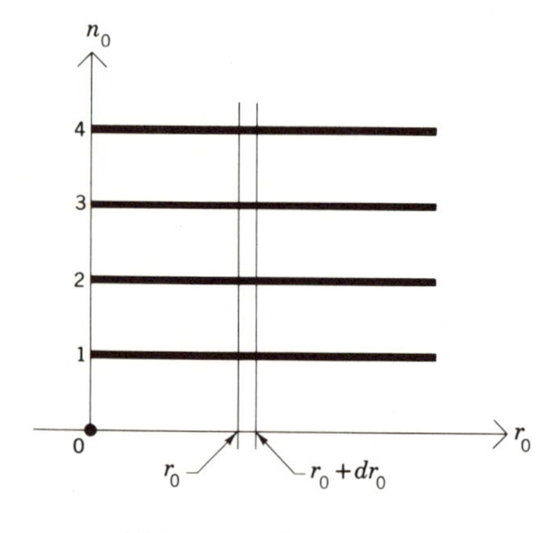

Every event point representing a possible outcome is either on one of these heavy lines or at the origin

Along each heavy line, there applies a conditional PDF $f_{r|n}(r_0 \mid n_0)$ which is the PDF for r given the experimental value of n. We know that $f_{r|n}(r_0 \mid n_0)$ is that PDF which describes the sum of n_0 independent experimental values of random variable x. As we noted at the start of this section, PDF $f_{r|n}(r_0 \mid n_0)$ has the s transform $[f_x^T(s)]^{n_0}$.

We use these observations to collect $f_r(r_0)$ as a summation in the

r_0,n_0 event space, and then we take the s transform on both sides of the equation.

$$f_r(r_0) = \sum_{n_0} p_n(n_0) f_{r|n}(r_0 \mid n_0)$$

$$f_r{}^T(s) = \int_{-\infty}^{\infty} e^{-sr_0} \sum_{n_0} p_n(n_0) f_{r|n}(r_0 \mid n_0)\, dr_0$$

$$= \sum_{n_0} p_n(n_0) \int_{-\infty}^{\infty} e^{-sr_0} f_{r|n}(r_0 \mid n_0)\, dr_0$$

$$= \sum_{n_0} p_n(n_0)[f_x{}^T(s)]^{n_0}$$

We recognize the last equation for $f_r{}^T(s)$ to be the z transform of PMF $p_n(n_0)$, with the transform evaluated at $z = f_x{}^T(s)$. We now restate this problem and its solution.

Let n and x be independent random variables, where n is described by the PMF $p_n(n_0)$ and x by the PDF $f_x(x_0)$. Define r to be the sum of n independent experimental values of random variable x. The s transform for the PDF $f_r(r_0)$ is $\quad f_r{}^T(s) = p_n{}^T[f_x{}^T(s)]$

We may use the chain rule for differentiation to obtain the expectation, second moment, and variance of r.

$$E(r) = -\left[\frac{df_r{}^T(s)}{ds}\right]_{s=0} = -\left\{\frac{dp_n{}^T[f_x{}^T(s)]}{d[f_x{}^T(s)]} \cdot \frac{df_x{}^T(s)}{ds}\right\}_{s=0}$$

To evaluate the first term in the right-hand brackets, we proceed,

$$\left\{\frac{dp_n{}^T[f_x{}^T(s)]}{d[f_x{}^T(s)]}\right\}_{s=0} = \left[\frac{dp_n{}^T(z)}{dz}\right]_{z=1} = E(n)$$

The first step in the equation immediately above made use of the fact that $[f_x{}^T(s)]_{s=0} = 1$. To solve for $E(r)$, we use $\left[\dfrac{df_x{}^T(s)}{ds}\right]_{s=0} = -E(x)$ in the expression for $E(r)$ to obtain

$$E(r) = E(n)E(x)$$

A second chain-rule differentiation of $f_r{}^T(s)$ and the use of the relation for $\sigma_r{}^2$ in terms of $f_r{}^T(s)$ leads to the further result

$$\sigma_r^2 = E(n)\sigma_x^2 + [E(x)]^2\sigma_n^2$$

We may note that this result checks out correctly for the case where n is deterministic ($\sigma_n^2 = 0$) and for the case where x is deterministic ($\sigma_x^2 = 0$, $[E(x)]^2 = x^2$).

If we had required that x, as well as n, be a discrete random variable which takes on only nonnegative integer experimental values, we could have worked with the PMF $p_x(x_0)$ to study a particular case of the above derivations. The resulting z transform of the PMF for discrete random variable r is

$$p_r^T(z) = p_n^T[p_x^T(z)]$$

and the above expressions for $E(r)$ and σ_r^2 still hold.

An example of the sum of a random number of independent identically distributed random variables is included in the following section.

3-8 An Example, with Some Notes on Inverse Transforms

As we undertake the study of some common probabilistic processes in the next chapter, our work will include numerous examples of applications of transform techniques. One problem is solved here to review some of the things we learned in this chapter, to indicate one new application, and [in part (e)] to lead us into a discussion of how we may attempt to go back from an s transform to its PDF. There are, of course, more general methods, which we shall not discuss.

Let discrete random variable k be described by the PMF

$$p_k(k_0) = \frac{8^{k_0}}{9^{k_0+1}} \qquad k_0 = 0, 1, 2, \ldots$$

(a) Determine the expected value and variance of random variable k.

(b) Determine the probability that an experimental value of k is even.

(c) Determine the probability that the sum of n independent experimental values of k is even.

(d) Let random variable k represent the number of light bulbs we happen to have on hand at time T_0. Furthermore, let x, the lifetime of each bulb, be an independent random variable with PDF

$$f_x(x_0) = \mu_{-1}(x_0 - 0)\lambda e^{-\lambda x_0}$$

We turn on one bulb at time T_0, replacing it immediately with another bulb as soon as it fails. This continues until the last of the k bulbs blows out. Determine the s transform, expectation, and variance for random variable τ, the time from T_0 until the last bulb dies.

(e) Determine the PDF $f_\tau(\tau_0)$ from the s transform $f_\tau{}^T(s)$ obtained in part (*d*).

a Rather than attempt to carry out some troublesome summations directly, it seems appropriate to employ the z transform.

$$p_k{}^T(z) = E(z^k) = \sum_{k_0=0}^{\infty} z^{k_0} p_k(k_0) = \frac{1}{9} \sum_{k_0=0}^{\infty} \left(\frac{8z}{9}\right)^{k_0} = (9 - 8z)^{-1}$$

Making a quick check, we note that $p_k{}^T(1)$ is equal to unity.

$$E(k) = \left[\frac{d}{dz} p_k{}^T(z)\right]_{z=1} = 8$$

$$\sigma_k{}^2 = \left[\frac{d^2}{dz^2} p_k{}^T(z) + \frac{d}{dz} p_k{}^T(z) - \left(\frac{d}{dz} p_k{}^T(z)\right)^2\right]_{z=1} = 72$$

b We shall do this part two ways. First, for our particular PMF we can evaluate the answer directly. Let A represent the event that the experimental value of k is even.

$$P(A) = \sum_{k_0 \text{ even}} p_k(k_0) = \frac{1}{9}\left(1 + \frac{8^2}{9^2} + \frac{8^4}{9^4} + \frac{8^6}{9^6} + \cdots\right)$$

$$P(A) = \frac{1/9}{1 - 64/81} = \frac{9}{17}$$

The monotonically decreasing PMF for random variable k makes it obvious that $P(A) > 0.5$, since $p_k(0) > p_k(1)$, $p_k(2) > p_k(3)$, etc.

Another approach, which is applicable to a more general problem where we may not be able to sum $\sum_{k_0 \text{ even}} p_k(k_0)$ directly, follows:

$$P(A) = \sum_{k_0 \text{ even}} p_k(k_0) = \frac{1}{2}\left[\sum_{k_0} p_k(k_0)(1)^{k_0} + \sum_{k_0} p_k(k_0)(-1)^{k_0}\right]$$

$$P(A) = \tfrac{1}{2}[1 + p_k{}^T(-1)]$$

For our example we have $p_k{}^T(z) = (9 - 8z)^{-1}$, $p_k{}^T(-1) = \frac{1}{17}$, resulting in $P(A) = 9/17$.

c Let r be the sum of n independent experimental values of random variable k. In Sec. 3-5, we learned that

$$p_r{}^T(z) = [p_k{}^T(z)]^n$$

which we simply substitute into the expression obtained in (*b*) above, to get

$$\text{Prob(exper. value of } r \text{ is even)} = \tfrac{1}{2}[1 + p_r{}^T(-1)] = \tfrac{1}{2}[1 + (\tfrac{1}{17})^n]$$

As we might expect on intuitive grounds, this probability rapidly approaches 0.5 as n grows.

d This part is concerned with the sum of a random number of independent identically distributed random variables. Continuous random variable τ is the sum of k independent experimental values of random variable x. From Sec. 3-7, we have

$$f_\tau^T(s) = p_k^T[f_x^T(s)] = [p_k^T(z)]_{z=f_x^T(s)}$$

For $f_x(x_0)$, the exponential PDF, we have

$$f_x^T(s) = \int_{x_0=0}^{\infty} e^{-sx_0}\lambda e^{-\lambda x_0}\,dx_0 = \frac{\lambda}{s+\lambda}$$

which results in

$$f_\tau^T(s) = \left(9 - \frac{8\lambda}{s+\lambda}\right)^{-1} = \frac{s+\lambda}{9s+\lambda} \qquad [f_\tau^T(s)]_{s=0} \overset{\checkmark}{=} 1$$

We may substitute $E(k)$, σ_k^2, $E(x)$, and σ_x^2 into the formulas of Sec. 3-7 to obtain $E(\tau)$ and σ_τ^2, or we may use the relations

$$E(\tau) = -\left[\frac{d}{ds}\left(\frac{s+\lambda}{9s+\lambda}\right)\right]_{s=0} \quad \text{and} \quad \sigma_\tau^2 = \left[\frac{d^2}{ds^2}\left(\frac{s+\lambda}{9s+\lambda}\right)\right]_{s=0} - \left[\frac{d}{ds}\left(\frac{s+\lambda}{9s+\lambda}\right)\right]_{s=0}^2$$

We'll use the former method, with

$$E(k) = 8$$

$$\sigma_k^2 = 72$$

from part (a) of this example and

$$E(x) = \frac{1}{\lambda}$$

$$\sigma_x^2 = \frac{1}{\lambda^2}$$

(from the example in Sec. 3-3), which, in the expressions of Sec. 3-7 for the expectation and variance of a sum of a random number of independent identically distributed random variables, yields

$$E(\tau) = E(k)E(x) = \frac{8}{\lambda} \qquad \sigma_\tau^2 = E(k)\sigma_x^2 + [E(x)]^2\sigma_k^2 = \frac{80}{\lambda^2}$$

The expected time until the last bulb dies is the same as it would be if we always had eight bulbs. But, because of the probabilistic behavior of k, the variance of this time is far greater than the value $8/\lambda^2$ which would describe σ_τ^2 if we always started out with exactly eight bulbs.

e Let $A_1, A_2, \ldots, A_k$ be a list of mutually exclusive collectively exhaustive events. Assume that there is a continuous random variable y which is not independent of the A_i's. Then it is useful to write

$$f_y(y_0) = \sum_i P(A_i) f_{y|A_i}(y_0 \mid A_i)$$

And, from the definition of the s transform, we note that $f_y{}^T(s)$ would be the weighted sum of the transforms of the conditional PDF's $f_{y|A_i}(y_0 \mid A_i)$. If we define

$$f_{y|A_i}^T(s) = \int_{y_0=-\infty}^{\infty} e^{-sy_0} f_{y|A_i}(y_0 \mid A_i)\, dy_0 = E(e^{-sy} \mid A_i)$$

we have

$$f_y{}^T(s) = \sum_i P(A_i) f_{y|A_i}^T(s)$$

When we wish to take inverse transforms (go back to a PDF from a transform), we shall try to express the s transform to be inverted, $f_y{}^T(s)$, in the above form such that we can recognize the inverse transform of each $f_{y|A_i}^T(s)$.

In our particular problem, where the PMF for k is of the form

$$p_k(k_0) = (1 - P)P^{k_0} \qquad k_0 = 0, 1\ 2, \ldots ; \quad 1 > P > 0$$

and the PDF for x is the exponential PDF

$$f_x(x_0) = \lambda \mu_{-1}(x_0 - 0) e^{-\lambda x_0}$$

it happens that we may obtain $f_\tau(\tau_0)$ from $f_\tau{}^T(s)$ by the procedure discussed above. We begin by using long division to obtain

$$f_\tau{}^T(s) = \frac{s + \lambda}{9s + \lambda} = \frac{1}{9} + \frac{8}{9} \frac{\lambda/9}{s + \lambda/9}$$

which is of the form

$$f_\tau{}^T(s) = \tfrac{1}{9} f_{\tau|A_1}^T(s) + \tfrac{8}{9} f_{\tau|A_2}^T(s)$$

From the examples carried out in Sec. 3-1, we note that

$$f_{\tau|A_1}^T(s) = 1$$

has the inverse transform $f_{\tau|A_1}(\tau_0 \mid A_1) = \mu_0(\tau_0 - 0)$, and also

$$f_{\tau|A_2}^T(s) = \frac{\lambda/9}{s + \lambda/9}$$

has the inverse transform $f_{\tau|A_2}(\tau_0 \mid A_2) = \mu_{-1}(\tau_0 - 0) \frac{\lambda}{9} e^{-\lambda\tau_0/9}$

and, finally, we have the PDF for the duration of the interval during which the lights are on.

$$f_\tau(\tau_0) = \tfrac{1}{9}\mu_0(\tau_0 - 0) + \tfrac{8}{9}\mu_{-1}(\tau_0 - 0) \frac{\lambda}{9} e^{-\lambda\tau_0/9}$$

The impulse at $\tau_0 = 0$ in this PDF is due to the fact that, with probability $\frac{1}{9}$, we start out with zero bulbs. Thus our PDF $f_\tau(\tau_0)$ is a *mixed* PDF, having both a discrete and a continuous component. We conclude with a sketch of the PDF $f_\tau(\tau_0)$

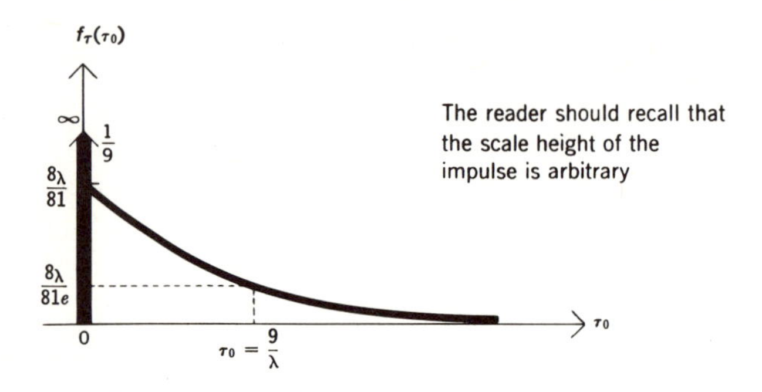

3-9 Where Do We Go from Here?

We are already familiar with most of the basic concepts, methods, and tools of applied probability theory. By always reasoning in an appropriate sample or event space, we have had little difficulty in going from concepts to applications.

Three main areas of study are to follow:

1. Probabilistic processes (Chaps. 4 and 5)
2. Limit theorems (Chap. 6)
3. Statistical reasoning (Chap. 7)

Although we shall consider these topics in the above order, this does not necessarily reflect their relative importance in the world of applied probability theory. Further study of the consequences of the summation of a large number of random variables (limit theorems) is indeed basic. Many probabilicists work solely at attempting to make reasonable inferences from actual physical data (statistical reasoning).

Our choice of the order of these topics is based on the contention that, if we first develop an understanding of several processes and their properties, we may then begin a more meaningful discussion of limit theorems and statistics. The following two chapters are concerned with those probabilistic processes which form the most basic building blocks from which models of actual physical processes are constructed.

PROBLEMS

3.01 A sufficient condition for the existence of an integral is that the integral of the *magnitude* of the integrand be finite. Show that, at least for purely imaginary values of s, $s = j\omega$, this condition is always satisfied by the s transform of a PDF.

3.02 If we allow s to be the complex quantity, $s = \alpha + j\omega$, determine for which values of s in the α,ω plane the s transforms of the following PDF's exist:

a $f_x(x_0) = \begin{cases} \lambda e^{-\lambda x_0} & x_0 \geq 0 \\ 0 & x_0 < 0 \end{cases}$ **b** $f_x(x_0) = \begin{cases} 0 & x_0 > 0 \\ \lambda e^{\lambda x_0} & x_0 \leq 0 \end{cases}$

c $f_x(x_0) = \begin{cases} 0.5\lambda e^{-\lambda x_0} & x_0 \geq 0 \\ 0.5\lambda e^{\lambda x_0} & x_0 < 0 \end{cases}$

3.03 Express the PMF $p_x(x_0) = (1 - P)P^{x_0}, x_0 = 0, 1, 2, \ldots$, as a PDF.

3.04 If z can be the complex number $z = \alpha + j\beta$, determine for which values of z in the α,β plane the z transform of the PMF of Prob. 3.03 will exist.

3.05 All parts of this problem require numerical answers.
a If $f_y{}^T(s) = K/(2 + s)$, evaluate K and $E(y^3)$.
b If $p_x{}^T(z) = (1 + z^2)/2$, evaluate $p_x[E(x)]$ and σ_x.
c If $f_x{}^T(s) = 2(2 - e^{-s/2} - e^{-s})/3s$, evaluate $E(e^{2x})$.
d If $p_x{}^T(z) = A(1 + 3z)^3$, evaluate $E(x^3)$ and $p_x(2)$.

3.06 Determine whether or not the following are valid z transforms of a PMF for a discrete random variable which can take on only nonnegative integer experimental values:
a $z^2 + 2z - 2$ **b** $2 - z$ **c** $(2 - z)^{-1}$

3.07 Show that neither of the following is an s transform of a PDF:
a $(1 - e^{-5s})/s$ **b** $7(4 + 3s)^{-1}$

3.08 Let l be a discrete random variable whose possible experimental values are all nonnegative integers. We are given

$$p_l{}^T(z) = K\left[\frac{14 + 5z - 3z^2}{8(2 - z)}\right]$$

Determine the numerical values of $E(l)$, $p_l(1)$ and of the conditional expected value of l given $l \neq 0$.

3.09 Use the expected-value definition of the s transform to prove that, if x and y are random variables with $y = ax + b$, $f_y{}^T(s) = e^{-sb}f_x{}^T(as)$. (This is a useful expression, and we shall use it in our proof of the central limit theorem in Chap. 6.)

3.10 For a particular batch of cookies, k, the number of nuts in any cookie is an independent random variable described by the probability mass function

$$p_k(k_0) = \tfrac{1}{3}(\tfrac{2}{3})^{k_0} \qquad k_0 = 0, 1, 2, 3, \ldots$$

Human tastes being what they are, assume that the cash value of a cookie is proportional to the third power of the number of nuts in the cookie. The cookie packers (they are chimpanzees) eat all the cookies containing exactly 0, 1, or 2 nuts. All series must be summed.

a What is the probability that a randomly selected cookie is eaten by the chimpanzees?

b What is the probability that a particular nut, chosen at random from the population of all nuts, is eaten by the chimpanzees?

c What is the fraction of the cash value which the chimpanzees consume?

d What is the probability that a random nut will go into a cookie containing exactly R nuts?

3.11 The hitherto uncaught burglar is hiding in city A (with a priori probability 0.3) or in city B (with a priori probability 0.6), or he has left the country. If he is in city A and N_A men are assigned to look for him there, he will be caught with probability $1 - f^{N_A}$. If he is in city B and N_B men are assigned to look for him there, he will be caught with probability $1 - f^{N_B}$. If he has left the country, he won't be captured.

Policemen's lives being as hectic as they are, N_A and N_B are independent random variables described by the probability mass functions

$$p_{N_A}(N) = \frac{2^N e^{-2}}{N!} \qquad N = 0, 1, 2, \ldots$$

$$p_{N_B}(N) = (\tfrac{1}{2})^N \qquad N = 1, 2, 3, \ldots$$

a What is the probability that a total of three men will be assigned to search for the burglar?

b What is the probability that the burglar will be caught? (All series are to be summed.)

c Given that he was captured in a city in which exactly K men had been assigned to look for him, what is the probability that he was found in city A?

3.12 The number of "leads" (contacts) available to a salesman on any given day is a Poisson random variable with probability mass function

$$p_k(k_0) = \frac{\mu^{k_0} e^{-\mu}}{k_0!} \qquad k_0 = 0, 1, 2, \ldots$$

The probability that any particular lead will result in a sale is 0.5. If your answers contain any series, the series must be summed.

a What is the probability that the salesman will make exactly one sale on any given day?

b If we randomly select a sales receipt from his file, what is the probability that it represents a sale made on a day when he had a total of R leads?

c What fraction of all his leads comes on days when he has exactly one sale?

d What is the probability that he has no sales on a given day?

3.13 The probability that a store will have exactly k_0 customers on any given day is

$$p_k(k_0) = \tfrac{1}{5}(\tfrac{4}{5})^{k_0} \qquad k_0 = 0, 1, 2, \ldots$$

On each day when the store has had at least one customer, one of the sales slips for that day is picked out of a hat, and a door prize is mailed to the corresponding customer. (No customer goes to this store more than once or buys more or less than exactly one item.)

a What is the probability that a customer selected randomly from the population of all customers will win a door prize?

b Given a customer who has won a door prize, what is the probability that he was in the store on a day when it had a total of exactly k_0 customers?

3.14 Independent random variables x and y have PDF's whose s transforms are $f_x{}^T(s)$ and $f_y{}^T(s)$. Random variable r is defined to be $r = x + y$. Use $f_r{}^T(s)$ and the moment generating properties of transforms to show that $E(r) = E(x) + E(y)$ and $\sigma_r{}^2 = \sigma_x{}^2 + \sigma_y{}^2$.

3.15 Let x and y be independent random variables with

$$f_x(x_0) = \begin{cases} \lambda e^{-\lambda x_0} & x_0 \geq 0 \\ 0 & x_0 < 0 \end{cases} \qquad f_y(y_0) = \begin{cases} 0 & y_0 > 0 \\ \lambda e^{\lambda y_0} & y_0 \leq 0 \end{cases}$$

Random variable r is defined by $r = x + y$.

Determine:

a $f_x{}^T(s)$, $f_y{}^T(s)$, and $f_r{}^T(s)$.

b $E(r)$ and σ_r^2.
c $f_r(r_0)$.
d Repeat the previous parts for the case $r = ax + by$.

3.16 Consider the PDF $f_x(x_0) = \mu_{-1}(x_0 - 0) - \mu_{-1}(x_0 - 1)$. Random variable y is defined to be the sum of two independent experimental values of x. Determine the PDF $f_y(y_0)$:
a In an appropriate two-dimensional event space
b By performing the convolution graphically
c By taking the inverse transform of $f_y^T(s)$ (if you can)

3.17 **a** If x and y are *any* independent discrete random variables with PMF's $p_x(x_0)$ and $p_y(y_0)$ and we define $r = x + y$, show that $p_r(r_0) = \sum_{x_0} p_x(x_0)p_y(r_0 - x_0) = \sum_{y_0} p_y(y_0)p_x(r_0 - y_0)$. These summations are said to represent the *discrete convolution*. Show how you would go about performing a discrete convolution graphically.
b For the case where x and y are discrete, independent random variables which can take on only nonnegative-integer experimental values, take the z transform of one of the above expressions for $p_r(r_0)$ to show that $p_r^T(z) = p_x^T(z)p_y^T(z)$.

3.18 Random variable x has the PDF $f_x(x_0)$, and we define the *Mellin transform* $f_x^M(s)$ to be

$$f_x^M(s) = E(x^s)$$

a Determine $E(x)$ and σ_x^2 in terms of $f_x^M(s)$.
b Let y be a random variable with

$$f_y(y_0) = Ky_0f_x(y_0)$$

i Determine K.
ii Determine $f_y^M(s)$ in terms of $f_x^M(s)$.
iii Evaluate $f_x^M(s)$ and $f_y^M(s)$ for

$$f_x(x_0) = \begin{cases} 1 & \text{if } 0 < x_0 \leq 1 \\ 0 & \text{otherwise} \end{cases}$$

and use your results to determine $E(y)$ and σ_y.
c Let w and r be independent random variables with PDF's $f_w(w_0)$ and $f_r(r_0)$ and Mellin transforms $f_w^M(s)$ and $f_r^M(s)$. If we define $l = wr$, find $f_l^M(s)$ in terms of the Mellin transforms for w and r.

3.19 A fair wheel of fortune, calibrated infinitely finely from zero to unity, is spun k times, and the resulting readings are summed to obtain an experimental value of random variable r. Discrete random variable k has the PMF $p_k(k_0) = \dfrac{\lambda^{k_0}e^{-\lambda}}{k_0!}$, $k_0 = 0, 1, 2, \ldots .$

Determine:

a The probability that at least one reading is larger than 0.3

b $f_r^T(s)$

c $E(r^2)$

3.20 Widgets are packed into cartons which are packed into crates. The weight (in pounds) of a widget is a continuous random variable with PDF

$$f_x(x_0) = \lambda e^{-\lambda x_0} \qquad x_0 \geq 0$$

The number of widgets in any carton, K, is a random variable with the PMF

$$p_K(K_0) = \frac{\mu^{K_0} e^{-\mu}}{K_0!} \qquad K_0 = 0, 1, 2, \ldots$$

The number of cartons in a crate, N, is a random variable with PMF

$$p_N(N_0) = P^{N_0-1}(1 - P) \qquad N_0 = 1, 2, 3, \ldots$$

Random variables x, K, and N are mutually independent.

Determine:

a The probability that a randomly selected crate contains exactly one widget

b The conditional PDF for the total weight of widgets in a carton given that the carton contains less than two widgets

c The s transform of the PDF for the total weight of the widgets in a crate

d The probability that a randomly selected crate contains an odd number of widgets

3.21 The number of customers who shop at a supermarket in a day has the PMF

$$p_k(k_0) = \frac{\lambda^{k_0} e^{-\lambda}}{k_0!} \qquad k_0 = 0, 1, 2, \ldots$$

and, independent of k, the number of items purchased by any customer has the PMF

$$p_l(l_0) = \frac{\mu^{l_0} e^{-\mu}}{l_0!} \qquad l_0 = 0, 1, 2, \ldots$$

Two ways the market can obtain a 10% increase in the expected value of the number of items sold are:

a To increase μ by 10%

b To increase λ by 10%

Which of these changes would lead to the smaller variance of the total items sold per day?

CHAPTER FOUR

some basic probabilistic processes

This chapter presents a few simple probabilistic processes and develops family relationships among the PMF's and PDF's associated with these processes.

Although we shall encounter many of the most common PMF's and PDF's here, it is not our purpose to develop a general catalogue. A listing of the most frequently occurring PMF's and PDF's and some of their properties appears as an appendix at the end of this book.

4-1 The Bernoulli Process

A single *Bernoulli trial* generates an experimental value of discrete random variable x, described by the PMF

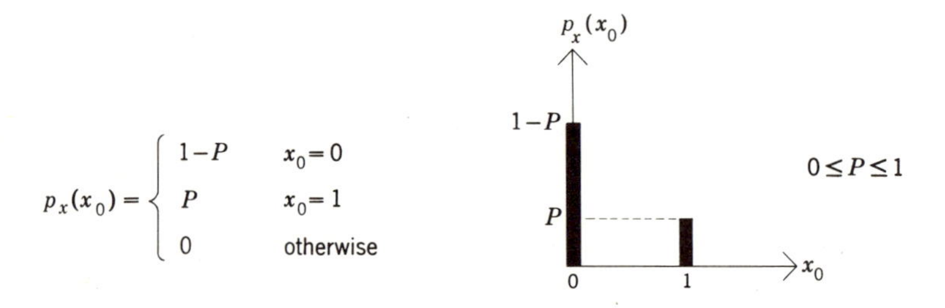

Random variable x, as described above, is known as a *Bernoulli random variable,* and we note that its PMF has the z transform

$$p_x{}^T(z) = \sum_{x_0} z^{x_0} p_x(x_0) = z^0(1 - P) + zP = 1 - P + zP$$

The sample space for each Bernoulli trial is of the form

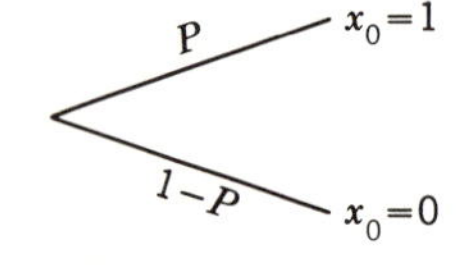

Either by use of the transform or by direct calculation we find

$$E(x) = P \qquad E(x^2) = P \qquad \sigma_x{}^2 = P(1 - P)$$

We refer to the outcome of a Bernoulli trial as a *success* when the experimental value of x is unity and as a *failure* when the experimental value of x is zero.

A *Bernoulli process* is a series of independent Bernoulli trials, each with the same probability of success. Suppose that n independent Bernoulli trials are to be performed, and define discrete random variable k to be the number of successes in the n trials. Random variable k is noted to be the sum of n independent Bernoulli random variables, so we must have

$$p_k{}^T(z) = [p_x{}^T(z)]^n = (1 - P + zP)^n$$

There are several ways to determine $p_k(k_0)$, the probability of exactly k_0 successes in n independent Bernoulli trials. One way would be to apply the binomial theorem

$$(a+b)^n = \sum_{l=0}^{n} \binom{n}{l} a^l b^{n-l}$$

to expand $p_k{}^T(z)$ in a power series and then note the coefficient of z^{k_0} in this expansion, recalling that any z transform may be written in the form

$$p_k{}^T(z) = p_k(0) + zp_k(1) + z^2p_k(2) + \cdots$$

This leads to the result known as the *binomial* PMF,

$$p_k(k_0) = \binom{n}{k_0} P^{k_0}(1-P)^{n-k_0} \qquad k_0 = 0, 1, 2, \ldots, n$$

where the notation is the common

$$\binom{n}{k_0} = \frac{n!}{(n-k_0)!k_0!}$$

discussed in Sec. 1-9.

Another way to derive the binomial PMF would be to work in a sequential sample space for an experiment which consists of n independent Bernoulli trials,

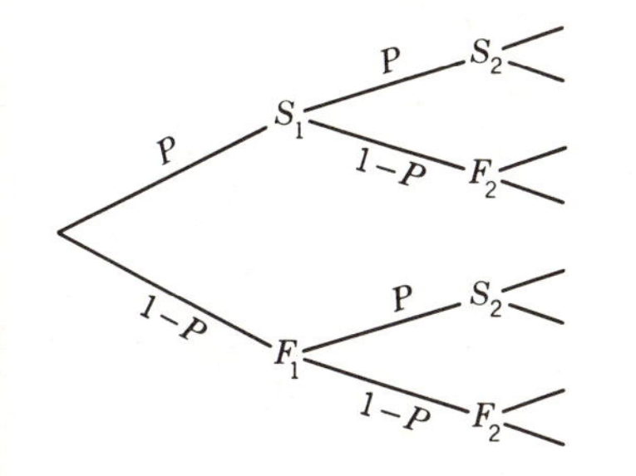

We have used the notation

$$\begin{Bmatrix} S_n \\ F_n \end{Bmatrix} = \begin{Bmatrix} \text{success} \\ \text{failure} \end{Bmatrix} \text{ on the } n\text{th trial}$$

Each sample point which represents an outcome of exactly k_0 successes in the n trials would have a probability assignment equal to $P^{k_0}(1-P)^{n-k_0}$. For each value of k_0, we use the techniques of Sec. 1-9 to determine that there are $\binom{n}{k_0}$ such sample points. Thus, we again obtain

$$p_k(k_0) = \binom{n}{k_0} P^{k_0}(1-P)^{n-k_0} \qquad k_0 = 0, 1, 2, \ldots, n$$

for the binomial PMF.

We can determine the expected value and variance of the *binomial random variable* k by any of three techniques. (One should always review his arsenal before selecting a weapon.) To evaluate $E(k)$ and $\sigma_k{}^2$ we may

1 Perform the expected value summations directly.
2 Use the moment-generating properties of the z transform, introduced in Sec. 3-3.
3 Recall that the expected value of a sum of random variables is *always* equal to the sum of their expected values and that the variance of a sum of *linearly independent* random variables is equal to the sum of their individual variances.

Since we know that binomial random variable k is the sum of n independent Bernoulli random variables, the last of the above methods is the easiest and we obtain

$$E(k) = nE(x) = nP \qquad \sigma_k^2 = n\sigma_x^2 = nP(1 - P)$$

Before moving on to other aspects of the Bernoulli process, let's look at a plot of a binomial PMF. The following plot presents $p_k(k_0)$ for a Bernoulli process, with $P = \frac{1}{3}$ and $n = 4$.

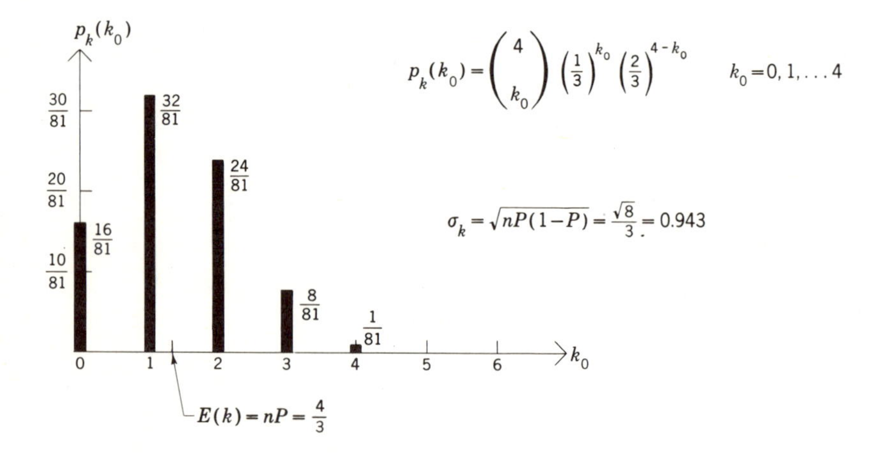

4-2 Interarrival Times for the Bernoulli Process

It is often convenient to refer to the successes in a Bernoulli process as *arrivals*. Let discrete random variable l_1 be the number of Bernoulli trials up to and including the first success. Random variable l_1 is known as the *first-order interarrival time*, and it can take on the experimental values 1, 2, ... We begin by determining the PMF $p_{l_1}(l)$. (Note that since we are subscripting the random variable there is no reason to use a subscripted dummy variable in the argument of the PMF.)

We shall determine $p_{l_1}(l)$ from a sequential sample space for the experiment of performing independent Bernoulli trials until we obtain our first success. Using the notation of the last section, we have

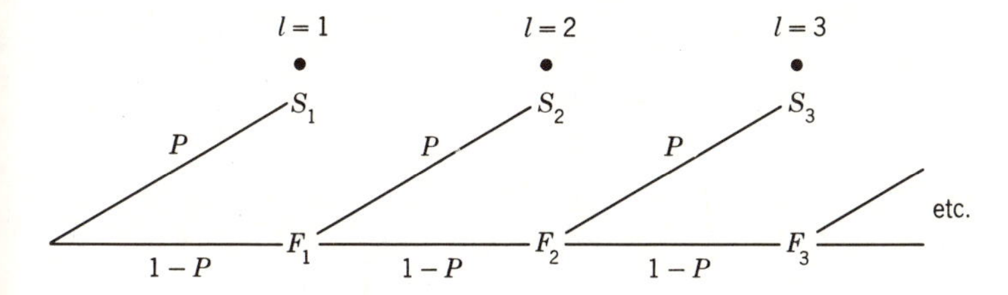

We have labeled each sample point with the experimental value of random variable l_1 associated with the experimental outcome represented by that point. From the above probability tree we find that

$$p_{l_1}(l) = P(1 - P)^{l-1} \qquad l = 1, 2, \ldots$$

and since its successive terms decrease in a geometric progression, this PMF for the first-order interarrival times is known as the *geometric* PMF. The z transform for the geometric PMF is

$$p_{l_1}{}^T(z) = \sum_{l=0}^{\infty} p_{l_1}(l)z^l = \sum_{l=1}^{\infty} P(1 - P)^{l-1}z^l = \frac{zP}{1 - z(1 - P)}$$

Since direct calculation of $E(l_1)$ and $\sigma_{l_1}{}^2$ in an l_1 event space involves difficult summations, we shall use the moment-generating property of the z transform to evaluate these quantities.

$$E(l_1) = \left[\frac{d}{dz} p_{l_1}{}^T(z)\right]_{z=1} = \frac{1}{P}$$

$$\sigma_{l_1}{}^2 = \left\{\frac{d^2}{dz^2} p_{l_1}{}^T(z) + \frac{d}{dz} p_{l_1}{}^T(z) - \left[\frac{d}{dz} p_{l_1}{}^T(z)\right]^2\right\}_{z=1} = \frac{1 - P}{P^2}$$

Suppose that we were interested in the conditional PMF for the *remaining number of trials* up to and including the next success, given that there were no successes in the first m trials. By conditioning the event space for l_1 we would find that the conditional PMF for $y = l_1 - m$, the remaining number of trials until the next success, is still a geometric random variable with parameter P (see Prob. 4.03). This is a result attributable to the "no-memory" property (independence of trials) of the Bernoulli process. The PMF $p_{l_1}(l)$ was obtained as the PMF for the number of trials up to and including the first success. Random variable l_1, the first-order interarrival time, represents both the waiting time (number of trials) from one success through the next success and the waiting time from *any* starting time through the next success.

Finally, we wish to consider the higher-order interarrival times for a Bernoulli process. Let random variable l_r, called the *rth-order interarrival time,* be the number of trials up to and including the rth success. Note that l_r is the sum of r independent experimental values of random variable l_1; so we must have

$$p_{l_r}{}^T(z) = [p_{l_1}{}^T(z)]^r = \left[\frac{zP}{1 - z(1 - P)}\right]^r$$

There are several ways we might attempt to take the inverse of this transform to obtain $p_{l_r}(l)$, the PMF for the rth-order interarrival time, but the following argument seems both more intuitive and more efficient. Since $p_{l_r}(l)$ represents the probability that the rth success in a Bernoulli process arrives on the lth trial, $p_{l_r}(l)$ may be expressed as

$$p_{l_r}(l) = \begin{pmatrix}\text{probability of having} \\ \text{exactly } r-1 \text{ successes} \\ \text{in the first } l-1 \text{ trials}\end{pmatrix} \times \begin{pmatrix}\text{conditional probability of hav-} \\ \text{ing } r\text{th success on the } l\text{th trial,} \\ \text{given exactly } r-1 \text{ successes} \\ \text{in the previous } l-1 \text{ trials}\end{pmatrix}$$

The first term in the above product is the binomial PMF evaluated for the probability of exactly $r - 1$ successes in $l - 1$ trials. Since the outcome of each trial is independent of the outcomes of all other trials, the second term in the above product is simply equal to P, the probability of success on any trial. We may now substitute for all the words in the above equation to determine the PMF for the rth-order interarrival time (the number of trials up to and including the rth success) for a Bernoulli process

$$\begin{aligned} p_{l_r}(l) &= \left[\binom{l-1}{r-1} P^{r-1}(1-P)^{l-1-(r-1)}\right] P \\ &= \binom{l-1}{r-1} P^r(1-P)^{l-r} \qquad l = r, r+1, r+2, \ldots; \quad r = 1, 2, 3, \ldots \end{aligned}$$

Of course, with $r = 1$, this yields the geometric PMF for l_1 and thus provides an alternative derivation of the PMF for the first-order interarrival times. The PMF $p_{l_r}(l)$ for the number of trials up to and including the rth success in a Bernoulli process is known as the *Pascal* PMF. Since l_r is the sum of r independent experimental values of l_1, we have

$$E(l_r) = rE(l_1) = \frac{r}{P} \qquad \sigma_{l_r}{}^2 = r\sigma_{l_1}{}^2 = \frac{r(1-P)}{P^2}$$

The *negative binomial PMF*, a PMF which is very closely related to the Pascal PMF, is noted in Prob. 4.01.

As one last note regarding the Bernoulli process, we recognize that the relation

$$\sum_{r=1}^{\infty} p_{l_r}(l) = P \qquad l = 1, 2, \ldots$$

is always true. The quantity $p_{l_r}(l)$, evaluated for any value of l, equals the probability that the rth success will occur on the lth trial. Any success on the lth trial must be the first, or second, or third, etc., success after $l = 0$. Therefore, the sum of $p_{l_r}(l)$ over r simply represents the probability of a success on the lth trial. From the definition of the Bernoulli process, this probability is equal to P.

Our results for the Bernoulli process are summarized in the following section.

4-3 Summary of the Bernoulli Process

Each performance of a Bernoulli trial generates an experimental value of the *Bernoulli* random variable x described by

$$p_x(x_0) = \begin{cases} 1 - P & x_0 = 0 \text{ (a "failure")} \\ P & x_0 = 1 \text{ (a "success")} \end{cases}$$

$$p_x^T(z) = 1 - P + zP \qquad E(x) = P \qquad \sigma_x^2 = P(1 - P)$$

A series of identical independent Bernoulli trials is known as a *Bernoulli process.* The number of successes in n trials, random variable k, is the sum of n independent Bernoulli random variables and is described by the *binomial* PMF

$$p_k(k_0) = \binom{n}{k_0} P^{k_0}(1 - P)^{n-k_0} \qquad k_0 = 0, 1, 2, \ldots, n$$

$$p_k^T(z) = (1 - P + zP)^n \qquad E(k) = nP \qquad \sigma_k^2 = nP(1 - P)$$

The number of trials up to and including the first success is described by the PMF for random variable l_1, called the *first-order interarrival (or waiting) time.* Random variable l_1 has a *geometric* PMF

$$p_{l_1}(l) = P(1 - P)^{l-1} \qquad l = 1, 2, \ldots$$

$$p_{l_1}^T(z) = \frac{zP}{1 - z(1 - P)} \qquad E(l_1) = \frac{1}{P} \qquad \sigma_{l_1}^2 = \frac{1 - P}{P^2}$$

The number of trials up to and including the rth success, l_r, is called the *rth-order interarrival time.* Random variable l_r, the sum of r independent experimental values of l_1, has the *Pascal* PMF

$$p_{l_r}(l) = \binom{l-1}{r-1} P^r(1-P)^{l-r} \qquad l = r, r+1, \ldots ; \quad r = 1, 2, \ldots$$

$$p_{l_r}{}^T(z) = \left[\frac{zP}{1 - z(1-P)}\right]^r \qquad E(l_r) = \frac{r}{P} \qquad \sigma_{l_r}{}^2 = \frac{r(1-P)}{P^2}$$

We conclude with one useful observation based on the definition of the Bernoulli process. *Any* events, defined on nonoverlapping sets of trials, are independent. If we have a list of events defined on a series of Bernoulli trials, but there is no trial whose outcome is relevant to the occurrence or nonoccurrence of more than one event in the list, the events in the list are mutually independent. This result, of course, is due to the independence of the individual trials and is often of value in the solution of problems.

4-4 An Example

We consider one example of the application of our results for the Bernoulli process. The first five parts are a simple drill, and part (*f*) will lead us into a more interesting discussion.

Fred is giving out samples of dog food. He makes calls door to door, but he leaves a sample (one can) only on those calls for which the door is answered *and* a dog is in residence. On any call the probability of the door being answered is 3/4, and the probability that any household has a dog is 2/3. Assume that the events "Door answered" and "A dog lives here" are independent and also that the outcomes of all calls are independent.

(a) Determine the probability that Fred gives away his first sample on his third call.

(b) Given that he has given away exactly four samples on his first eight calls, determine the conditional probability that Fred will give away his fifth sample on his eleventh call.

(c) Determine the probability that he gives away his second sample on his fifth call.

(d) Given that he did not give away his second sample on his second call, determine the conditional probability that he will leave his second sample on his fifth call.

(e) We shall say that Fred "needs a new supply" immediately *after* the call on which he gives away his last can. If he starts out with two cans, determine the probability that he completes at least five calls before he needs a new supply.

(f) If he starts out with exactly m cans, determine the expected value and variance of d_m, the number of homes with dogs which he passes up (because of no answer) before he needs a new supply.

We begin by sketching the event space for each call.

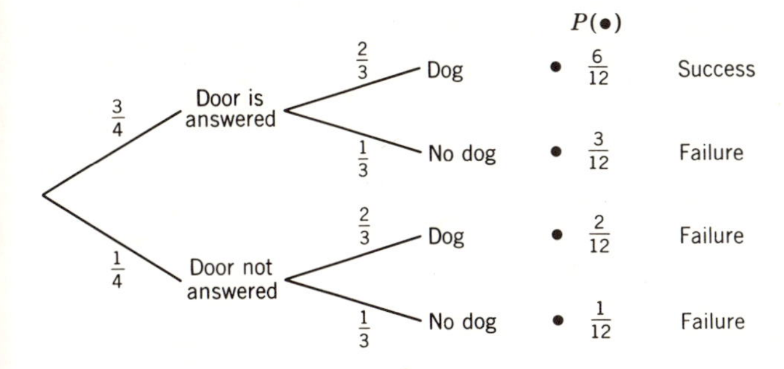

For all but the last part of this problem, we may consider each call to be a Bernoulli trial where the probability of success (door answered and dog in residence) is given by $P = \frac{3}{4} \cdot \frac{2}{3} = \frac{1}{2}$.

a Fred will give away his first sample on the third call if the first two calls are failures and the third is a success. Since the trials are independent, the probability of this sequence of events is simply $(1 - P)(1 - P)P = 1/8$. Another way to obtain this answer is to realize that, in the notation of the previous section, we want $p_{l_1}(3)$ which is $(1 - P)^2P = 1/8$.

b The event of interest requires failures on the ninth and tenth trials and a success on the eleventh trial. For a Bernoulli process, the outcomes of these three trials are independent of the results of any other trials, and again our answer is $(1 - P)(1 - P)P = 1/8$.

c We desire the probability that l_2, the second-order interarrival time, is equal to five trials. We know that $p_{l_2}(l)$ is a Pascal PMF, and we have

$$p_{l_2}(5) = \binom{5-1}{2-1} P^2(1 - P)^{5-2} = \frac{4}{32} = \frac{1}{8}$$

d Here we require the conditional probability that the experimental value of l_2 is equal to 5, given that it is greater than 2.

$$p_{l_2|l_2>2}(5 \mid l_2 > 2) = \frac{p_{l_2}(5)}{\text{Prob}(l_2 > 2)} = \frac{p_{l_2}(5)}{1 - p_{l_2}(2)} \qquad l = 3, 4, 5, \ldots$$

$$= \frac{\binom{5-1}{2-1} P^2(1 - P)^{5-2}}{1 - \binom{2-1}{2-1} P^2(1 - P)^0} = \frac{1/8}{3/4} = \frac{1}{6}$$

As we would expect, by excluding the possibility of one particular experimental value of l_2, we have increased the probability that the experimental value of l_2 is equal to 5. The PMF for the total number of trials up to and including the rth success (since the process began) does, of course, depend on the past history of the process.

e The probability that Fred will complete at least five calls before he needs a new supply is equal to the probability that the experimental value of l_2 is greater than or equal to 5.

$$\text{Prob}(l_2 \geq 5) = 1 - \text{Prob}(l_2 \leq 4) = 1 - \sum_{l=2}^{4} \binom{l-1}{2-1} P^2(1-P)^{l-2} = \frac{5}{16}$$

f Let discrete random variable f_m represent the number of failures before Fred runs out of samples on his mth successful call. Since l_m is the number of trials up to and including the mth success, we have $f_m = l_m - m$. Given that Fred makes l_m calls before he needs a new supply, we can regard each of the f_m unsuccessful calls as trials in another Bernoulli process where P', the probability of a success (a disappointed dog), is found from the above event space to be

$$P' = \text{Prob(dog lives there} \mid \text{Fred did not leave a sample)}$$
$$= \frac{(1/4)(2/3)}{(3/4)(1/3) + (1/4)(2/3) + (1/4)(1/3)} = \frac{1}{3}$$

We define x to be a Bernoulli random variable with parameter P'.

The number of dogs passed up before Fred runs out, d_m, is equal to the sum of f_m (a random number) Bernoulli random variables each with $P' = 1/3$. From Sec. 3-7, we know that the z transform of $p_{d_m}(d)$ is equal to the z transform of $p_{f_m}(f)$, with z replaced by the z transform of Bernoulli random variable x. Without formally obtaining $p_{d_m}{}^T(z)$, we may use the results of Sec. 3-7 to evaluate $E(d_m)$ and $\sigma_{d_m}{}^2$.

$$E(d_m) = E(f_m)E(x) \qquad \text{from Sec. 3-7}$$

$$E(f_m) = E(l_m - m) = \frac{m}{P} - m = m\frac{1-P}{P} \qquad E(x) = P'$$

We substitute these expected values into the above equation for $E(d_m)$, the expected value of the number of dogs passed up.

$$E(d_m) = m\frac{1-P}{P}P' = m\frac{\frac{1}{2}}{\frac{1}{2}}\frac{1}{3} = \frac{m}{3} = \text{expected value of no. of dogs passed up before Fred gives away } m\text{th sample}$$

We find the variance of d_m by

$$\sigma_{d_m}{}^2 = E(f_m)\sigma_x{}^2 + [E(x)]^2\sigma_{f_m}{}^2 \qquad \text{from Sec. 3-7}$$

$$\sigma_{f_m}{}^2 = \sigma_{l_m}{}^2$$

Since $f_m = l_m - m$, the PMF for f_m is simply the PMF for l_m shifted to the left by m. Such a shift doesn't affect the spread of the PMF about its expected value.

$$\sigma_{l_m}{}^2 = m\frac{(1-P)}{P^2} \qquad \text{from properties of Pascal PMF noted in previous section}$$

We may now substitute into the above equation for $\sigma_{d_m}{}^2$, the variance of the number of dogs passed up.

$$\sigma_{d_m}{}^2 = m\frac{1-P}{P}P'(1-P') + (P')^2 m\frac{1-P}{P^2}$$
$$= m\frac{\frac{1}{2}}{\frac{1}{2}}\cdot\frac{1}{3}\cdot\frac{2}{3} + \left(\frac{1}{3}\right)^2 m\frac{\frac{1}{2}}{\frac{1}{4}} = \frac{4m}{9}$$

Although the problem did not require it, let's obtain the z transform of $p_{d_m}(d)$, which is to be obtained by substitution into

$$p_{f_m}{}^T[p_x{}^T(z)]$$

We know that $p_x{}^T(z) = 1 - P' + P'z = \frac{2}{3} + \frac{1}{3}z$, and, using the fact that $f_m = l_m - m$, we can write out $p_{l_m}{}^T(z)$ and $p_{f_m}{}^T(z)$ and note a simple relation to obtain the latter from the former.

$$p_{l_m}{}^T(z) = p_{l_m}(m)z^m + p_{l_m}(m+1)z^{m+1} + p_{l_m}(m+2)z^{m+2} + \cdots$$
$$p_{f_m}{}^T(z) = p_{l_m}(m)z^0 + p_{l_m}(m+1)z^1 + p_{l_m}(m+2)z^2 + \cdots$$

From these expansions and our results from the Pascal process we have

$$p_{f_m}{}^T(z) = z^{-m}p_{l_m}{}^T(z) = P^m[1 - z(1-P)]^{-m}$$

and, finally,

$$p_{d_m}{}^T(z) = p_{f_m}{}^T[p_x{}^T(z)] = P^m[1 - (1 - P' + zP')(1-P)]^{-m}$$
$$= \left(\frac{1}{2}\right)^m\left(\frac{2}{3} - \frac{z}{6}\right)^{-m}$$

Since the z transform for the PMF of the number of dogs passed up happened to come out in such a simple form, we can find the PMF $p_{d_m}(d)$ by applying the inverse transform relationship from Sec. 3-2. We omit the algebraic work and present the final form of $p_{d_m}(d)$.

$$p_{d_m}(d) = \frac{1}{d!}\left[\frac{d^d}{dz^d}p_{d_m}{}^T(z)\right]_{z=0}$$
$$= \binom{d+m-1}{d}\left(\frac{3}{4}\right)^m\left(\frac{1}{4}\right)^d \qquad m = 1, 2, 3, \ldots; \quad d = 0, 1, 2, \ldots$$

For instance, if Fred starts out with only one sample, we have $m = 1$ and

$$p_{d_1}(d) = (\tfrac{3}{4})(\tfrac{1}{4})^d \qquad d = 0, 1, 2, \ldots$$

is the PMF for the number of dogs who were passed up (Fred called but door not answered) while Fred was out making calls to try and give away his one sample.

4-5 The Poisson Process

We defined the Bernoulli process by a particular probabilistic description of the "arrivals" of successes in a series of independent identical discrete trials. The Poisson process will be defined by a probabilistic description of the behavior of arrivals at *points* on a continuous line.

For convenience, we shall generally refer to this line as if it were a time (t) axis. From the definition of the process, we shall see that a Poisson process may be considered to be the limit, as $\Delta t \to 0$ of a series of identical independent Bernoulli trials, one every Δt, with the probability of a success on any trial given by $P = \lambda \, \Delta t$.

For our study of the Poisson process we shall adopt the somewhat improper notation:

$\mathcal{P}(k,t)$ = the probability that there are exactly k arrivals during any interval of duration t

This notation, while not in keeping with our more aesthetic habits developed earlier, is compact and particularly convenient for the types of equations to follow. We observe that $\mathcal{P}(k,t)$ is a PMF for random variable k for any fixed value of parameter t. In any interval of length t, with $t \geq 0$, we must have exactly zero, or exactly one, or exactly two, etc., arrivals. Thus we have

$$\sum_{k=0}^{\infty} \mathcal{P}(k,t) = 1$$

We also note that $\mathcal{P}(k,t)$ is *not* a PDF for t. Since $\mathcal{P}(k,t_1)$ and $\mathcal{P}(k,t_2)$ are not mutually exclusive events, we can state only that

$$0 \leq \int_{t=0}^{\infty} \mathcal{P}(k,t) \, dt \leq \infty$$

The use of random variable k to count arrivals is consistent with our notation for counting successes in a Bernoulli process.

There are several equivalent ways to define the Poisson process. We shall define it directly in terms of those properties which are most useful for the analysis of problems based on physical situations.

Our definition of the Poisson process is as follows:

1 *Any* events defined on nonoverlapping time intervals are mutually independent.

2 The following statements are correct for suitably small values of Δt:

$$\mathcal{P}(k,\Delta t) = \begin{cases} 1 - \lambda \, \Delta t & k = 0 \\ \lambda \, \Delta t & k = 1 \\ 0 & k > 1 \end{cases}$$

The first of the above two defining properties establishes the no-memory attribute of the Poisson process. As an example, for a

Poisson process, events A, B, and C, defined on the intervals shown below, are mutually independent.

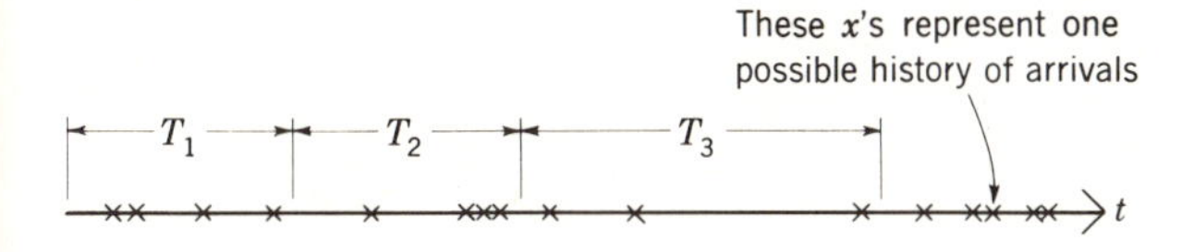

Event A: Exactly k_1 arrivals in interval T_1 and exactly k_3 arrivals in interval T_3
Event B: More than k_2 arrivals in interval T_2
Event C: No arrivals in the hour which begins 10 minutes after the third arrival following the end of interval T_3

The second defining property for the Poisson process states that, *for small enough intervals*, the probability of having exactly one arrival within one such interval is proportional to the duration of the interval and that, to the first order, the probability of more than one arrival within one such interval is zero. This simply means that $\mathcal{P}(k,\Delta t)$ can be expanded in a Taylor series about $\Delta t = 0$, and when we neglect terms of order $(\Delta t)^2$ or higher, we obtain the given expressions for $\mathcal{P}(k,\Delta t)$.

Among other things, we wish to determine the expression for $\mathcal{P}(k,t)$ for $t \geq 0$ and for $k = 0, 1, 2, \ldots$. Before doing the actual derivation, let's reason out how we would expect the result to behave. From the definition of the Poisson process and our interpretation of it as a series of Bernoulli trials in incremental intervals, we expect that

$\mathcal{P}(0,t)$ as a function of t should be unity at $t = 0$ and decrease monotonically toward zero as t increases. (The event of exactly zero arrivals in an interval of length t requires more and more successive failures in incremental intervals as t increases.)

$\mathcal{P}(k,t)$ as a function of t, for $k > 0$, should start out at zero for $t = 0$, increase for a while, and then decrease toward zero as t gets very large. [The probability of having exactly k arrivals (with $k > 0$) should be very small for intervals which are too long or too short.]

$\mathcal{P}(k,0)$ as a function of k should be a bar graph with only one nonzero bar; there will be a bar of height unity at $k = 0$.

We shall use the defining properties of the Poisson process to relate $\mathcal{P}(k, t + \Delta t)$ to $\mathcal{P}(k,t)$ and then solve the resulting differential equations to obtain $\mathcal{P}(k,t)$.

For a Poisson process, if Δt is small enough, we need consider only the possibility of zero or one arrivals between t and $t + \Delta t$. Taking advantage also of the independence of events in nonoverlapping time intervals, we may write

$$\mathcal{P}(k, t + \Delta t) = \mathcal{P}(k,t)\mathcal{P}(0,\Delta t) + \mathcal{P}(k - 1, t)\mathcal{P}(1,\Delta t)$$

The two terms summed on the right-hand side are the probabilities of the only two (mutually exclusive) histories of the process which may lead to having exactly k arrivals in an interval of duration $t + \Delta t$. Our definition of the process specified $\mathcal{P}(0,\Delta t)$ and $\mathcal{P}(1,\Delta t)$ for small enough Δt. We substitute for these quantities to obtain

$$\mathcal{P}(k, t + \Delta t) = \mathcal{P}(k,t)(1 - \lambda\,\Delta t) + \mathcal{P}(k - 1, t)\lambda\,\Delta t$$

Collecting terms, dividing through by Δt, and taking the limit as $\Delta t \to 0$, we find

$$\frac{d}{dt}\,\mathcal{P}(k,t) + \lambda\mathcal{P}(k,t) = \lambda\mathcal{P}(k - 1, t)$$

which may be solved iteratively for $k = 0$ and then for $k = 1$, etc., subject to the initial conditions

$$\mathcal{P}(k,0) = \begin{cases} 1 & k = 0 \\ 0 & k \neq 0 \end{cases}$$

The solution for $\mathcal{P}(k,t)$, which may be verified by direct substitution, is

$$\mathcal{P}(k,t) = \frac{(\lambda t)^k e^{-\lambda t}}{k!} \qquad t \geq 0; \quad k = 0, 1, 2, \ldots$$

And we find that $\mathcal{P}(k,t)$ does have the properties we anticipated earlier.

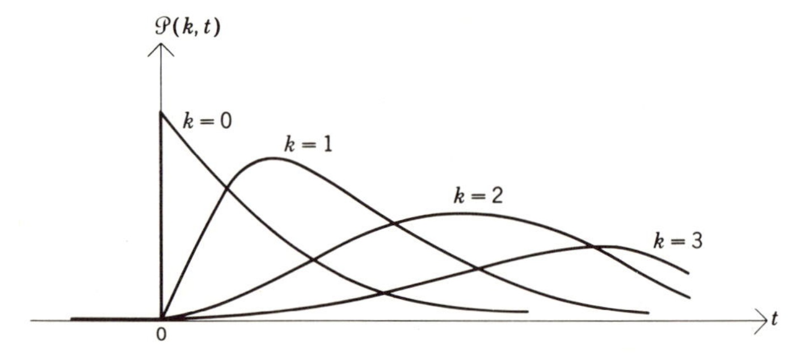

Letting $\mu = \lambda t$, we may write this result in the more proper notation for a PMF as

$$p_k(k_0) = \frac{(\lambda t)^{k_0} e^{-\lambda t}}{k_0!} = \frac{\mu^{k_0} e^{-\mu}}{k_0!} \qquad \mu = \lambda t; \quad k_0 = 0, 1, 2, \ldots$$

This is known as the *Poisson* PMF. Although we derived the Poisson PMF by considering the number of arrivals in an interval of length t for a certain process, this PMF arises frequently in many other situations.

To obtain the expected value and variance of the Poisson PMF, we'll use the z transform

$$p_k{}^T(z) = \sum_{k_0=0}^{\infty} p_k(k_0)z^{k_0} = e^{-\mu} \sum_{k_0=0}^{\infty} \frac{(\mu z)^{k_0}}{k_0!} = e^{\mu(z-1)}$$

$$E(k) = \left[\frac{d}{dz} p_k{}^T(z)\right]_{z=1} = \mu$$

$$\sigma_k{}^2 = \left\{\frac{d^2}{dz^2} p_k{}^T(z) + \frac{d}{dz} p_k{}^T(z) - \left[\frac{d}{dz} p_k{}^T(z)\right]^2\right\}_{z=1} = \mu$$

Thus the expected value and variance of Poisson random variable k are both equal to μ.

We may also note that, since $E(k) = \lambda t$, we have an interpretation of the constant λ used in

$$\mathcal{P}(k,\Delta t) = \begin{cases} 1 - \lambda\,\Delta t & k = 0 \\ \lambda\,\Delta t & k = 1 \\ 0 & k = 2, 3, \ldots \end{cases}$$

as part of the definition of the Poisson process. The relation $E(k) = \lambda t$ indicates that λ is the expected number of arrivals per unit time in a Poisson process. The constant λ is referred to as the *average arrival rate* for the process.

Incidentally, another way to obtain $E(k) = \lambda t$ is to realize that, for sufficiently short increments, the expected number of arrivals in a time increment of length Δt is equal to $0 \cdot (1 - \lambda\,\Delta t) + 1 \cdot \lambda\,\Delta t = \lambda\,\Delta t$. Since an interval of length t is the sum of $t/\Delta t$ such increments, we may determine $E(k)$ by summing the expected number of arrivals in each such increment. This leads to $E(k) = \lambda\,\Delta t \cdot \dfrac{t}{\Delta t} = \lambda t$.

4-6 Interarrival Times for the Poisson Process

Let l_r be a continuous random variable defined to be the interval of time between any arrival in a Poisson process and the rth arrival after it. Continuous random variable l_r, the *rth-order interarrival time*, has the same interpretation here as discrete random variable l_r had for the Bernoulli process.

We wish to determine the PDF's

$$f_{l_r}(l) \qquad l \geq 0; \quad r = 1, 2, 3, \ldots$$

And we again use an argument similar to that for the derivation of the Pascal PMF,

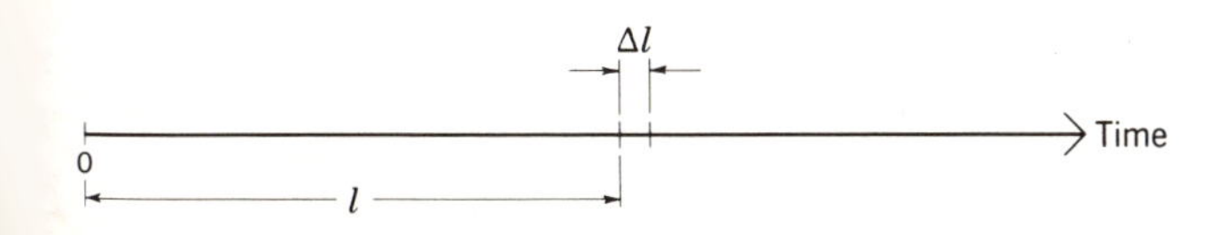

For small enough Δl we may write

$$\text{Prob}(l < l_r \leq l + \Delta l) = f_{l_r}(l)\,\Delta l$$

$$f_{l_r}(l)\,\Delta l = \underbrace{\mathcal{P}(r-1, l)}_{A}\,\underbrace{\lambda\,\Delta l}_{B} = \frac{(\lambda l)^{r-1}e^{-\lambda l}}{(r-1)!}\,\lambda\,\Delta l \qquad l \geq 0;\quad r = 1, 2, \ldots$$

where A = probability that there are exactly $r - 1$ arrivals in an interval of duration l

B = conditional probability that rth arrival occurs in next Δl, given exactly $r - 1$ arrivals in previous interval of duration l

Thus we have obtained the PDF for the rth-order interarrival time

$$f_{l_r}(l) = \frac{\lambda^r l^{r-1} e^{-\lambda l}}{(r-1)!} \qquad l \geq 0;\quad r = 1, 2, \ldots$$

which is known as the *Erlang* family of PDF's. (Random variable l_r is said to be an *Erlang random variable of order r*.)

The first-order interarrival times, described by random variable l_1, have the PDF

$$f_{l_1}(l) = \mu_{-1}(l - 0)\lambda e^{-\lambda l}$$

which is the *exponential* PDF. We may obtain its mean and variance by use of the s transform.

$$f_{l_1}{}^T(s) = \int_{l=-\infty}^{\infty} e^{-sl} f_{l_1}(l)\,dl = \frac{\lambda}{s + \lambda}$$

$$E(l_1) = -\left[\frac{d}{ds} f_{l_1}{}^T(s)\right]_{s=0} = \frac{1}{\lambda}$$

$$\sigma_{l_1}{}^2 = \left\{\frac{d^2}{ds^2} f_{l_1}{}^T(s) - \left[\frac{d}{ds} f_{l_1}{}^T(s)\right]^2\right\}_{s=0} = \frac{1}{\lambda^2}$$

Suppose we are told that it has been τ units of time since the last arrival and we wish to determine the conditional PDF for the duration of the remainder $(l_1 - \tau)$ of the present interarrival time. By conditioning the event space for l_1, we would learn that the PDF for the remaining time until the next arrival is still an exponential random variable with parameter λ (see Prob. 4.06). This result is due to the no-memory (independence of events in nonoverlapping intervals) property of the Poisson process; we discussed a similar result for the Bernoulli process in Sec. 4-2.

Random variable l_r is the sum of r independent experimental values of random variable l_1. Therefore we have

$$E(l_r) = rE(l_1) = \frac{r}{\lambda} \qquad \sigma_{l_r}{}^2 = r\sigma_{l_1}{}^2 = \frac{r}{\lambda^2}$$

$$f_{l_r}{}^T(s) = [f_{l_1}{}^T(s)]^r = \left(\frac{\lambda}{s + \lambda}\right)^r$$

The following is a sketch of some members of Erlang family of PDF's:

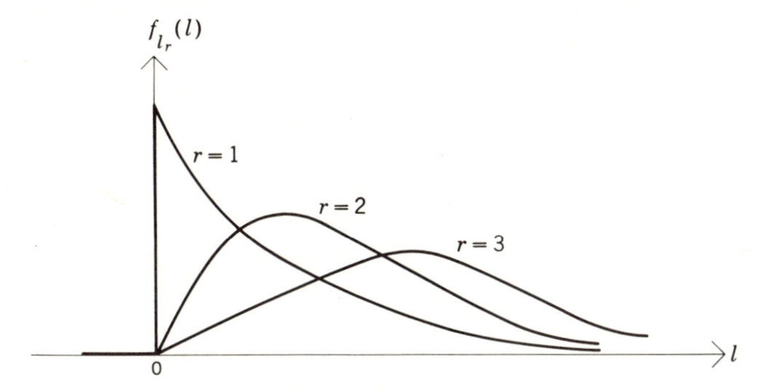

We established that the first-order interarrival times for a Poisson process are exponentially distributed mutually independent random variables. Had we taken this to be our definition of the Poisson process, we would have arrived at identical results. The usual way of determining whether it is reasonable to model a physical process as a Poisson process involves checking whether or not the first-order interarrival times are approximately independent exponential random variables.

Finally, we realize that the relation

$$\sum_{r=1}^{\infty} f_{l_r}(l)\, dl = \lambda\, dl \qquad l \geq 0$$

holds for reasons similar to those discussed at the end of Sec. 4-2.

4-7 Some Additional Properties of Poisson Processes and Poisson Random Variables

Before summarizing our results for the Poisson process, we wish to note a few additional properties.

Consider discrete random variable w, the sum of two *independent* Poisson random variables x and y, with expected values $E(x)$ and $E(y)$. There are at least three ways to establish that $p_w(w_0)$ is also a Poisson PMF. One method involves direct summation in the x_0,y_0 event space (see Prob. 2.03). Or we may use z transforms as follows,

$$p_x{}^T(z) = e^{E(x)(z-1)} \qquad p_y{}^T(z) = e^{E(y)(z-1)}$$

$$w = x + y \qquad x, y \text{ independent}$$

$$p_w{}^T(z) = p_x{}^T(z)p_y{}^T(z) = e^{[E(x)+E(y)](z-1)}$$

which we recognize to be the z transform of the Poisson PMF

$$p_w(w_0) = \frac{[E(x) + E(y)]^{w_0}e^{-[E(x)+E(y)]}}{w_0!} \qquad w_0 = 0, 1, \ldots$$

A third way would be to note that $w = x + y$ could represent the total number of arrivals for two independent Poisson processes within a certain interval. A new process which contains the arrivals due to both of the original processes would still satisfy our definition of the Poisson process with $\lambda = \lambda_1 + \lambda_2$ and would generate experimental values of random variable w for the total number of arrivals within the given interval.

We have learned that the arrival process representing all the arrivals in several independent Poisson processes is also Poisson.

Furthermore, suppose that a new arrival process is formed by performing an independent Bernoulli trial for each arrival in a Poisson process. With probability P, any arrival in the Poisson process is also considered an arrival at the same time in the new process. With probability $1 - P$, any particular arrival in the original process does not appear in the new process. The new process formed in this manner (by "independent random erasures") still satisfies the definition of a Poisson process and has an average arrival rate equal to λP and the expected value of the first-order interarrival time is equal to $(\lambda P)^{-1}$.

If the erasures are not independent, then the derived process has memory. For instance, if we erase alternate arrivals in a Poisson process, the remaining arrivals do not form a Poisson process. It is clear that the resulting process violates the definition of the Poisson process, since, given that an arrival in the new process just occurred, the probability of another arrival in the new process in the next Δt is zero (this would require two arrivals in Δt in the underlying Poisson process). This particular derived process is called an *Erlang process* since the first-order interarrival times are independent and have (second-order) Erlang PDF's. This derived process is one example of how we can use the memoryless Poisson process to model more involved situations with memory.

4-8 Summary of the Poisson Process

For convenience, assume that we are concerned with arrivals which occur at points on a continuous time axis. Quantity $\mathcal{P}(k,t)$ is defined to be the probability that any interval of duration t will contain exactly k arrivals. A process is said to be a *Poisson* process if and only if

1 *For suitably small* Δt, $\mathcal{P}(k,\Delta t)$ *satisfies*

$$\mathcal{P}(k,\Delta t) = \begin{cases} 1 - \lambda\,\Delta t & k = 0 \\ \lambda\,\Delta t & k = 1 \\ 0 & k > 1 \end{cases}$$

2 *Any events defined on nonoverlapping intervals of time are mutually independent.*

An alternative definition of a Poisson process is the statement that the first-order interarrival times be independent identically distributed exponential random variables.

Random variable k, the number of arrivals in an interval of duration t, is described by the *Poisson* PMF

$$p_k(k_0) = \frac{(\lambda t)^{k_0} e^{-\lambda t}}{k_0!} \qquad t \geq 0; \quad k_0 = 0, 1, 2, \ldots$$

$$p_k{}^T(z) = e^{\lambda t(z-1)} \qquad E(k) = \lambda t \qquad \sigma_k{}^2 = \lambda t$$

The first-order interarrival time l_1 is an *exponential* random variable with the PDF

$$f_{l_1}(l) = \lambda e^{-\lambda l} \qquad l \geq 0$$

$$f_{l_1}{}^T(s) = \frac{\lambda}{s + \lambda} \qquad E(l_1) = \frac{1}{\lambda} \qquad \sigma_{l_1}{}^2 = \frac{1}{\lambda^2}$$

The time until the rth arrival, l_r, is known as the *rth-order waiting time*, is the sum of r independent experimental values of l_1, and is described by the *Erlang* PDF

$$f_{l_r}(l) = \frac{\lambda^r l^{r-1} e^{-\lambda l}}{(r-1)!} \qquad l \geq 0; \quad r = 1, 2, \ldots$$

$$f_{l_r}{}^T(s) = \left(\frac{\lambda}{s + \lambda}\right)^r \qquad E(l_r) = rE(l_1) = \frac{r}{\lambda} \qquad \sigma_{l_r}{}^2 = r\sigma_{l_1}{}^2 = \frac{r}{\lambda^2}$$

The sum of several independent Poisson random variables is also a random variable described by a Poisson PMF. If we form a new process by including all arrivals due to several independent Poisson processes, the new process is also Poisson. If we perform Bernoulli trials to make independent random erasures from a Poisson process, the remaining arrivals also form a Poisson process.

4-9 Examples

The Poisson process finds wide application in the modeling of probabilistic systems. We begin with a simple example and proceed to consider some rather structured situations. Whenever it seems informative, we shall solve these problems in several ways.

example 1 The PDF for the duration of the (independent) interarrival times between successive cars on the Trans-Australian Highway is given by

$$f_t(t_0) = \begin{cases} \frac{1}{12}e^{-t_0/12} & t_0 \geq 0 \\ 0 & t_0 < 0 \end{cases}$$

where these durations are measured in seconds.

(a) An old wombat requires 12 seconds to cross the highway, and he starts out immediately after a car goes by. What is the probability that he will survive?

(b) Another old wombat, slower but tougher, requires 24 seconds to cross the road, but it takes two cars to kill him. (A single car won't even slow him down.) If he starts out at a random time, determine the probability that he survives.

(c) If both these wombats leave at the same time, immediately after a car goes by, what is the probability that exactly one of them survives?

a Since we are given that the first-order interarrival times are independent exponentially distributed random variables, we know that the vehicle arrivals are Poisson, with

$$\mathcal{P}(k,t) = \frac{(t/12)^k e^{-t/12}}{k!} \qquad k = 0, 1, 2, \ldots ; \quad t \geq 0$$

Since the car-arrival process is memoryless, the time since the most recent car went by until the wombat starts to cross is irrelevant. The fast wombat will survive only if there are exactly zero arrivals in the first 12 seconds after he starts to cross.

$$\mathcal{P}(0,12) = \frac{1^0 e^{-1}}{0!} = e^{-1} = 0.368$$

Of course, this must be the same as the probability that the wait until the next arrival is longer than 12 seconds.

$$\mathcal{P}(0,12) = \int_{t=12}^{\infty} \frac{1}{12} e^{-t/12}\, dt = 0.368$$

b The slower but tougher wombat will survive only if there is exactly zero or one car in the first 24 seconds after he starts to cross.

$$\mathcal{P}(0,24) + \mathcal{P}(1,24) = \frac{2^0 e^{-2}}{0!} + \frac{2^1 e^{-2}}{1!} = 3e^{-2} = 0.406$$

c Let $\begin{Bmatrix} N_1 \\ N_2 \end{Bmatrix}$ be the number of cars in the $\begin{Bmatrix} \text{first} \\ \text{second} \end{Bmatrix}$ 12 seconds after the wombats start out. It will be helpful to draw a sequential event space for the experiment.

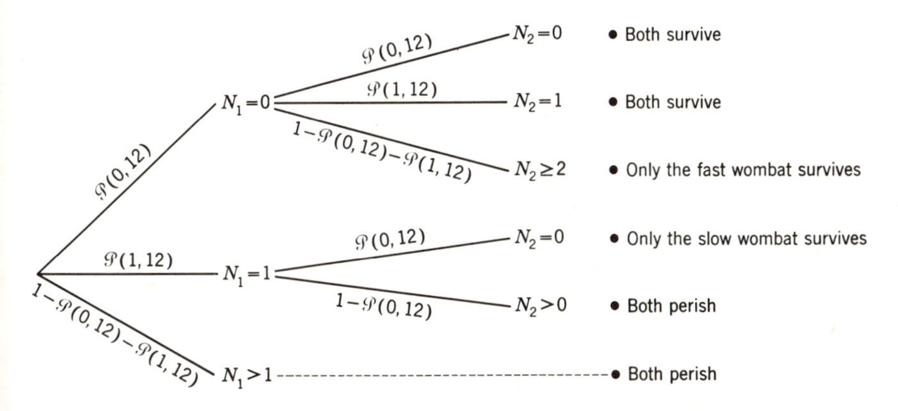

$$\text{Prob(exactly one wombat survives)} = \text{Prob}(N_1 = 0, N_2 \geq 2) + \text{Prob}(N_1 = 1, N_2 = 0)$$

Quantities N_1 and N_2 are independent random variables because they are defined on nonoverlapping intervals of a Poisson process. We may now collect the probability of exactly one survival from the above event space.

$$\text{Prob(exactly 1 wombat survives)} = \mathcal{P}(0,12)[1 - \mathcal{P}(0,12) - \mathcal{P}(1,12)] + \mathcal{P}(1,12)\mathcal{P}(0,12)$$

$$\text{Prob(exactly 1 wombat survives)} = e^{-1}(1 - 2e^{-1}) + e^{-2} = e^{-1} - e^{-2} = 0.233$$

example 2 Eight light bulbs are turned on at $t = 0$. The lifetime of any particular bulb is independent of the lifetimes of all other bulbs and is described by the PDF

$$f_t(t_0) = \begin{cases} \lambda e^{-\lambda t_0} & \text{if } t_0 \geq 0 \\ 0 & \text{otherwise} \end{cases}$$

Determine the mean, variance, and s transform of random variable y, the time until the third failure.

We define t_{ij} to be a random variable representing the time from the ith failure until the jth failure, where t_{01} is the duration from $t = 0$ until the first failure. We may write

$$y = t_{03} = t_{01} + t_{12} + t_{23}$$

The length of the time interval during which exactly $8 - i$ bulbs are on is equal to $t_{i(i+1)}$. While $8 - i$ bulbs are on, we are dealing with the sum of $8 - i$ independent Poisson processes and the probability of a failure in the next Δt is equal to $(8 - i)\lambda \Delta t$. Thus, from the properties of the Poisson process, we have

$$E(y) = E(t_{01}) + E(t_{12}) + E(t_{23}) = \frac{1}{8\lambda} + \frac{1}{7\lambda} + \frac{1}{6\lambda}$$

Knowledge of the experimental value of, for instance, t_{01} does not tell us anything about t_{12}. Random variable t_{12} would still be an exponential random variable representing the time until the next arrival for a Poisson process with an average arrival rate of 7λ. Random variables t_{01}, t_{12}, and t_{23} are mutually independent (why?), and we have

$$\sigma_y{}^2 = \sigma_{t_{01}}^2 + \sigma_{t_{12}}^2 + \sigma_{t_{23}}^2 = \frac{1}{(8\lambda)^2} + \frac{1}{(7\lambda)^2} + \frac{1}{(6\lambda)^2}$$

$$f_y{}^T(s) = f_{t_{01}}^T(s) \times f_{t_{12}}^T(s) \times f_{t_{23}}^T(s) = \frac{8\lambda}{s + 8\lambda}\,\frac{7\lambda}{s + 7\lambda}\,\frac{6\lambda}{s + 6\lambda}$$

This has been one example of how easily we can obtain answers for many questions related to Poisson models. A harder way to go about it would be to determine first the PDF for y, the third smallest of eight independent identically distributed exponential random variables.

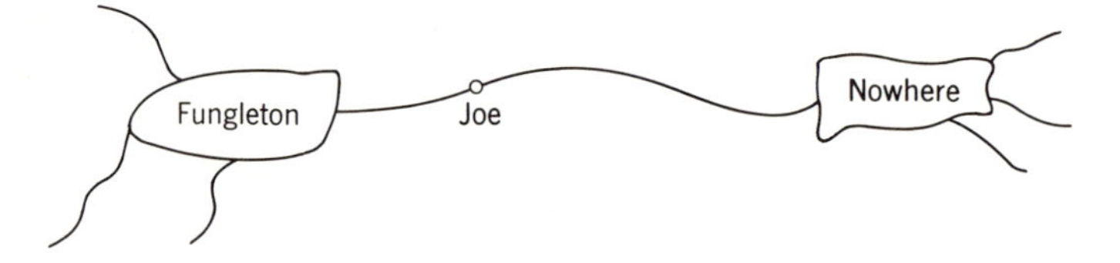

example 3 Joe is waiting for a Nowhere-to-Fungleton (NF) bus, and he knows that, out where he is, arrivals of $\begin{Bmatrix} FN \\ NF \end{Bmatrix}$ buses may be considered independent Poisson processes with average arrival rates of $\begin{Bmatrix} \lambda_{FN} \\ \lambda_{NF} \end{Bmatrix}$ buses per hour. Determine the PMF and the expectation for random variable K, the number of "wrong-way" buses he will see arrive before he boards the next NF bus.

We shall do this problem in several ways.

Method A

We shall obtain the compound PDF for the amount of time he waits and the number of wrong-way buses he sees. Then we determine $p_K(K_0)$ by integrating out over the other random variable. We know the marginal PDF for his waiting time, and it is simple to find the PMF for K conditional on his waiting time. The product of these probabilities tells us all there is to know about the random variables of interest.

The time Joe waits until the first right-way (NF) bus is simply the waiting time until the first arrival in a Poisson process with average arrival rate λ_{NF}. The probability that his total waiting time t will be between t_0 and $t_0 + dt_0$ is

$$f_t(t_0)\,dt_0 = \lambda_{NF}e^{-\lambda_{NF}t_0}\,dt_0$$

Given that the experimental value of Joe's waiting time is exactly t_0 hours, the conditional PMF for K is simply the probability of exactly K_0 arrivals in an interval of duration t_0 for a Poisson process with average arrival rate λ_{FN}.

$$p_{K|t}(K_0 \mid t_0) = \frac{(\lambda_{FN}t_0)^{K_0}e^{-\lambda_{FN}t_0}}{K_0!} \qquad K_0 = 0, 1, 2, \ldots$$

The experiment of Joe's waiting for the next NF bus and observing the number of wrong-way buses while he waits has a two-dimensional event space which is discrete in K and continuous in t.

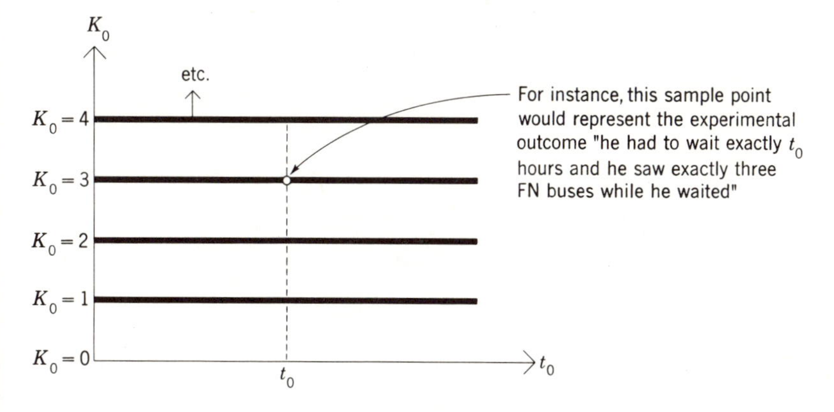

We obtain the probability assignment in this event space, $f_{t,k}(t_0,k_0)$.

$$\begin{aligned} f_{t,K}(t_0,K_0) &= f_t(t_0)p_{K|t}(K_0 \mid t_0) \\ &= \frac{\lambda_{NF}e^{-\lambda_{NF}t_0}(\lambda_{FN}t_0)^{K_0}e^{-\lambda_{FN}t_0}}{K_0!} \qquad t_0 \geq 0; \quad K_0 = 0, 1, 2, \ldots \end{aligned}$$

The marginal PMF $p_K(K_0)$ may be found from

$$p_K(K_0) = \int_{t_0=0}^{\infty} f_{t,K}(t_0,K_0)\,dt_0 = \frac{\lambda_{NF}\lambda_{FN}^{K_0}}{K_0!}\int_{t_0=0}^{\infty} t_0^{K_0}e^{-(\lambda_{FN}+\lambda_{NF})t_0}\,dt_0$$

By noting that

$$\frac{(\lambda_{FN} + \lambda_{NF})^{K_0+1}t_0^{K_0}e^{-(\lambda_{FN}+\lambda_{NF})t_0}}{K_0!}$$

would integrate to unity over the range $0 \leq t_0 \leq \infty$ (since it is an Erlang PDF of order $K_0 + 1$), we can perform the above integration by inspection to obtain (with $\lambda_{NF}/\lambda_{FN} = \rho$),

$$p_K(K_0) = \frac{\rho}{(1+\rho)^{K_0+1}} \qquad K_0 = 0, 1, 2, \ldots$$

If the average arrival rates λ_{NF} and λ_{FN} are equal ($\rho = 1$), we note that the probability that Joe will see a total of exactly K_0 wrong-way buses before he boards the first right-way bus is equal to $(\frac{1}{2})^{K_0+1}$. For this case, the probability is 0.5 that he will see no wrong way buses while he waits.

The expected value of the number of FN buses he will see arrive may be obtained from the z transform.

$$p_K{}^T(z) = \frac{\rho}{1+\rho} \sum_{K_0=0}^{\infty} \frac{z^{K_0}}{(1+\rho)^{K_0}} = \rho(1+\rho-z)^{-1}$$

$$E(K) = \left[\frac{d}{dz} p_K{}^T(z)\right]_{z=1} = \rho^{-1}$$

This answer seems reasonable for the cases $\rho >> 1$ and $\rho << 1$.

Method B

Regardless of when Joe arrives, the probability that the next bus is a wrong bus is simply the probability that an experimental value of a random variable with PDF

$$f_x(x_0) = \lambda_{FN} e^{-\lambda_{FN} x_0} \qquad x_0 \geq 0$$

is smaller than an experimental value of another, independent, random variable with PDF

$$f_y(y_0) = \lambda_{NF} e^{-\lambda_{NF} y_0} \qquad y_0 \geq 0$$

So, working in the x_0, y_0 event space

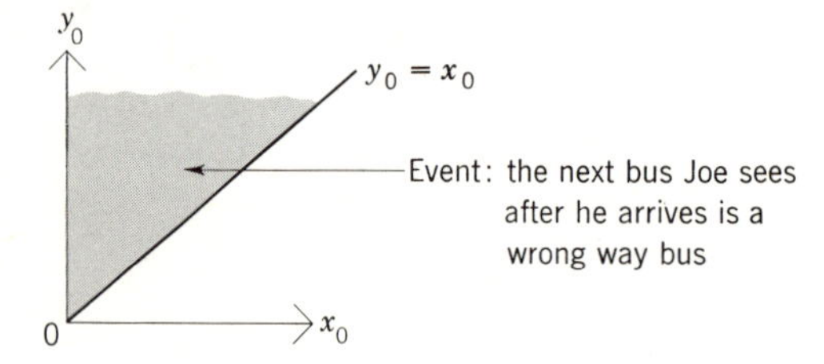

$$\text{Prob (next bus is an } FN \text{ bus)} = \int_{x_0=0}^{\infty} dx_0 \int_{y_0=x_0}^{\infty} dy_0\, \lambda_{FN}\lambda_{NF} e^{-\lambda_{FN}x_0} e^{-\lambda_{NF}y_0}$$

$$= \frac{\lambda_{FN}}{\lambda_{FN} + \lambda_{NF}} = \frac{1}{1+\rho} \qquad \text{with } \rho = \frac{\lambda_{NF}}{\lambda_{FN}}$$

As soon as the next bus does come, *the same result holds for the following bus;* so we can draw out the sequential event space where each trial corresponds to the arrival of another bus, and the experiment terminates with the arrival of the first NF bus.

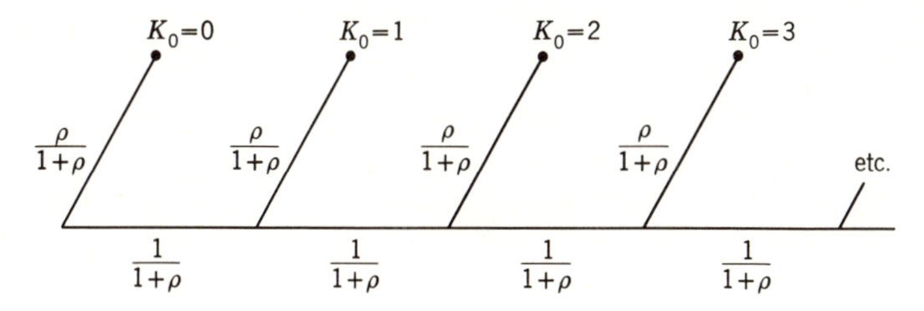

This again leads to

$$p_K(K_0) = \frac{\rho}{(1+\rho)^{K_0+1}} \qquad K_0 = 0, 1, 2, \ldots$$

Method C

Consider the event space for *any* adequately small Δt,

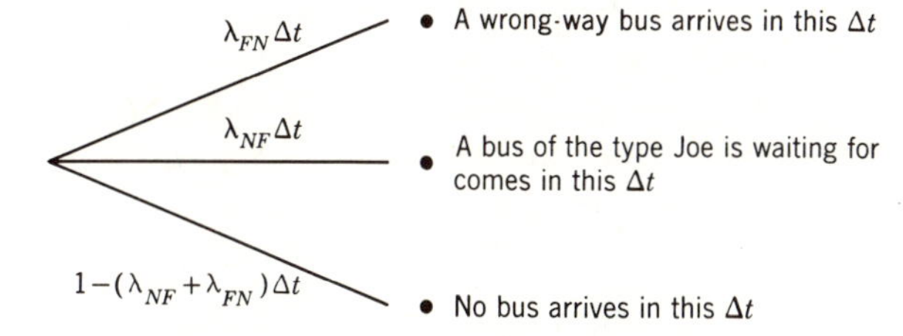

We need be interested in a Δt only if a bus arrives during that Δt; so we may work in a conditional space containing only the upper two event points to obtain

$$\text{Prob (any particular bus is } FN) = \frac{\lambda_{FN}}{\lambda_{FN} + \lambda_{NF}}$$

$$\text{Prob (any particular bus is } NF) = \frac{\lambda_{NF}}{\lambda_{FN} + \lambda_{NF}}$$

This approach replaces the integration in the x,y event space for the previous solution and, of course, leads to the same result.

As a final point, note that N, the *total* number of buses Joe would see if he waited until the Rth NF bus, would have a Pascal PMF. The arrival of each bus would be a Bernoulli trial, and a success is represented by the arrival of an NF bus. Thus, we have

$$p_N(N_0) = \binom{N_0 - 1}{R - 1}\left(\frac{\lambda_{NF}}{\lambda_{FN} + \lambda_{NF}}\right)^R \left(\frac{\lambda_{FN}}{\lambda_{FN} + \lambda_{NF}}\right)^{N_0 - R}$$
$$N_0 = R, R+1, \ldots; \quad R = 1, 2, 3, \ldots$$

where N is the total number of buses (including the one he boards) seen by Joe if his policy is to board the Rth right-way bus to arrive after he gets to the bus stop.

4-10 Renewal Processes

Consider a somewhat more general case of a random process in which arrivals occur at points in time. Such a process is known as a *renewal process* if its first-order interarrival times are mutually independent random variables described by the same PDF. The Bernoulli and Poisson processes are two simple examples of the renewal process. In this and the following section, we wish to study a few basic aspects of the general renewal process.

To simplify our discussion, we shall assume in our formal work that the PDF for the first-order interarrival times (*gaps*) $f_x(x_0)$ is a continuous PDF which does not contain any impulses. [A notational change from $f_{l_1}(l)$ to $f_x(x_0)$ will also simplify our work.]

We begin by determining the conditional PDF for the time until the next arrival when we know how long ago the most recent arrival occurred. In the next section, we develop the consequences of beginning to observe a renewal process at a *random* time.

If it is known that the most recent arrival occurred exactly τ units of time ago, application of the definition of conditional probability results in the following conditional PDF for x, the *total* duration of the present interarrival gap:

$$f_{x|x>\tau}(x_0 \mid x > \tau) = \frac{f_x(x_0)}{\int_\tau^\infty f_x(x_0)\,dx_0} = \frac{f_x(x_0)}{1 - p_{x\leq}(\tau)} \qquad x_0 > \tau$$

If we let random variable y represent the *remaining* time in the present gap, $y = x - \tau$, we obtain the conditional PDF for y,

$$f_{y|x>\tau}(y_0 \mid x > \tau) = \frac{f_x(y_0 + \tau)}{\int_\tau^\infty f_x(x_0)\,dx_0} = \frac{f_x(y_0 + \tau)}{1 - p_{x\leq}(\tau)} \qquad y_0 > 0$$

As an example, suppose that we are burning light bulbs one at a time and replacing each bulb the instant it fails. If the lifetimes of the bulbs are independent random variables with PDF $f_x(x_0)$, we have a renewal process in which the points in time at which bulb replacements occur are the arrivals. Let's use the results obtained above to work out one example with a particularly simple form for $f_x(x_0)$.

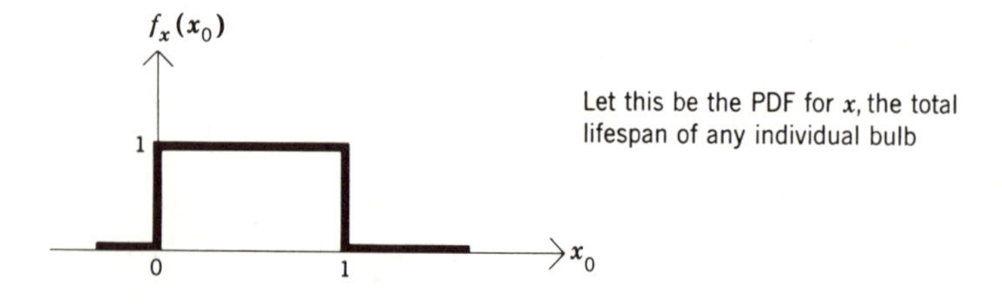

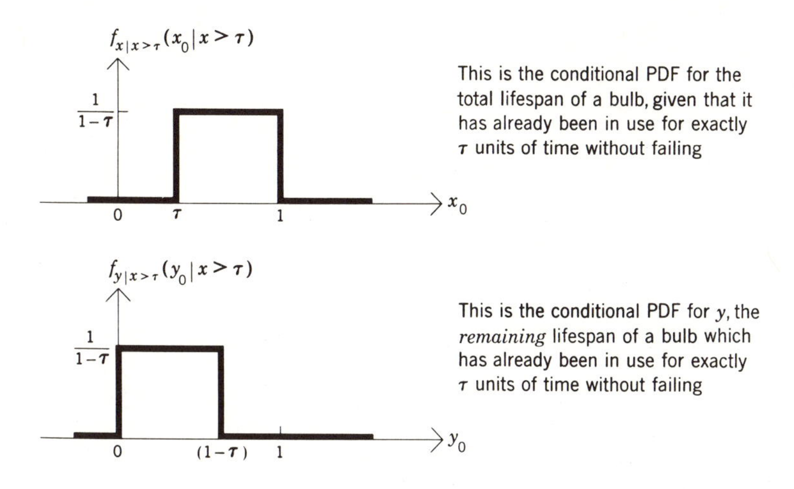

We learned earlier that the first-order interarrival times for a Poisson process are independent random variables with the PDF $f_x(x_0) = \lambda e^{-\lambda x_0}$ for $x_0 \geq 0$. For a Poisson process we can show by direct substitution that the conditional PDF for the remaining time until the next arrival, $f_{y|x>\tau}(y_0 \mid x > \tau)$, *does not depend on* τ (Prob. 4.06) and is equal to $f_x(y_0)$, the original unconditional PDF for the first-order interarrival times. For the Poisson process (but *not* for the more general renewal process) the time until the next arrival is independent of when we start waiting. If the arrivals of cars at a line across a street constituted a Poisson process, it would be just as safe to start crossing the street at a random time as it would be to start crossing immediately after a car goes by.

4-11 Random Incidence

Assume that a renewal process, characterized by the PDF of its first-order interarrival times, $f_x(x_0)$, has been in progress for a long time. We are now interested in *random incidence*. The relevant experiment is *to pick a time randomly* (for instance, by spinning the hands of a clock) *and then wait until the first arrival in the renewal process after our randomly selected entry time.* The instant of the random entry must always be chosen in a manner which is independent of the actual arrival history of the process.

We wish to determine the PDF for random variable y, the waiting time until the next arrival (or the remaining gap length) following random entry. Several intermediate steps will be required to obtain the unconditional PDF $f_y(y_0)$.

First we shall obtain the PDF for random variable w, the *total* duration of the interarrival gap into which we enter by random incidence. Random variable w describes the duration of an interval which begins with the most recent arrival in the renewal process prior to the instant of random incidence and which terminates with the first arrival in the process after the instant of random incidence.

Note that random variables w and x both refer to total interarrival-gap durations for the renewal process, but the experiments on which they are defined are different. An experimental value of w is obtained by determining the total duration of the interarrival gap into which a randomly selected instant falls. An experimental value of x is obtained by noting the duration from any arrival in the renewal process until the next arrival.

After obtaining $f_w(w_0)$, we shall then find the conditional PDF for the remaining time in the gap, y, given the experimental value of the total duration of the gap, w. Thus, our procedure is to work in a w_0,y_0 event space, first obtaining $f_w(w_0)$ and $f_{y|w}(y_0 \mid w_0)$. We then use the relations

$$f_{w,y}(w_0,y_0) = f_w(w_0)f_{y|w}(y_0 \mid w_0) \qquad \text{and} \qquad f_y(y_0) = \int_{w_0} f_{w,y}(w_0,y_0)\, dw_0$$

to obtain the unconditional PDF $f_y(y_0)$ for the waiting time from our randomly selected instant until the next arrival in the renewal process.

To determine the PDF $f_w(w_0)$, let's begin by considering an example where the first-order interarrival times of the renewal process have the discrete PMF

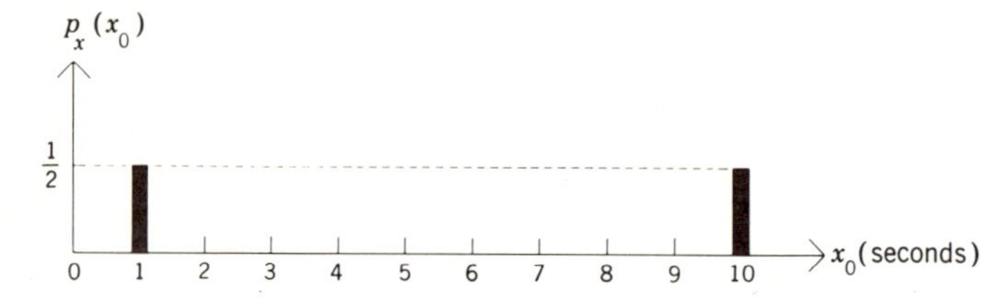

Although any interarrival time is equally likely to be either 1 or 10 seconds long, note that each 10-second gap consumes 10 times as much time as each 1-second gap. The probability that a randomly selected instant of time falls into a 10-second gap is proportional to the fraction of all time which is included in 10-second gaps.

The fraction of all time which is included in gaps of duration x_0 should be, in general, proportional to $p_x(x_0)$ weighted by x_0, since $p_x(x_0)$ is the fraction of the gaps which are of duration x_0 and each such gap consumes x_0 seconds. Recalling that random variable w is to be the total duration of the interarrival gap into which our randomly selected instant falls, we have argued that

$$p_w(w_0) = \frac{w_0 p_x(w_0)}{\sum_{w_0} w_0 p_x(w_0)} = \frac{w_0 p_x(w_0)}{E(x)}$$

where the denominator is the required normalization factor.

For the particular example given above, we obtain the PMF for the total duration of the gap into which a random entry falls,

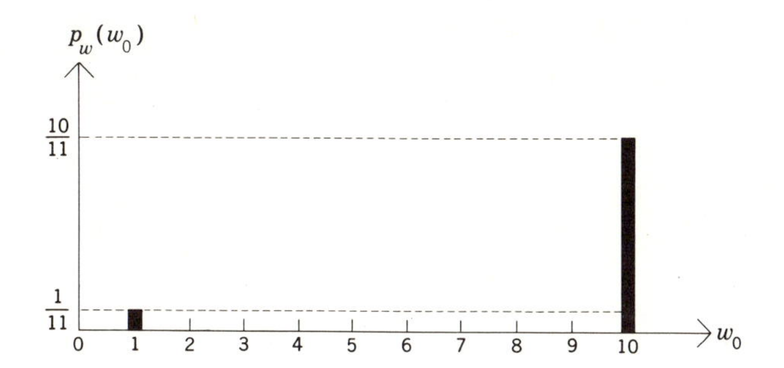

A random entry, for this example, is ten times as likely to fall into a ten-second gap as a one-second gap, even though a gap length is equally likely to be of either kind.

Extending the general form of $p_w(w_0)$ to the continuous case, we have the desired $f_w(w_0)$

$$f_w(w_0) = \frac{w_0 f_x(w_0)}{\int_{w_0} w_0 f_x(w_0)\, dw_0} = \frac{w_0 f_x(w_0)}{E(x)}$$

where $f_x(\cdot)$ is the PDF for the first-order interarrival times for the renewal process and $f_w(w_0)$ is the PDF for the total duration of the interarrival gap entered by random incidence.

In reasoning our way to this result, we have made certain assumptions about the relation between the probability of an event and the fraction of a large number of trials on which the event will occur. We speculated on the nature of this relation in Sec. 3-6, and the proof will be given in Chap. 6.

Given that we have entered into a gap of total duration w_0 by random incidence, the remaining time in the gap, y, is uniformly distributed between 0 and w_0 with the conditional PDF

$$f_{y|w}(y_0 \mid w_0) = \begin{cases} \dfrac{1}{w_0} & \text{if } 0 \le y_0 \le w_0 \\ 0 & \text{otherwise} \end{cases}$$

because a random instant is as likely to fall within any increment of a w_0-second gap as it is to fall within any other increment of equal duration within the w_0-second gap.

Now we may find the joint PDF for random variables w and y,

$$f_{w,y}(w_0,y_0) = f_w(w_0)f_{y|w}(y_0 \mid w_0) = \frac{w_0 f_x(w_0)}{E(x)} \cdot \frac{1}{w_0} \qquad 0 \le y_0 \le w_0 \le \infty$$

To determine $f_y(y_0)$, the PDF for the time until the next arrival after the instant of random incidence, we need only integrate (carefully) over w_0 in the w_0,y_0 event space. Note that w, the total length of the gap entered by random incidence, must be greater than or equal to y, the remaining time in that gap; so we have

$$f_y(y_0) = \int_{w_0} f_{w,y}(w_0,y_0)\,dw_0 = \int_{w_0=y_0}^{\infty} \frac{f_x(w_0)\,dw_0}{E(x)}$$

$$f_y(y_0) = \frac{1 - p_{x\le}(y_0)}{E(x)}$$

where $f_y(y_0)$ is the PDF for the duration of the interval which begins at a "random" time and terminates with the next arrival for a renewal process with first-order interarrival times described by random variable x.

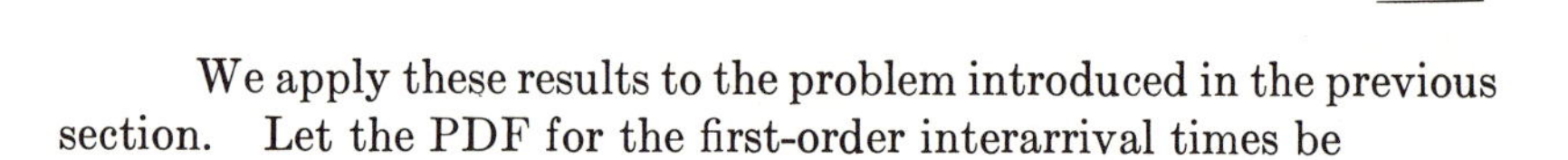

We apply these results to the problem introduced in the previous section. Let the PDF for the first-order interarrival times be

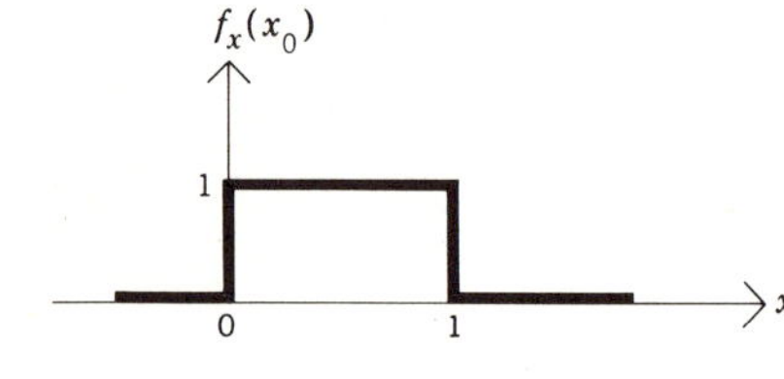

Let this be the PDF for the first-order interarrival times of a renewal process

Now, first we apply the relation

$$f_w(w_0) = \frac{w_0 f_x(w_0)}{E(x)}$$

to obtain

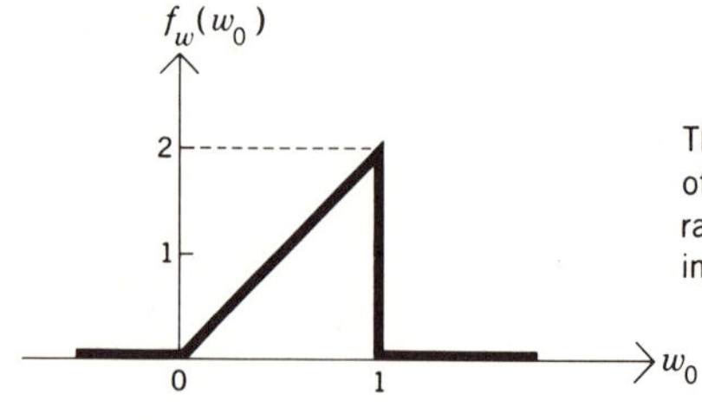

This is the PDF for the total duration of the interarrival gap entered by random incidence. Obviously, random incidence favors entry into longer gaps and we use

$$f_y(y_0) = \frac{1 - p_{x\leq}(y_0)}{E(x)}$$

to obtain

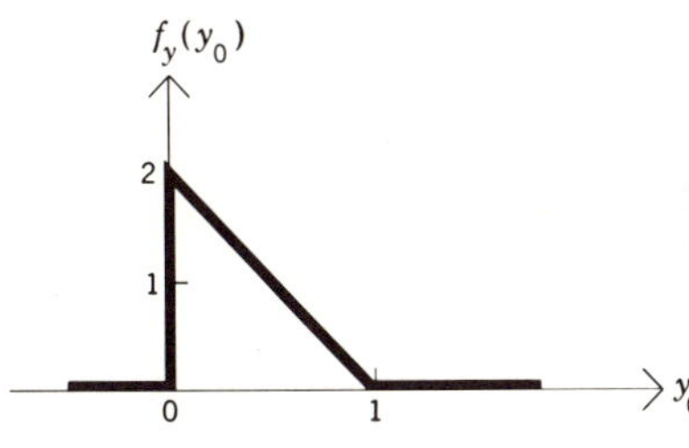

This is the PDF for the remaining duration of the interarrival gap entered by random incidence

It is interesting to note that the expected value of the remaining duration of the gap entered by random incidence, $E(y)$, may be greater than, equal to, or less than the "expected gap length" given by $E(x)$. In fact, we have already learned that $E(x) = E(y)$ for a Poisson process.

Thus, for instance, if car interarrival times are independent random variables described by $f_x(x_0)$, the expected waiting time until the next car arrives may be greater if we start to cross at a random time than if we start right after a car goes by! If we understand the different experiments which give rise to $E(x)$ and $E(y)$, this seems entirely reasonable, since we realize that random incidence favors entry into large gaps.

We should realize that statements about average values or expected values of a random variable are meaningless unless we have a full description of the experiment on whose outcomes the random variable is defined. In the above discussion, $E(x)$ and $E(y)$ are generally different, but each is the "expected value of the time until the next arrival." The experiments of "picking a gap" and "picking an instant of time" may lead to distinctly different results. (Probs. 2.08, 2.11, and 3.10 have already introduced similar concepts.)

PROBLEMS

4.01 The PMF for the number of failures before the rth success in a Bernoulli process is sometimes called the *negative binomial* PMF. Derive it and explain its relation to the Pascal PMF.

4.02 A channel contains a series flow of objects, each of fixed length L. All objects travel at constant velocity V. Each separation S between successive objects is some integral multiple of L, $S = nL$, where the n

for each separation is an independent random variable described by the probability mass function

$$p_n(n_0) = a(1 - a)^{n_0 - 1} \qquad n_0 = 1, 2, 3, \ldots ; \quad 0 < a < 1$$

a Find the average flow rate, in objects per unit time, observable at some point in the channel.

b Calculate what additional flow can exist under a rule that the resulting arrangements of objects must have at least a separation of L from adjacent objects.

c As seen by an electric eye across the channel, what fraction of all the gap time is occupied by gaps whose total length is greater than $2L$? A numerical answer is required.

4.03 Let x be a discrete random variable described by a geometric PMF. Given that the experimental value of random variable x is greater than integer y, show that the conditional PMF for $x - y$ is the same as the original PMF for x. Let $r = x - y$, and sketch the following PMF's:

a $p_x(x_0)$ **b** $p_{x|x>y}(x_0 \mid x > y)$ **c** $p_r(r_0)$

4.04 We are given two independent Bernoulli processes with parameters P_1 and P_2. A new process is defined to have a success on its kth trial ($k = 1, 2, 3, \ldots$) only if *exactly* one of the other two processes has a success on its kth trial.

a Determine the PMF for the number of trials up to and including the rth success in the new process.

b Is the new process a Bernoulli process?

4.05 Determine the expected value, variance, and z transform for the total number of trials from the start of a Bernoulli process up to and including the nth success after the mth failure.

4.06 Let x be a continuous random variable whose PDF $f_x(x_0)$ contains no impulses. Given that $x > T$, show that the conditional PDF for $r = x - T$ is equal to $f_x(r_0)$ if $f_x(x_0)$ is an exponential PDF.

4.07 To cross a single lane of moving traffic, we require at least a duration T. Successive car interarrival times are independently and identically distributed with probability density function $f_t(t_0)$. If an interval between successive cars is longer than T, we say that the interval represents a single opportunity to cross the lane. Assume that car lengths are small relative to intercar spacing and that our experiment begins the instant after the zeroth car goes by.

Determine, in as simple a form as possible, expressions for the probability that:

a We can cross for the first time just before the Nth car goes by.

b We shall have had exactly n opportunities by the instant the Nth car goes by.

c The occurrence of the nth opportunity is immediately followed by the arrival of the Nth car.

4.08 Consider the manufacture of Grandmother's Fudge Nut Butter Cookies. Grandmother has noted that the number of nuts in a cookie is a random variable with a Poisson mass function and that the average number of nuts per cookie is 1.5.

a What is the numerical value of the probability of having at least one nut in a randomly selected cookie?

b Determine the numerical value of the variance of the number of nuts per cookie.

c Determine the probability that a box of exactly M cookies contains exactly the expected value of the number of nuts for a box of N cookies. ($M = 1, 2, 3, \ldots$; $N = 1, 2, 3, \ldots$)

d What is the probability that a nut selected at random goes into a cookie containing exactly K nuts?

e The customers have been getting restless; so grandmother has instructed her inspectors to discard each cookie which contains exactly zero nuts. Determine the mean and variance of the number of nuts per cookie for the remaining cookies.

4.09 A woman is seated beside a conveyer belt, and her job is to remove certain items from the belt. She has a narrow line of vision and can get these items only when they are right in front of her.

She has noted that the probability that exactly k of her items will arrive in a minute is given by

$$p_k(k_0) = \frac{2^{k_0}e^{-2}}{k_0!} \qquad k_0 = 0, 1, 2, 3, \ldots$$

and she assumes that the arrivals of her items constitute a Poisson process.

a If she wishes to sneak out to have a beer but will not allow the expected value of the number of items she misses to be greater than 5, how much time may she take?

b If she leaves for two minutes, what is the probability that she will miss exactly two items the first minute and exactly one item the second minute?

c If she leaves for two minutes, what is the probability that she will miss a total of exactly three items?

d The union has installed a bell which rings once a minute with precisely one-minute intervals between gongs. If, between two successive gongs, more than three items come along the belt, she will handle only three of them properly and will destroy the rest. Under this system, what is the probability that any particular item will be destroyed?

4.10 Arrivals of certain events at points in time are known to constitute a Poisson process, but it is not known which of two possible values of λ, the average arrival rate, describes the process. Our a priori estimate is that $\lambda = 2$ or $\lambda = 4$ with equal probability.

We observe the process for T units of time and observe exactly K arrivals. Given this information, determine the conditional probability that $\lambda = 2$. Check to see whether or not your answer is reasonable for some simple limiting values for K and T.

4.11 Independent experimental values of a geometric random variable are obtained, and we label these values $K_1, K_2, K_3, \ldots$. Random variable r_i is defined by

$$r_i = \sum_{j=1}^{i} K_j \qquad i = 1, 2, \ldots$$

If we eliminate arrivals number $r_1, r_2, r_3, \ldots$ in a Poisson process, do the remaining arrivals constitute a Poisson process?

4.12

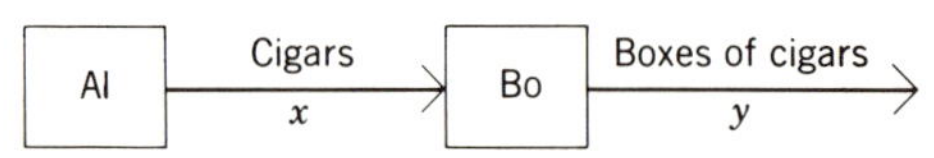

Al makes cigars, placing each cigar on a constant-velocity conveyer belt as soon as it is finished. Bo packs the cigars into boxes of four cigars each, placing each box back on the belt as soon as it is filled. The time Al takes to construct any particular cigar is, believe it or not, an independent exponential random variable with an expected value of five minutes.

a Determine $\mathcal{P}_A(k,T)$, the probability that Al makes exactly k cigars in T minutes. Determine the mean and variance of k as a function of T. $k = 0, 1, 2, \ldots; \quad 0 \leq T \leq \infty$.

b Determine the probability density function $f_\tau(\tau_0)$, where τ is the interarrival time (measured in minutes) between successive cigars at point x.

c Determine $\mathcal{P}_B(r,T)$, the probability that Bo places exactly r boxes of cigars back on the belt during an interval of T minutes.

d Determine the probability density function $f_t(t_0)$, where t is the interarrival time (measured in minutes) between successive boxes of cigars at point y.

e If we arrive at point y at a random instant, long after the process began, determine the PDF $f_r(r_0)$, where r is the duration of our wait until we see a box of cigars at point y.

4.13 Dave is taking a multiple-choice exam. You may assume that the number of questions is infinite. *Simultaneously, but independently,* his conscious and subconscious facilities are generating answers for him, each in a Poisson manner. (His conscious and subconscious are always working on different questions.)

Average rate at which conscious responses are generated
$= \lambda_c$ responses/min

Average rate at which subconscious responses are generated
$= \lambda_s$ responses/min

Each conscious response is an independent Bernoulli trial with probability p_c of being correct. Similarly, each subconscious response is an independent Bernoulli trial with probability p_s of being correct.

Dave responds only once to each question, and you can assume that his time for recording these conscious and subconscious responses is negligible.

a Determine $p_k(k_0)$, the probability mass function for the number of *conscious responses* Dave makes in an interval of T minutes.

b If we pick any question to which Dave has responded, what is the probability that his answer to that question:

i Represents a conscious response

ii Represents a conscious correct response

c If we pick an interval of T minutes, what is the probability that in that interval Dave will make exactly R_0 conscious responses *and* exactly S_0 subconscious responses?

d Determine the s transform for the probability density function for random variable x, where x is the time from the start of the exam until Dave makes his first conscious response which is preceded by at least one subconscious response.

e Determine the probability mass function for the total number of responses up to and including his third conscious response.

f The papers are to be collected as soon as Dave has completed exactly N responses. Determine:

i The expected number of questions he will answer correctly

ii The probability mass function for L, the number of questions he answers correctly

g Repeat part (f) for the case in which the exam papers are to be collected at the end of a fixed interval of T minutes.

4.14 Determine, in an efficient manner, the fourth moment of a continuous random variable described by the probability density function

$$f_x(x_0) = \begin{cases} \dfrac{4^3 x_0^2 e^{-4x_0}}{2} & x_0 \geq 0 \\ 0 & x_0 < 0 \end{cases}$$

4.15 The probability density function for L, the length of yarn purchased by any particular customer, is given by

$$f_L(L_0) = \frac{\lambda^3 L_0^2 e^{-\lambda L_0}}{2} \qquad L_0 \geq 0$$

A single dot is placed on the yarn at the mill. Determine the expected value of r, where r is the length of yarn purchased by that customer whose purchase included the dot.

4.16 A communication channel fades (degrades beyond use) in a random manner. The length of any fade is an exponential random variable with expected value λ^{-1}. The duration of the interval between the end of any fade and the start of the next fade is an Erlang random variable with PDF

$$f_t(t_0) = \frac{\mu^4 t_0^3 e^{-\mu t_0}}{3!} \qquad t_0 \geq 0$$

a If we observe the channel at a randomly selected instant, what is the probability that it will be in a fade at that time? Would you expect this answer to be equal to the fraction of all time for which the channel is degraded beyond use?

b A device can be built to make the communication system continue to operate during the first T units of time in any fade. The cost of the device goes up rapidly with T. What is the smallest value of T which will reduce by 90% the amount of time the system is out of service?

4.17 The random variable t corresponds to the interarrival time between consecutive events and is specified by the probability density function

$$f_t(t_0) = 4t_0^2 e^{-2t_0} \qquad t_0 \geq 0$$

Interarrival times are independent.

a Determine the expected value of the interarrival time x between the 11th and 13th events.

b Determine the probability density function for the interarrival time y between the 12th and 16th events.

c If we arrive at the process at a random time, determine the probability density function for the total length of the interarrival gap which we shall enter.

d Determine the expected value and the variance of random variable r, defined by $r = x + y$.

4.18 Bottles arrive at the Little Volcano Bottle Capper (LVBC) in a Poisson manner, with an average arrival rate of λ bottles per minute. The LVBC works instantly, but we also know that it destroys any bottles which arrive within $1/5\lambda$ minutes of the most recent successful capping operation.

a A long time after the process began, what is the probability that a randomly selected arriving bottle (marked at the bottle factory) will be destroyed?

b What is the probability that neither the randomly selected bottle nor any of the four bottles arriving immediately after it will be destroyed?

4.19 In the diagram below, each ⊣⊢ represents a communication link. Under the present maintenance policy, link failures may be considered independent events, and one can assume that, at any time, the probability that any link is working properly is p.

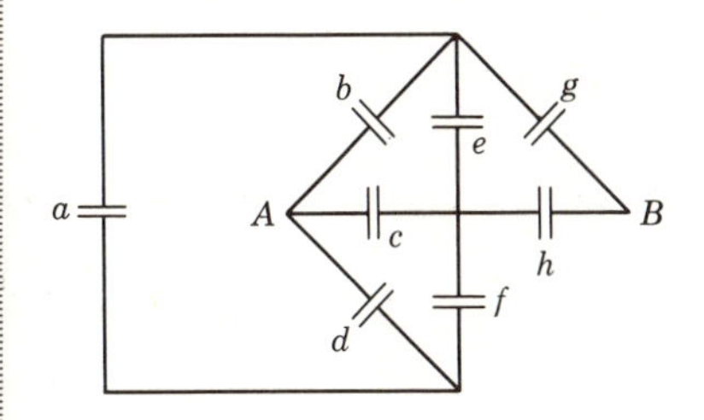

a If we consider the system at a random time, what is the probability that:

i A total of exactly two links are operating properly?

ii Link g and exactly one other link are operating properly?

b Given that exactly six links are not operating properly at a particular time, what is the probability that A can communicate with B?

c *Under a new maintenance policy*, the system was put into operation in perfect condition at $t = 0$, and the PDF for the time until failure of any link is

$$f_t(t_0) = \lambda e^{-\lambda t_0} \qquad t_0 \geq 0$$

Link failures are still independent, but no repairs are to be made until the third failure occurs. At the time of this third failure, the system is shut down, fully serviced, and then "restarted" in perfect order. The down time for this service operation is a random variable with probability density function

$$f_x(x_0) = \mu^2 x_0 e^{-\mu x_0} \qquad x_0 \geq 0$$

i What is the probability that link g will fail before the first service operation?

ii Determine the probability density function for random variable y, the time until the first link failure after $t = 0$.

iii Determine the mean and variance and s transform for w, the time from $t = 0$ until the end of the first service operation.

4.20 The interarrival times (gaps) between the arrivals of successive events at points in time are independent random variables with PDF,

$$f_t(t_0) = \begin{cases} Kt_0(1 - t_0) & \text{if } 0 \leq t_0 \leq 1 \\ 0 & \text{otherwise} \end{cases}$$

a What fraction of time is spent in gaps longer than the average gap?

b If we come along at a random instant after the process has been proceeding for a long time, determine

i The probability we shall see an arrival in the next (small) Δt

ii The PDF for l, the time we wait until the next arrival

c Find any $f_t(t_0)$ for which, in the notation of this problem, there would result $E(l) > E(t)$.

4.21 Two types of tubes are processed by a certain machine. Arrivals of type I tubes and of type II tubes form independent Poisson processes with average arrival rates of λ_1 and λ_2 tubes per hour, respectively. The processing time required for any type I tube, x_1, is an independent random variable with PDF

$$f_{x_1}(x) = \begin{cases} 1 & \text{if } 0 \leq x \leq 1 \\ 0 & \text{otherwise} \end{cases}$$

The processing time required for any type II tube, x_2, is also a uniformly distributed independent random variable

$$f_{x_2}(x) = \begin{cases} 0.5 & \text{if } 0 \leq x \leq 2 \\ 0 & \text{otherwise} \end{cases}$$

The machine can process only one tube at a time. If any tube arrives while the machine is occupied, the tube passes on to another machine station.

a Let y be the time between successive tube arrivals (regardless of type and regardless of whether the machine is free). Determine $f_y(y_0)$, $E(y)$, and $\sigma_y{}^2$.

b Given that a tube arrives when the machine is free, what is the probability that the tube is of type I?

c Given that the machine starts to process a tube at time T_0, what is the PDF for the time required to process the tube?

d If we inspect the machine at a random time and find it processing a tube, what is the probability that the tube we find in the machine is type I?

e Given that an idle period of the machine was exactly T hours long, what is the probability that this particular idle period was terminated by the arrival of a type I tube?

4.22 The first-order interarrival times for cars passing a checkpoint are independent random variables with PDF

$$f_t(t_0) = \begin{cases} 2e^{-2t_0} & t_0 > 0 \\ 0 & t_0 \leq 0 \end{cases}$$

where the interarrival times are measured in minutes. The successive experimental values of the durations of these first-order interarrival times are recorded on small computer cards. The recording operation occupies a negligible time period following each arrival. Each card has space for three entries. As soon as a card is filled, it is replaced by the next card.

a Determine the mean and the third moment of the first-order interarrival times.

b Given that no car has arrived in the last four minutes, determine the PMF for random variable K, the number of cars to arrive in the next six minutes.

c Determine the PDF, the expected value, and the s transform for the total time required to use up the first dozen computer cards.

d Consider the following two experiments:

i Pick a card at random from a group of completed cards and note the total time, t_i, the card was in service. Find $E(t_i)$ and $\sigma_{t_i}{}^2$.

ii Come to the corner at a random time. When the card in use at the time of your arrival is completed, note the total time it was in service (the time from the start of its service to its completion). Call this time t_j. Determine $E(t_j)$, and $\sigma_{t_j}{}^2$.

e Given that the computer card presently in use contains exactly two entries and also that it has been in service for exactly 0.5 minute, determine and sketch the PDF for the remaining time until the card is completed.

CHAPTER FIVE

discrete-state Markov processes

The Bernoulli and Poisson processes are defined by probabilistic descriptions of series of *independent* trials. The Markov process is one type of characterization of a series of *dependent* trials.

We have emphasized the no-memory properties of the Bernoulli and Poisson processes. Markov processes do have memory (events in nonoverlapping intervals of time need not be independent), but the dependence of future events on past events is of a particularly simple nature.

5-1 Series of Dependent Trials for Discrete-state Processes

Consider a system which may be described at any time as being in one of a set of mutually exclusive collectively exhaustive *states* S_1, S_2, $\ldots$, S_m. According to a set of probabilistic rules, the system may, at certain discrete instants of time, undergo *changes of state* (or *state transitions*). We number the particular instants of time at which transitions may occur, and we refer to these instants as the *first trial*, the *second trial*, etc.

Let $S_i(n)$ be the event that the system is in state S_i immediately after the nth trial. The probability of this event may be written as $P[S_i(n)]$. Each trial in the general process of the type (*discrete state, discrete transition*) introduced in the above paragraph may be described by transition probabilities of the form

$$P[S_j(n) \mid S_a(n-1)S_b(n-2)S_c(n-3) \ldots]$$
$$1 \le j, a, b, c, \ldots, \le m; \quad n = 1, 2, 3, \ldots$$

These transition probabilities specify the probabilities associated with each trial, and they are conditional on the entire past history of the process. The above quantity, for instance, is the conditional probability that the system will be in state S_j immediately after the nth trial, given that the previous state history of the process is specified by the event $S_a(n-1)S_b(n-2)S_c(n-3) \cdots$.

We note some examples of series of dependent trials in discrete-state discrete-transition processes. The states might be nonnegative integers representing the number of people on a bus, and each bus stop might be a probabilistic trial at which a change of state may occur. Another example is a process in which one of several biased coins is flipped for each trial and the selection of the coin for each trial depends in some manner on the outcomes of the previous flips. The number of items in a warehouse at the start of each day is one possible state description of an inventory. For this process, the state transition due to the total transactions on any day could be considered to be the result of one of a continuing series of dependent trials.

5-2 Discrete-state Discrete-transition Markov Processes

If the transition probabilities for a series of dependent trials satisfy the

Markov condition:

$$P[S_j(n) \mid S_a(n-1)S_b(n-2)S_c(n-3) \cdots] = P[S_j(n) \mid S_a(n-1)] \qquad \text{for all } n, j, a, b, c, \ldots$$

the system is said to be a *discrete-state discrete-transition Markov process.*

If the state of the system immediately prior to the nth trial is known, the Markov condition requires that the conditional transition probabilities describing the nth trial do not depend on any additional past history of the process. The present state of the system specifies all historical information relevant to the future behavior of a Markov process.

We shall not consider processes for which the conditional transition probabilities

$$P[S_j(n) \mid S_i(n-1)]$$

depend on the number of the trial. Thus we may define the *state transition probabilities* for a discrete-transition Markov process to be

$$p_{ij} = P[S_j(n) \mid S_i(n-1)] \qquad 1 \leq i, j \leq m; \quad p_{ij} \text{ independent of } n$$

Quantity p_{ij} is the conditional probability that the system will be in state S_j immediately after the next trial, given that the present state of the process is S_i. We always have $0 \leq p_{ij} \leq 1$, and, because the list of states must be mutually exclusive and collectively exhaustive, it must also be true that

$$\sum_j p_{ij} = 1 \qquad \text{for } i = 1, 2, 3, \ldots, m$$

It is often convenient to display these transition probabilities as members of an $m \times m$ *transition matrix* $[p]$, for which p_{ij} is the entry in the ith row and jth column

$$[p] = \begin{bmatrix} p_{11} & p_{12} & \cdots & p_{1m} \\ p_{21} & p_{22} & \cdots & p_{2m} \\ \cdots & \cdots & \cdots & \cdots \\ p_{m1} & p_{m2} & \cdots & p_{mm} \end{bmatrix}$$

We also define the *k-step state transition probability* $p_{ij}(k)$,

$$p_{ij}(k) = \begin{pmatrix} \text{conditional probability that proc-} \\ \text{ess will be in state } S_j \text{ after exactly} \\ k \text{ more trials, given that present} \\ \text{state of process is } S_i \end{pmatrix} = P[S_j(n+k) \mid S_i(n)]$$

$$p_{ij}(0) = \begin{cases} 1 & i = j \\ 0 & i \neq j \end{cases} \qquad\qquad p_{ij}(1) = p_{ij}$$

Consider any integer l, subject to $0 \leq l \leq k$. We may always write

$$p_{ij}(k) = P[S_j(n+k) \mid S_i(n)] = \sum_{x=1}^{m} P[S_j(n+k)S_x(n+k-l) \mid S_i(n)]$$

which simply notes that the process had to be in *some* state immediately after the $(n+k-l)$th trial. From the definition of conditional probability we have

$$P[S_j(n+k)S_x(n+k-l) \mid S_i(n)] = P[S_x(n+k-l) \mid S_i(n)] \cdot P[S_j(n+k) \mid S_x(n+k-l)S_i(n)]$$

For a discrete-state discrete-transition Markov process we may use the Markov condition on the right-hand side of this equation to obtain

$$P[S_j(n+k) \mid S_x(n+k-l)S_i(n)] = P[S_j(n+k) \mid S_x(n+k-l)]$$

$$P[S_j(n+k)S_x(n+k-l) \mid S_i(n)] = p_{ix}(k-l)p_{xj}(l)$$

which may be substituted in the above equation for $p_{ij}(k)$ to obtain the result

$$p_{ij}(k) = \sum_{x=1}^{m} p_{ix}(k-l)p_{xj}(l) \qquad k = 1, 2, 3, \ldots ; \quad 0 \leq l \leq k; \quad 1 \leq i, j \leq m$$

This relation is a simple case of the *Chapman-Kolmogorov* equation, and it may be used as an alternative definition for the discrete-state discrete-transition Markov process with constant transition probabilities. This equation need *not* apply to the more general process described in Sec. 5-1.

Note that the above relation, with $l = 1$,

$$p_{ij}(k) = \sum_{x=1}^{m} p_{ix}(k-1)p_{xj}$$

provides a means of calculation of the k-step transition probabilities which is more efficient than preparing a probability tree for k trials and then collecting the probabilities of the appropriate events (see Prob. 5.02).

We consider one example. Suppose that a series of dependent-coin flips can be described by a model which assigns to any trial conditional probabilities which depend only on the outcome of the previous trial. In particular, we are told that any flip immediately following an experimental outcome of a head has probability 0.75 of also resulting in a head and that any flip immediately following a tail is a fair toss. Using the most recent outcome as the state description, we have the two-state Markov process

S_1: heads

S_2: tails

$$[p] = \begin{bmatrix} 0.75 & 0.25 \\ 0.50 & 0.50 \end{bmatrix}$$

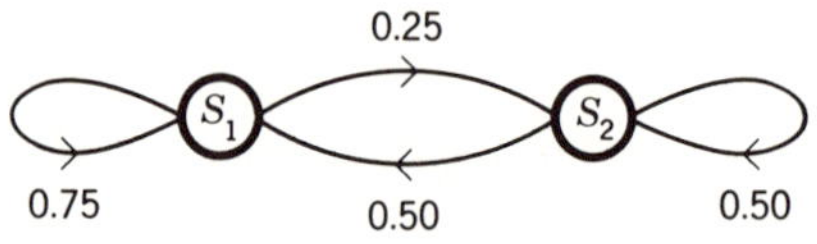

In the *state-transition diagram* shown above, we have made a picture of the process in which the states are circles and the trial transition probabilities are labeled on the appropriate arrowed branches.

We may use the relation

$$p_{ij}(k) = \sum_{x=1}^{m} p_{ix}(k-1)p_{xj}$$

first for $k = 2$, then for $k = 3$, etc., to compute the following table (in which we round off to three significant figures):

	$k = 1$	$k = 2$	$k = 3$	$k = 4$	$k = 5$	$k = 6$	$k = 7$	$k = 8$
$p_{11}(k)$	0.750	0.688	0.672	0.668	0.667	0.667	0.667	0.667
$p_{12}(k)$	0.250	0.312	0.328	0.332	0.333	0.333	0.333	0.333
$p_{21}(k)$	0.500	0.625	0.656	0.664	0.666	0.667	0.667	0.667
$p_{22}(k)$	0.500	0.375	0.344	0.336	0.334	0.333	0.333	0.333

Our table informs us, for instance, that, given the process is in state S_1 at any time, the conditional probability that the process will be in state S_2 exactly three trials later is equal to 0.328. In this example it appears that the k-step transition probabilities $p_{ij}(k)$ reach a limiting value as k increases and that these limiting values do not depend on i. We shall study this important property of *some* Markov processes in the next few sections.

If the probabilities describing each trial had depended on the results of the previous C flips, the resulting sequence of dependent trials could still be represented as a Markov process. However, the state description might require as many as 2^C states (for an example, see Prob. 5.01).

It need not be obvious whether or not a particular physical system can be modeled accurately by a Markov process with a finite number of states. Often this turns out to depend on how resourceful we are in suggesting an appropriate state description for the physical system.

5-3 State Classification and the Concept of Limiting-state Probabilities

We observed one interesting result from the dependent coin-flip example near the end of Sec. 5-2. As $k \to \infty$, the k-step state transition probabilities $p_{ij}(k)$ appear to depend neither on k nor on i.

If we let $P[S_i(0)]$ be the probability that the process is in state S_i just before the first trial, we may use the definition of $p_{ij}(k)$ to write

$$P[S_j(k)] = \sum_{i=1}^{m} P[S_i(0)]p_{ij}(k)$$

The quantities $P[S_i(0)]$ are known as the *initial conditions* for the process. If it is the case that, as $k \to \infty$, the quantity $p_{ij}(k)$ depends neither on k nor on i, then we would conclude from the above equation that $P[S_j(k)]$ approaches a constant as $k \to \infty$ and this constant is independent of the initial conditions.

Many (*but not all*) Markov processes do in fact exhibit this behavior. For processes for which the *limiting-state probabilities*

$$\lim_{k \to \infty} P[S_j(k)] = P_j \qquad j = 1, 2, \ldots, m$$

exist and are independent of the initial conditions, many significant questions may be answered with remarkable ease. A correct discussion of this matter requires several definitions.

State S_i is called *transient* if there exists a state S_j and an integer l such that $p_{ij}(l) \neq 0$ and $p_{ji}(k) = 0$ for $k = 0, 1, 2, \ldots$. This simply states that S_i is a transient state if there exists any state to which the system (in some number of trials) can get to from S_i but from which it can never return to S_i. For a Markov process with a finite number of states, we might expect that, after very many trials, the probability that the process is in any transient state approaches zero, no matter what the initial state of the process may have been.

As an example, consider the process shown below,

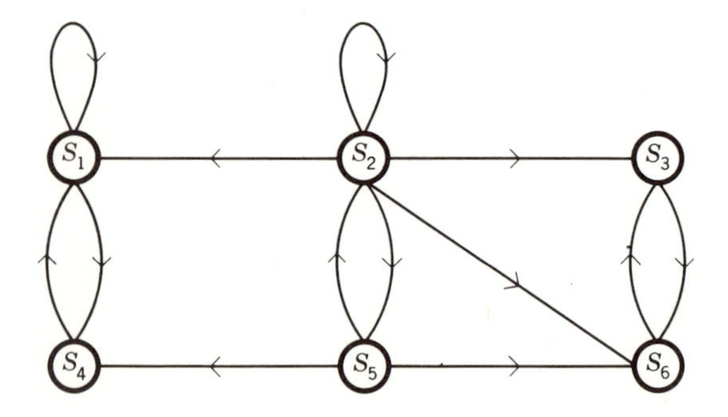

for which we have indicated branches for all state transitions which are to have nonzero transition probabilities. States S_2 and S_5 are the only states which the process can leave in some manner such that it may never return to them; so S_2 and S_5 are the only transient states in this example.

State S_i is called *recurrent* if, for every state S_j, the existence of an integer r_j such that $p_{ij}(r_j) > 0$ implies the existence of an integer r_i such that $p_{ji}(r_i) > 0$. From this definition we note that, no matter what state history may occur, once the process enters a recurrent state it will *always* be possible, in some number of transitions, to return to that state. Every state must be either recurrent or transient. In the above example, states S_1, S_3, S_4, and S_6 are recurrent states.

The fact that each of two states is recurrent does not necessarily require that the process can ever get from one of these states to the other. One example of two recurrent states with $p_{ij}(k) = p_{ji}(k) = 0$

for all k is found by considering the pair of states S_1 and S_3 in the above diagram.

Recurrent state S_i is called *periodic* if there exists an integer d, with $d > 1$, such that $p_{ii}(k)$ is equal to zero for all values of k other than $d, 2d, 3d, \ldots$. In our example above, recurrent states S_3 and S_6 are the only periodic states. (For our purposes, there is no reason to be concerned with periodicity for transient states.)

A set W of recurrent states forms one *class* (or a *single chain*) if, for every pair of recurrent states S_i and S_j of W, there exists an integer r_{ij} such that $p_{ij}(r_{ij}) > 0$. Each such set W includes all its possible members. The members of a class of recurrent states satisfy the condition that it is possible for the system (eventually) to get from any member state of the class to any other member state. In our example, there are two single chains of recurrent states. One chain is composed of states S_1 and S_4, and the other chain includes S_3 and S_6. Note that the definition of a single chain is concerned only with the properties of the recurrent states of a Markov process.

After informally restating these four definitions for m-state Markov processes ($m < \infty$), we indicate why they are of interest.

TRANSIENT STATE S_i: From at least one state which may be reached eventually from S_i, system can never return to S_i.

RECURRENT STATE S_i: From every state which may be reached eventually from S_i, system can eventually return to S_i.

PERIODIC STATE S_i: A recurrent state for which $p_{ii}(k)$ may be nonzero only for $k = d, 2d, 3d, \ldots$, with d an integer greater than unity.

SINGLE CHAIN W: A set of recurrent states with the property that the system can eventually get from any member state to any other state which is also a member of the chain. All possible members of each such chain are included in the chain.

For a Markov process with a finite number of states *whose recurrent states form a single chain* and *which contains no periodic states*, we might expect that the k-step transition probabilities $p_{ij}(k)$ become independent of i and k as k approaches infinity. We might argue that such a process has "limited memory." Although successive trials are strongly dependent, it is hard to see how $P[S_i(k)]$ should be strongly influenced by either k or the initial state after a large number of trials. In any case, it should be clear that, for either of the following processes,

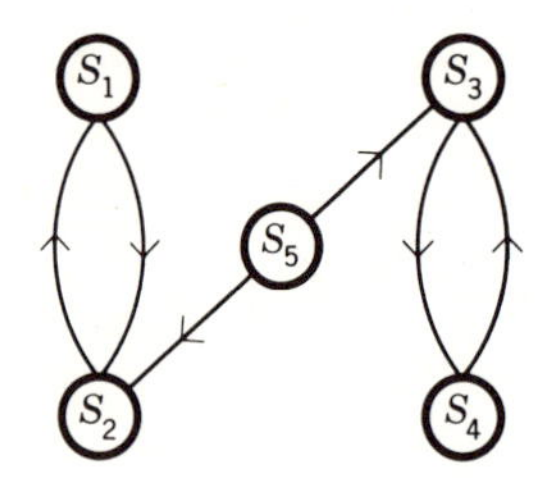

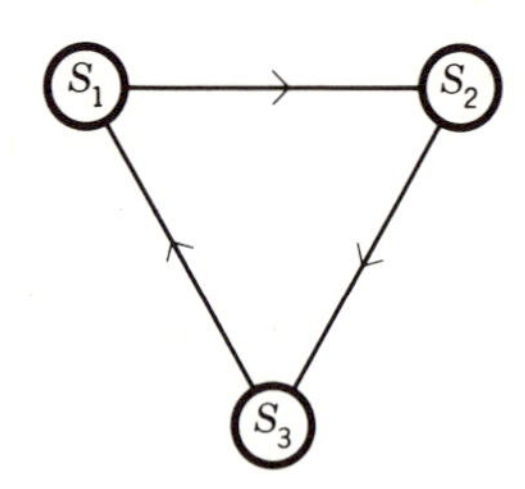

we would certainly *not* expect *any* $p_{ij}(k)$ to become independent of i and k as k gets very large.

We speculated that, for a Markov process with a finite number of states, whose recurrent states form a single chain, and which contains no periodic states, we might expect that

$$\lim_{k \to \infty} p_{ij}(k) = P_j \qquad \sum_{j=1}^{m} P_j = 1$$

where P_j depends neither on k nor on i. In fact this result is established by a simplified form of the *ergodic theorem*, which we shall state without proof in the following section. The P_j's, known as the *limiting-state probabilities*, represent the probabilities that a single-chain process with no periodic states will be in state S_j after very many trials, no matter what the initial conditions may have been.

Since our example of the dependent coin flips in the previous section satisfies these restrictions, the ergodic theorem states, for example, that quantity $P[S_1(n)]$ = Prob(heads on nth toss) will approach a constant as $n \to \infty$ and that this constant will not depend on the initial state of the process.

5-4 The Ergodic Theorem

In this section we shall present and discuss a formal statement of a simple form of the ergodic theorem for a discrete-state discrete-transition Markov process. The ergodic theorem is as follows:

Let M_k be the matrix of k-step transition probabilities of a Markov process with a finite number of states $S_1, S_2, \ldots, S_m$. If there exists an integer k such that the terms $p_{ij}(k)$ of the matrix M_k satisfy the relation

$$\min_{1 \le i \le m} p_{ij}(k) = \delta > 0$$

for at least one column of M_k, then the equalities

$$\lim_{n \to \infty} p_{ij}(n) = P_j \quad j = 1, 2, \ldots, m \quad i = 1, 2, \ldots, m; \qquad \sum_j P_j = 1$$

are satisfied.

The restriction

$$\min_{1 \le i \le m} p_{ij}(k) = \delta > 0$$

for at least one column of M_k simply requires that there be at least one state S_j and some number k such that it be possible to get to S_j from every state in exactly k transitions. This requirement happens to correspond to the conditions that the recurrent states of the system form a single chain and that there be no periodic states.

When the above restriction on the $p_{ij}(k)$ is satisfied for some value of k, the ergodic theorem states that, as $n \to \infty$, the n-step transition probabilities $p_{ij}(n)$ approach the limiting, or "steady-state," probabilities P_j. A formal test of whether this restriction does in fact hold for a given process requires certain matrix operations not appropriate to the mathematical background assumed for our discussions. We shall work with the "single chain, finite number of states, and no periodic states" restriction as being equivalent to the restriction in the ergodic theorem. (The single-chain and no-periodic-states restrictions are necessary conditions for the ergodic theorem; the finite-number-of-states restriction is not a necessary condition.) For the representative Markov systems to be considered in this book, we may test for these properties by direct observation.

5-5 The Analysis of Discrete-state Discrete-transition Markov Processes

We begin this section with a review of some of the things we already know about discrete-state discrete-transition Markov processes. We then write the general difference equations which describe the behavior of the state probabilities, the $P[S_j(n)]$'s, as the process operates over a number of trials. For processes to which the ergodic theorem applies, we also consider the solution of these difference equations as $n \to \infty$ to obtain the limiting-state probabilities. Finally, we note how our results simplify for the important class of Markov processes known as *birth-and-death* processes.

As we did at the beginning of our study of the Poisson process in Chap. 4, let us make use of an efficient but somewhat improper notation to suit our purposes. We define

$P_j(n) = P[S_j(n)] =$ probability process is in state S_j immediately after nth trial

From the definition of $p_{ij}(n)$ we may write

$$P_j(n) = \sum_i P_i(0)p_{ij}(n) \qquad \sum_j P_j(n) = 1 \qquad \text{for } n = 0, 1, 2, \ldots$$

where the $P_i(0)$'s, the *initial conditions*, represent the probabilities of the process being in its various states prior to the first trial. Because of the Markov condition, we also have

$$p_{ij}(n) = \sum_k p_{ik}(l)p_{kj}(n-l) \qquad l = 0, 1, 2, \ldots, n$$

and when the ergodic theorem applies, we know that

$$\lim_{n\to\infty} P_j(n) = P_j \qquad j = 1, 2, \ldots, m$$

and these limits do not depend on the initial conditions.

One expresses his "state of knowledge" about the state of a Markov process at any time by specifying the *probability state vector*

$$\lfloor P(n) \rfloor = \lfloor P_1(n) \quad P_2(n) \quad \cdots \quad P_m(n) \rfloor$$

Now we shall write the describing equations for the state probabilities of a Markov process as it continues to undergo trials. There are efficient transform and flow-graph techniques for the general solution of these equations. Here we shall develop only the describing equations and consider their solution as n goes to infinity. Even this limited tour will prepare us to study many important practical systems.

The equations of interest will relate the $P_j(n+1)$'s to the state probabilities one trial earlier, the $P_i(n)$'s. Each of the first $m-1$ following relations sums the probabilities of the mutually exclusive and collectively exhaustive ways the event $S_j(n+1)$ can occur, in terms of the state probabilities immediately after the nth trial. We start with the appropriate event space,

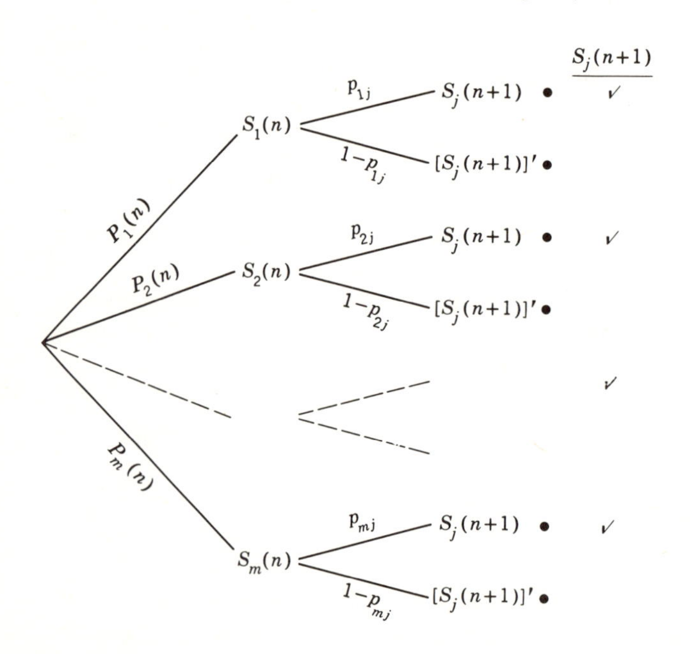

We collect the probabilities of the events $S_j(n+1)$ for $j = 1$ (first equation), for $j = 2$ (second equation), etc., up through $j = m - 1$. Thus, for any discrete-state discrete-transition Markov process

$$
m-1 \text{ eqs.:} \begin{cases} P_1(n+1) = P_1(n)p_{11} + P_2(n)p_{21} + \cdots + P_m(n)p_{m1} \\ P_2(n+1) = P_1(n)p_{12} + P_2(n)p_{22} + \cdots + P_m(n)p_{m2} \\ \cdots\cdots\cdots\cdots\cdots\cdots\cdots\cdots \\ P_{m-1}(n+1) = P_1(n)p_{1(m-1)} + P_2(n)p_{2(m-1)} + \cdots + P_m(n)p_{m(m-1)} \end{cases}
$$
$$
m\text{th eq.:} \quad 1 = P_1(n+1) + P_2(n+1) + \cdots + P_m(n+1)
$$

where the mth equation states the dependence of the $P_j(n+1)$'s. Given a set of initial conditions $P_1(0), P_2(0), \ldots, P_m(0)$, these equations may be solved iteratively for $n = 0, 1, 2,$ etc., to determine the state probabilities as a function of n. More advanced methods which allow one to obtain the probability state vector $\lfloor P(n) \rfloor$ in closed form will not be considered here.

For the remainder of this section, we consider only processes *with no periodic states* and *whose recurrent states form a single chain.* The ergodic theorem applies to such processes; so we may let

$$
\lim_{n\to\infty} P_i(n+1) = \lim_{n\to\infty} P_i(n) = P_i
$$

and, rewriting the above equations for the limit as $n \to \infty$, we have

$$
m-1 \text{ eqs.:} \begin{cases} 0 = P_1(p_{11} - 1) + P_2p_{21} + \cdots + P_mp_{m1} \\ 0 = P_1p_{12} + P_2(p_{22} - 1) + \cdots + P_mp_{m2} \\ \cdots\cdots\cdots\cdots\cdots\cdots\cdots\cdots \\ 0 = P_1p_{1(m-1)} + P_2p_{2(m-1)} + \cdots + P_mp_{m(m-1)} \end{cases}
$$
$$
m\text{th eq.:} \quad 1 = P_1 + P_2 + \cdots + P_m
$$

The solution to these m simultaneous equations determines the limiting-state probabilities for those processes which meet the restrictions for the ergodic theorem. Examples of the use of our two sets of simultaneous equations are given in the following section. (By writing the mth equation in the form used above, we avoid certain problems with this set of simultaneous equations.)

Happily, there is a special and very practical type of Markov process for which the equations for the limiting-state probabilities may be solved on sight. We further limit our discussion of single-chain processes with no periodicities to the case of *birth-and-death* processes.

A discrete-state discrete-transition birth-and-death process is a Markov process whose transition probabilities obey

$$p_{ij} = 0 \qquad \text{if } j \neq i - 1, \quad i, \quad i + 1$$

and for these processes it is advantageous to adopt the *b*irth-and-*d*eath notation

$$p_{i(i+1)} = b_i \qquad p_{i(i-1)} = d_i$$

One example of a birth-and-death process is

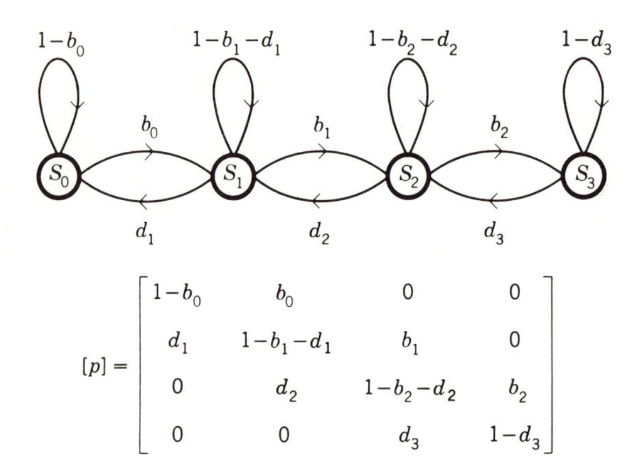

Many practical instances of this type of process are mentioned in the problems at the end of this chapter. Note that our definition of the birth-and-death process (and the method of solution for the limiting-state probabilities to follow) does not include the process pictured below:

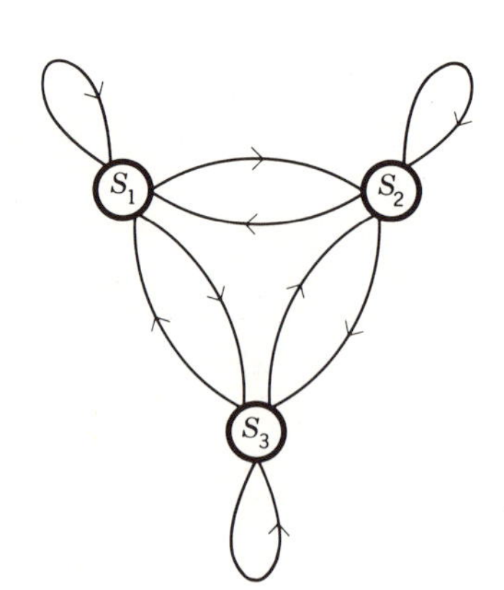

For the given assignment of state labels, this process will violate the definition of the birth and death process if either p_{13} or p_{31} is nonzero

We shall now demonstrate one argument for obtaining the

limiting-state probabilities for a single-chain birth-and-death process. We begin by choosing any *particular* state S_K and noting that, at any time in the history of a birth-and-death process, the total number of $S_K \to S_{K+1}$ transitions made so far must either be one less than, equal to, or one greater than the total number of $S_{K+1} \to S_K$ transitions made so far. (Try to trace out a possible state history which violates this rule.)

Consider the experiment which results if we approach a birth-and-death process after it has undergone a great many transitions and our state of knowledge about the process is given by the limiting-state probabilities. The probability that the *first* trial after we arrive will result in an $S_K \to S_{K+1}$ transition is $P_K b_K$; the probability that it will result in an $S_{K+1} \to S_K$ transition is $P_{K+1} d_{K+1}$.

Since, over a long period of time, the fractions of all trials which have these two outcomes must be equal and we are simply picking a trial at random, we must have (for a single-chain birth-and-death process with no periodic states)

$$P_K b_K = P_{K+1} d_{K+1}$$

and thus the limiting-state probabilities may be obtained by finding all P_i's in terms of P_0 from

$$P_{i+1} = \frac{P_i b_i}{d_{i+1}} \qquad i = 0, 1, 2, \ldots$$

and then solving for P_0 by using $\sum_i P_i = 1$

Another way to derive this result would be to notice that, for a birth-and-death process, many of the coefficients in the simultaneous equations for the P_i's for the more general single-chain Markov process are equal to zero. The resulting equations may easily be solved by direct substitution to obtain the solution stated above.

The first paragraph of this section may now serve as a road map for the above work. Several examples are discussed and solved in the following section.

5-6 Examples Involving Discrete-transition Markov Processes

example 1 Experience has shown that the general mood of Herman may be realistically modeled as a three-state Markov process with the mutually exclusive collectively exhaustive states

S_1: Cheerful $\quad$ S_2: So-so $\quad$ S_3: Glum

His mood can change only overnight, and the following transition probabilities apply to each night's trial:

$$[p] = \begin{bmatrix} 0.6 & 0.2 & 0.2 \\ 0.3 & 0.4 & 0.3 \\ 0.0 & 0.3 & 0.7 \end{bmatrix}$$

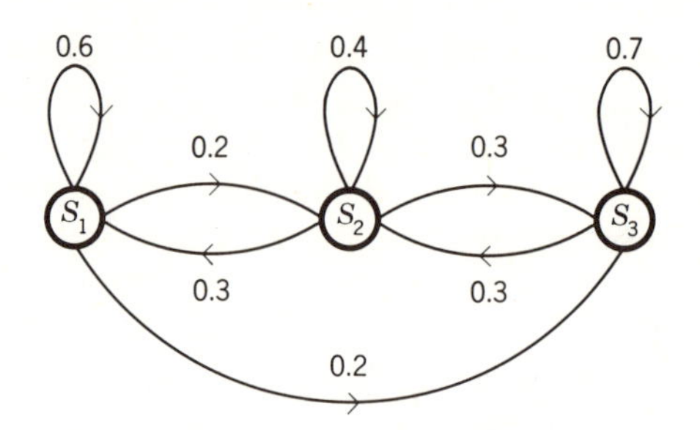

We are told that Herman's mood today is so-so.

(a) Determine the components of the probability state vector, the $P_i(n)$'s, for Herman's mood for the next few days.

(b) Determine this probability state vector for a day a few months hence. Is the answer dependent on the initial conditions?

(c) Determine the PMF for the number of trials until Herman's mood undergoes its first change of state.

(d) What is the probability that Herman will become glum before he becomes cheerful?

a We are given $\lfloor P(0) \rfloor = \lfloor P_1(0) \quad P_2(0) \quad P_3(0) \rfloor = \lfloor 0 \quad 1 \quad 0 \rfloor$, and we may use the original set of difference equations for the $P_j(n+1)$'s,

$$P_j(n+1) = \sum_i P_i(n) p_{ij} \qquad \text{for } j = 1, 2, \ldots, m-1$$

$$\sum_j P_j(n+1) = 1$$

first with $n = 0$, then with $n = 1$, etc. For instance, with $n = 0$ we find

$$P_1(1) = \sum_i P_i(0) p_{i1} = (0)(0.6) + (1)(0.3) + (0)(0.0) = 0.3$$

$$P_2(1) = \sum_i P_i(0) p_{i2} = (0)(0.2) + (1)(0.4) + (0)(0.3) = 0.4$$

$$1 = \sum_j P_j(1) = 0.3 + 0.4 + P_3(1) \qquad \therefore P_3(1) = 0.3$$

And thus we have obtained

$$\lfloor P(1) \rfloor = \lfloor P_1(1) \quad P_2(1) \quad P_3(1) \rfloor = \lfloor 0.3 \quad 0.4 \quad 0.3 \rfloor$$

Further iterations using the difference equations allow us to generate the following table:

	$n = 0$	$n = 1$	$n = 2$	$n = 3$	$n = 4$	$n = 5$	$n = 6$
$P_1(n)$	0.000	0.300	0.300	0.273	0.254	0.243	0.237
$P_2(n)$	1.000	0.400	0.310	0.301	0.303	0.305	0.306
$P_3(n)$	0.000	0.300	0.390	0.426	0.443	0.452	0.457

All entries have been rounded off to three significant figures. The difference equations apply to *any* discrete-state discrete-transition Markov process.

b Since our Markov model for this process has no periodic states and its recurrent states form a single chain, the limiting-state probabilities are independent of the initial conditions. (The limiting-state probabilities *would* depend on the initial conditions if, for instance, we had $p_{12} = p_{32} = p_{13} = p_{31} = 0$.) We shall assume that the limiting-state probabilities are excellent approximations to what we would get by carrying out the above table for about 60 more trials (two months). Thus we wish to solve the simultaneous equations for the limiting-state probabilities,

$$0 = \sum_i P_i p_{ij} - P_j \qquad j = 1, 2, \ldots, m-1$$

$$1 = \sum_j P_j$$

which, for our example, are

$$0 = P_1(0.6 - 1.0) + P_2(0.3) \qquad + P_3(0.0)$$
$$0 = P_1(0.2) \qquad + P_2(0.4 - 1.0) + P_3(0.3)$$
$$1 = P_1 + P_2 + P_3$$

which may be solved to obtain

$$P_1 = 3/13 \approx 0.231 \qquad P_2 = 4/13 \approx 0.308 \qquad P_3 = 6/13 \approx 0.461$$

These values seem consistent with the behavior displayed in the above table. The probability that Herman will be in a glum mood 60 days hence is very close to 6/13. In fact, for this example, the limiting-state probabilities are excellent approximations to the actual-state probabilities 10 or so days hence. Note also that this is *not* a birth-and-death process ($p_{13} \neq 0$) and, therefore, we may not use the more rapid method of solution for the P_j's which applies only to birth-and-death processes.

c Given that Herman is still in state S_2, the conditional probability that he will undergo a change of state (of mind) at the next transition is given by $1 - p_{22}$. Thus the PMF for l, the number of (Bernoulli) trials up to and including his first change of mood, is the geometric PMF with parameter P equal to $1 - p_{22}$.

$$p_l(l_0) = (1 - p_{22})p_{22}^{l_0 - 1} = (0.6)(0.4)^{l_0 - 1} \qquad l_0 = 1, 2, 3, \ldots$$

We would obtain a similar result for the conditional PMF for the number of trials up to and including the next actual change of state for any discrete-transition Markov process, given the present state of the process. For this reason, one may say that such a process is characterized by *geometric holding times.* A similar phenomenon will be discussed in the introduction to the next section.

d Given that Herman's mood is so-so, the following event space describes any trial while he is in this state:

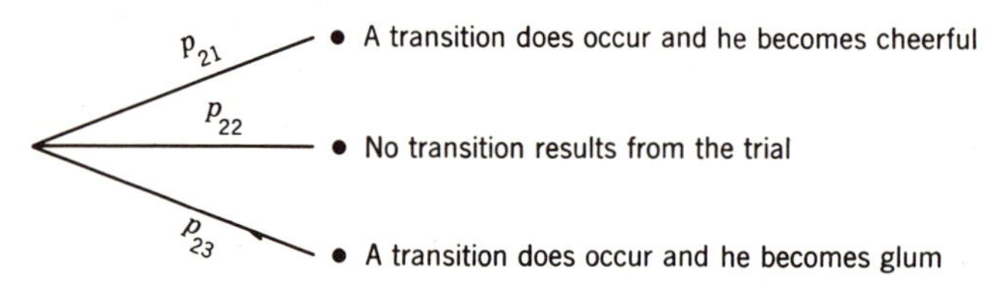

Thus we may calculate that the conditional probability he becomes glum given that a transition does occur is equal to $p_{23}/(p_{21} + p_{23})$. This is, of course, equal to the probability that he becomes glum before he becomes cheerful, and its numerical value is $0.3/(0.3 + 0.3) = 0.5$.

example 2 Roger Yogi Mantle, an exceptional baseball player who tends to have streaks, hit a home run during the first game of this season. The conditional probability that he will hit at least one homer in a game is 0.4 if he hit at least one homer in the previous game, but it is only 0.2 if he didn't hit any homers in the previous game. We assume this is a complete statement of the dependence. Numerical answers are to be correct within $\pm 2\%$.

(a) What is the probability that Roger hit at least one home run during the third game of this season?
(b) If we are told that he hit a homer during the third game, what is the probability that he hit at least one during the second game?
(c) If we are told that he hit a homer during the ninth game, what is the probability that he hit at least one during the tenth game?
(d) What is the probability that he will get at least one home run during the 150th game of this season?
(e) What is the probability that he will get home runs during both the 150th and 151st games of this season?
(f) What is the probability that he will get home runs during both the 3d and 150th games of this season?
(g) What is the probability that he will get home runs during both the 75th and 150th games of this season?

This situation may be formulated as a two-state Markov process. A game is type H if Roger hits at least one homer during the game; otherwise it is type N. For our model, we shall consider the trials to occur between games.

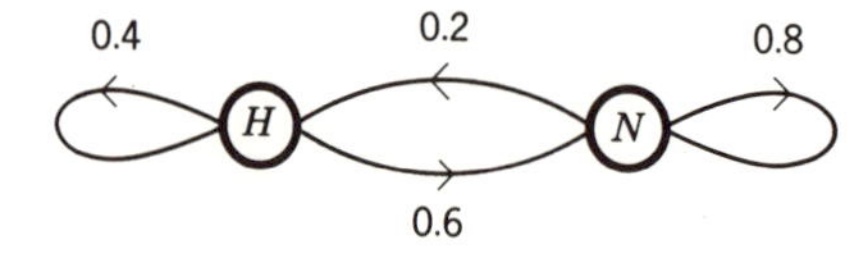

For this example we shall not go to an S_i description for each state, but we shall work directly with H and N, using the notation

$P(H_n)$ = probability Roger is in state H during nth game

$P(N_n)$ = probability Roger is in state N during nth game

We are given the initial condition $P(H_1) = 1$. We also note that this is a single-chain process with no periodic states, and it also happens to be a birth-and-death process.

a We wish to determine $P(H_3)$. One method would be to use the sequential sample space

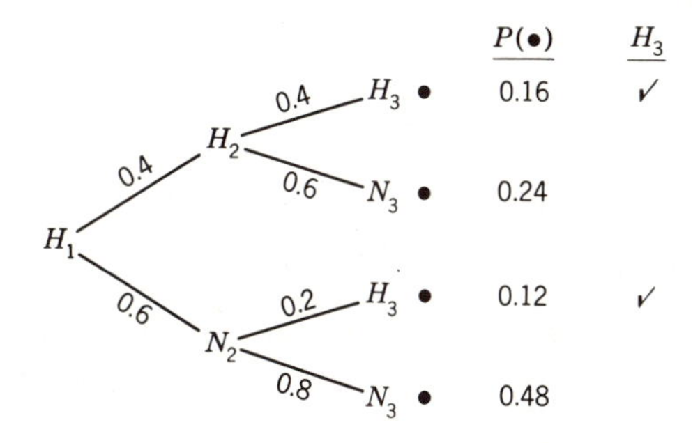

to find $P(H_3) = 0.16 + 0.12 = 0.28$. Since the conditional branch traversal probabilities for the tree of a Markov process depend only on the most recent node, it is usually more efficient to solve for such state probabilities as a function of n from the difference equations, which, for this example, are

$$\left.\begin{aligned} P(H_{n+1}) &= 0.4P(H_n) + 0.2P(N_n) \\ 1 &= P(H_{n+1}) + P(N_{n+1}) \end{aligned}\right\} \qquad n = 1, 2, \ldots$$

and which lead, of course, to the same result.

b The desired conditional probability is easily calculated from the above sequential sample space,

$$P(H_2 \mid H_3) = \frac{P(H_2H_3)}{P(H_3)} = \frac{0.16}{0.28} = \frac{4}{7}$$

We have chosen to write $P(H_2 \mid H_3)$ rather than $P(H_2 \mid H_3H_1)$ because the event H_1 is given as part of the overall problem statement.

c The conditional probability that Roger hits at least one homer in the 10th game, given he hit at least one in the 9th game (and given no information about later games), is simply p_{HH}, which is given to be 0.4 in the problem statement.

d If we carry out a few iterations using the difference equations given after the solution to part (*a*) we find, working to three significant figures,

	$n = 1$	$n = 2$	$n = 3$	$n = 4$	$n = 5$	$n = 6$	$n = 7$	$n = 8$
$P(H_n)$	1.000	0.400	0.280	0.256	0.251	0.250	0.250	0.250
$P(N_n)$	0.000	0.600	0.720	0.744	0.749	0.750	0.750	0.750

Thus it is conservative to state that, for all practical purposes, his performances in games separated by more than 10 games may be considered independent events. $P(H_{150})$ is just the limiting-state probability P_H, which, taking advantage of our method for birth-and-death processes, is determined by

$$0.6P_H = 0.2P_N \qquad P_H + P_N = 1.0$$

resulting in $P_H = 0.25$, which checks with the result obtained by iteration above.

e The desired quantity is simply $P_H p_{HH} = \frac{1}{4} \cdot \frac{4}{10} = 0.1$. Note that the strong dependence of results in successive or nearly successive games must always be considered and that the required answer is certainly *not* P_H^2.

f $P(H_3H_{150}) = P(H_3)P(H_{150} \mid H_3) \approx P(H_3)P_H = (0.28)(\frac{1}{4}) = 0.07$

g Roger's performances on games this far apart may be considered independent events, and we have

$$P(H_{75}H_{150}) = P(H_{75})P(H_{150} \mid H_{75}) \approx P_H^2 = 1/16$$

The reader is reminded that these have been two elementary problems, intended to further our understanding of results obtained earlier in this chapter. Some more challenging examples will be found at the end of the chapter.

Some questions concerned with random incidence (Sec. 4-11) for a Markov process, a situation which did not arise here, will be introduced in the examples in Sec. 5-8.

5-7 Discrete-state Continuous-transition Markov Processes

Again we are concerned with a system which may be described at any time as being in one of a set of mutually exclusive collectively exhaustive discrete states $S_1, S_2, S_3, \ldots, S_m$. For a *continuous-transition* process, the probabilistic rules which describe the transition behavior allow changes of state to occur at any instants on a continuous time axis. If an observer knows the present state of *any* Markov process, any other information about the past state history of the process is irrelevant to his probabilistic description of the future behavior of the process.

In this section, we consider Markov processes for which, given that the present state is S_i, the conditional probability that an $S_i \rightarrow S_j$

transition will occur in the next Δt is given by $\lambda_{ij}\,\Delta t$ (for $j \neq i$ and suitably small Δt). Thus, each incremental Δt represents a trial whose outcome may result in a change of state, and the transition probabilities which describe these trials depend only on the present state of the system. *We shall not allow* λ_{ij} *to be a function of time;* this restriction corresponds to our not allowing p_{ij} to depend on the number of the trial in our discussion of the discrete-transition Markov process.

We begin our study of these discrete-state continuous-transition Markov processes by noting some consequences of the above description of the state transition behavior and by making some comparisons with discrete-transition processes. (All the following statements need hold only for suitably small Δt.)

The conditional probability that *no* change of state will occur in the next Δt, given that the process is at present in state S_i, is

$$\text{Prob(no change of state in next } \Delta t, \text{ given present state is } S_i) = 1 - \sum_{\substack{j \\ j \neq i}} \lambda_{ij}\,\Delta t$$

Although p_{ii} was a meaningful parameter for the discrete-transition process, a quantity λ_{ii} has no similar interpretation in the continuous-transition process. This is one reason why our equations for the state probabilities as a function of time will be somewhat different in form from those describing the state probabilities for the discrete-transition process. (For reasons outside the scope of this text, it is preferable that we let λ_{ii} remain undefined rather than define λ_{ii} to be equal to zero.)

Given that the system is at present in state S_i, the probability of leaving this state in the next Δt, no matter how long the system has already been in state S_i, is equal to

$$\sum_{\substack{j \\ j \neq i}} \lambda_{ij}\,\Delta t$$

and, from our earlier study of the Poisson process, we realize that the remaining time until the next departure from the present state is an exponentially distributed random variable with expected value

$$\Big(\sum_{\substack{j \\ j \neq i}} \lambda_{ij}\Big)^{-1}$$

For this reason, the type of continuous process we have described is said to have *exponential holding times.* Surprisingly general physical systems, many of which do not have exponential holding times, may be modeled realistically by the resourceful use of such a Markov model.

For the continuous-transition process, we shall again define a transient state S_i to be one from which it is possible for the process eventually to get to some other state from which it can never return to S_i. A recurrent state S_i is one to which the system can return from any state which can eventually be reached from S_i. No concept of periodicity is required, and a single class (or chain) of recurrent states again includes all its possible members and has the property that it is possible eventually to get from any state which is a member of the class to any other member state.

A useful compact notation, similar to that used in Sec. 5-5, is $P_j(t) = P[S_j(t)]$ = probability process is in state S_j at time t. $P_j(t)$ must have the properties

$$0 \leq P_j(t) \leq 1 \qquad \sum_j P_j(t) = 1$$

We would expect, at least for a process with a finite number of states, that the probability of the process being in a transient state goes to zero as $t \to \infty$. For a recurrent state S_i in a single chain with a finite number of states we might expect

$$\int_0^\infty P_i(t)\,dt = \infty \qquad \begin{array}{l}\text{if } S_i \text{ is a recurrent state in a single-chain process}\\ \text{with a finite number of states}\end{array}$$

since we expect $P_i(t)$ to approach a nonzero limit as $t \to \infty$.

We shall now develop the equations which describe the behavior of the state probabilities, the $P_i(t)$'s, as the process operates over time for any m-state continuous-transition Markov process with exponential holding times. The formulation is very similar to that used earlier for discrete-transition Markov processes. We shall write $m - 1$ incremental relations relating $P_j(t + \Delta t)$ to the $P_i(t)$'s, for $j = 1, 2, \ldots, m - 1$. Our mth equation will be the constraint that $\sum_j P_j(t + \Delta t) = 1$.

To express each $P_j(t + \Delta t)$, we sum the probabilities of all the mutually exclusive ways that the process could come to be in state S_j at $t + \Delta t$, in terms of the state probabilities at time t,

$m - 1$ eqs.:

$$\begin{cases} P_1(t + \Delta t) = P_1(t)\left(1 - \sum\limits_{\substack{j\\ j\neq 1}} \lambda_{1j}\,\Delta t\right) + \sum\limits_{\substack{j\\ j\neq 1}} P_j(t)\lambda_{j1}\,\Delta t \\ P_2(t + \Delta t) = P_2(t)\left(1 - \sum\limits_{\substack{j\\ j\neq 2}} \lambda_{2j}\,\Delta t\right) + \sum\limits_{\substack{j\\ j\neq 2}} P_j(t)\lambda_{j2}\,\Delta t \\ \cdots\cdots\cdots\cdots\cdots\cdots\cdots\cdots\cdots\cdots \\ P_{m-1}(t + \Delta t) = P_{m-1}(t)\left(1 - \sum\limits_{\substack{j\\ j\neq m-1}} \lambda_{(m-1)j}\,\Delta t\right) + \sum\limits_{\substack{j\\ j\neq m-1}} P_j(t)\lambda_{j(m-1)}\,\Delta t \end{cases}$$

mth eq.: $\qquad 1 = \sum_j P_j(t + \Delta t)$

On the right-hand side of the ith of the first $m - 1$ equations, the first term is the probability of the process being in state S_i at time

t and not undergoing a change of state in the next Δt. The second term is the probability that the process entered state S_i from some other state during the incremental interval between t and $t + \Delta t$. We can simplify these equations by multiplying, collecting terms, dividing through by Δt, and taking the limit as $\Delta t \to 0$ to obtain, for *any* discrete-state continuous-transition Markov process with exponential holding times

$$
m - 1 \text{ eqs.:} \begin{cases} \dfrac{dP_1(t)}{dt} = \sum\limits_{\substack{j \\ j \neq 1}} P_j(t)\lambda_{j1} - P_1(t) \sum\limits_{\substack{j \\ j \neq 1}} \lambda_{1j} \\ \dfrac{dP_2(t)}{dt} = \sum\limits_{\substack{j \\ j \neq 2}} P_j(t)\lambda_{j2} - P_2(t) \sum\limits_{\substack{j \\ j \neq 2}} \lambda_{2j} \\ \cdots\cdots\cdots\cdots\cdots\cdots\cdots\cdots \\ \dfrac{dP_{m-1}(t)}{dt} = \sum\limits_{\substack{j \\ j \neq m-1}} P_j(t)\lambda_{j(m-1)} - P_{(m-1)}(t) \sum\limits_{\substack{j \\ j \neq m-1}} \lambda_{(m-1)j} \end{cases}
$$

$$
m\text{th eq.:} \qquad 1 = \sum_j P_j(t)
$$

Each of the first $m - 1$ equations above relates the rate of change of a state probability to the probability of being elsewhere (in the first term) and to the probability of being in that state (in the second term). The solution of the above set of simultaneous differential equations, subject to a given set of initial conditions, would provide the state probabilities, the $P_i(t)$'s, for $i = 1, 2, \ldots, m$ and $t \geq 0$. Effective flow-graph and transform techniques for the solution of these equations exist but are outside the scope of our discussion. For some simple cases, as in Example 1 in Sec. 5-8, the direct solution of these equations presents no difficulties.

For the remainder of this section we limit our discussion to *processes whose recurrent states form a single chain.* We might expect for such processes that the effects of the initial conditions vanish as $t \to \infty$ and that $P_i(t + \Delta t) \to P_i(t)$ $\left(\text{or } \dfrac{dP_i(t)}{dt} \to 0\right)$ as $t \to \infty$. We define the limiting-state (or steady-state) probabilities by

$$
\lim_{t \to \infty} P_i(t) = P_i
$$

And we comment without proof that a suitable ergodic theorem does exist to establish the validity of the above speculations.

To obtain the equations for the limiting-state probabilities, we need only rewrite the simultaneous differential equations for the limiting case of $t \to \infty$. There results

$$0 = \sum_{\substack{j \\ j \neq 1}} P_j \lambda_{j1} - P_1 \sum_{\substack{j \\ j \neq 1}} \lambda_{1j}$$

$$0 = \sum_{\substack{j \\ j \neq 2}} P_j \lambda_{j2} - P_2 \sum_{\substack{j \\ j \neq 2}} \lambda_{2j}$$

. . . .

$$0 = \sum_{\substack{j \\ j \neq m-1}} P_j \lambda_{j(m-1)} - P_{m-1} \sum_{\substack{j \\ j \neq m-1}} \lambda_{(m-1)j}$$

$$1 = \sum_j P_j$$

If the process has been in operation for a long time, this term, multiplied by Δt, is the probability that the process will enter S_2 from elsewhere in a randomly selected Δt.

If the process has been in operation for a long time, this term, multiplied by Δt, is the probability that, in a randomly selected Δt, the process will leave S_2 to enter another state.

Thus, for any Markov process with exponential holding times whose recurrent states form a single chain, we may obtain the limiting-state probabilities by solving these m simultaneous equations.

Again, there exists the important case of birth-and-death processes for which the equations for the limiting-state probabilities are solved with particular ease.

A continuous birth-and-death process is a discrete-state continuous-transition Markov process which obeys the constraint

$$\lambda_{ij} = 0 \qquad \text{if } j \neq i - 1, \quad i + 1$$

(Recall that λ_{ii} has not been defined for continuous-transition processes.) The parameters of the process may be written in a *b*irth-and-*d*eath notation and interpreted as follows

$$\lambda_{i(i+1)} = b_i = \textit{average birth rate} \text{ when process is in state } S_i$$

$$\lambda_{i(i-1)} = d_i = \textit{average death rate} \text{ when process is in state } S_i$$

Either by direct substitution into the simultaneous equations or by arguing that, when a birth-and-death process is in the steady state (i.e., the limiting-state probabilities apply), the process must be as likely, in any randomly selected Δt, to undergo an $S_i \to S_{i+1}$ transition as to undergo the corresponding $S_{i+1} \to S_i$ transition, we obtain

$$P_i b_i = P_{i+1} d_{i+1}$$

Thus, for a continuous birth-and-death process, the limiting-state probabilities are found from the simple relations

$$P_{i+1} = \frac{b_i}{d_{i+1}} P_i \qquad \text{and} \qquad \sum_i P_i = 1$$

Several examples with solutions are presented in Sec. 5-8. A wider range of applications is indicated by the set of problems at the end of this chapter. Example 2 in Sec. 5-8 introduces some elementary topics from the theory of queues.

5-8 Examples Involving Continuous-transition Processes

example 1 A young brown-and-white rabbit, named Peter, is hopping about his newly leased two-room apartment. From all available information, we conclude that with probability $\mathcal{P}$ he was in room 1 at $t = 0$. Whenever Peter is in room 1, the probability that he will enter room 2 in the next Δt is known to be equal to $\lambda_{12}\,\Delta t$. At any time when he is in room 2, the probability that he will enter room 1 in the next Δt is $\lambda_{21}\,\Delta t$. It is a bright, sunny day, the wind is 12 mph from the northwest (indoors!), and the inside temperature is 70°F.

(a) Determine $P_1(t)$, the probability that Peter is in room 1 as a function of time for $t \geq 0$.

(b) If we arrive at a random time with the process in the steady state:

(i) What is the probability the first transition we see will be Peter entering room 2 from room 1?

(ii) What is the probability of a transition occurring in the first Δt after we arrive?

(iii) Determine the PDF $f_x(x_0)$, where x is defined to be the waiting time from our arrival until Peter's next change of room.

(iv) If we observe no transition during the first T units of time after we arrive, what is then the conditional probability that Peter is in room 1?

Let state S_n represent the event "Peter is in room n." We have a two-state Markov process with exponential holding times. We sketch a transition diagram of the process, labeling the branches with the conditional-transition probabilities for trials in an incremental interval Δt.

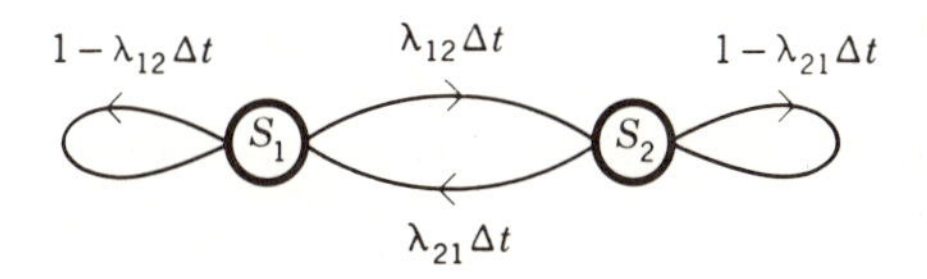

a It happens that, in this example, the recurrent states form a single chain. However, since we shall solve the general differential equations, we are not taking any steps which require this condition.

$m - 1$ eqs.: $P_1(t + \Delta t) = P_1(t)(1 - \lambda_{12}\,\Delta t) + P_2(t)\lambda_{21}\,\Delta t$

mth eq.: $1 = P_1(t) + P_2(t)$

and we have the initial conditions $\underbrace{P(0)} = \underbrace{P_1(0)\ P_2(0)} = \underbrace{\mathcal{P}\ 1-\mathcal{P}}$.
After collecting terms, dividing both sides of the first equation by Δt, taking the limit as $\Delta t \to 0$, and substituting the second equation into the first equation, we have

$$\frac{dP_1(t)}{dt} + (\lambda_{21} + \lambda_{12})P_1(t) = \lambda_{21} \qquad P_1(0) = \mathcal{P}$$

which is a first-order linear differential equation which has the complete solution

$$P_1(t) = \left(\mathcal{P} - \frac{\lambda_{21}}{\lambda_{21} + \lambda_{12}}\right) e^{-(\lambda_{12}+\lambda_{21})t} + \frac{\lambda_{21}}{\lambda_{12} + \lambda_{21}} \qquad t \geq 0$$

(For readers unfamiliar with how such an equation is solved, this knowledge is not requisite for any other work in this book.) We sketch $P_1(t)$ for a case where $\mathcal{P}$ is greater than $\dfrac{\lambda_{21}}{\lambda_{12} + \lambda_{21}}$.

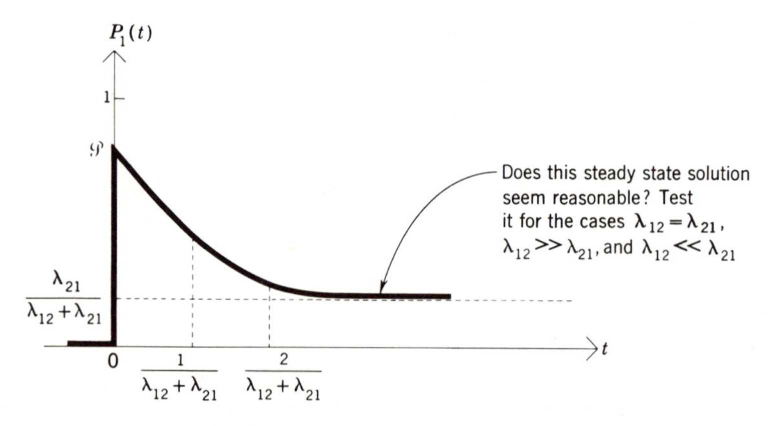

Since we do happen to be dealing with a birth-and-death process which satisfies the ergodicity condition, we may also obtain the limiting-state probabilities from

$$P_2\lambda_{21} = P_1\lambda_{12} \qquad \text{and} \qquad P_1 + P_2 = 1$$

which does yield the same values for the limiting-state probabilities as those obtained above.

bi The first transition we see will be an $S_1 \to S_2$ transition if and only if Peter happened to be in room 1 when we arrive. Thus, our answer is simply

$$P_1 = \lambda_{21}(\lambda_{12} + \lambda_{21})^{-1}$$

Note that, although we are equally likely to observe an $S_1 \to S_2$ or an $S_2 \to S_1$ transition in the *first* Δt after we arrive at a random time, it need not follow that the *first transition* we observe is equally likely to be either type. Outcomes of trials in successive Δt's after we arrive at a random time are not independent. For instance, if $\lambda_{12} > \lambda_{21}$ and we arrive at a random instant and wait a very long time without noting any transitions, the conditional probability that the process is in state S_2 approaches unity. We'll demonstrate this phenomenon in the last part of this problem.

bii The probability of a transition in the first Δt after we arrive is simply

$$P_1\lambda_{12}\,\Delta t + P_2\lambda_{21}\,\Delta t = 2\left(\frac{\lambda_{12}\lambda_{21}}{\lambda_{12} + \lambda_{21}}\right)\Delta t$$

which is the sum of the probabilities of the two mutually exclusive ways Peter may make a transition in this first Δt. The quantity $2\lambda_{12}\lambda_{21}(\lambda_{12} + \lambda_{21})^{-1}$ may be interpreted as the average rate at which Peter changes rooms. Of course, the two terms added above are equal, since the average rate at which he is making room $1 \to$ room 2 transitions must equal the average rate at which he makes the only other possible type of transitions. If $\lambda_{12} > \lambda_{21}$, it is true that Peter makes transitions more frequently when he is in room 1 than he does when he is in room 2, but the average transition rates over all time come out equal because he would be in room 1 much less often than he would be in room 2.

biii
$$\begin{aligned} f_x(x_0) &= P_1 f_{x|S_1}(x_0 \mid S_1) + P_2 f_{x|S_2}(x_0 \mid S_2). \\ &= \frac{\lambda_{21}}{\lambda_{12} + \lambda_{21}}\lambda_{12}e^{-\lambda_{12}x_0} + \frac{\lambda_{12}}{\lambda_{12} + \lambda_{21}}\lambda_{21}e^{-\lambda_{21}x_0} \qquad x_0 \geq 0 \\ &= \frac{\lambda_{12}\lambda_{21}}{\lambda_{12} + \lambda_{21}}(e^{-\lambda_{12}x_0} + e^{-\lambda_{21}x_0}) \qquad x_0 \geq 0 \end{aligned}$$

This answer checks out if $\lambda_{12} = \lambda_{21}$, and furthermore we note that, if $\lambda_{21} \gg \lambda_{12}$, we are almost certain to find the process in state 1 in the steady state and to have the PDF until the next transition be

$$\lambda_{12}e^{-\lambda_{12}x_0} \qquad x_0 \geq 0$$

and the above answer does exhibit this behavior.

biv Define event A: "No transition in first T units of time after we arrive." We may now sketch a sequential event space for the experiment in which we arrive at a random time, given the process is in the steady state.

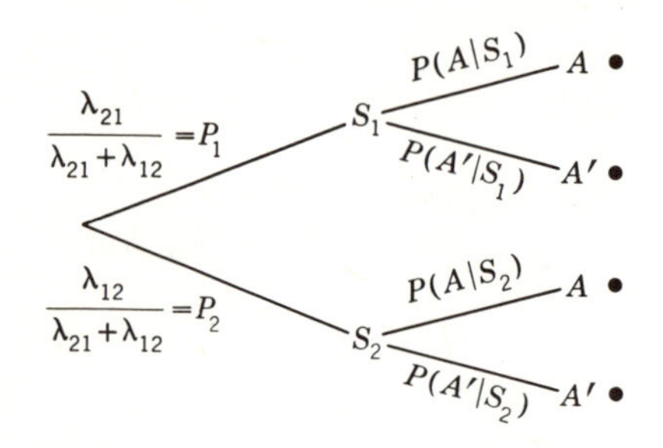

The quantity $P(A \mid S_1)$ is the conditional probability that we shall see no transitions in the first T units of time after we arrive, given that the process is in state S_1. This is simply the probability that the holding time in S_1 after we arrive is greater than T.

$$P(A \mid S_1) = \int_{t_0=T}^{\infty} \lambda_{12}e^{-\lambda_{12}t_0}\,dt_0 = e^{-\lambda_{12}T} \qquad P(A \mid S_2) = e^{-\lambda_{21}T}$$

We wish to obtain the conditional probability that we found the process in state S_1, given there were no changes of state in the first T units of time after we arrived at a random instant.

$$P(S_1 \mid A) = \frac{P(S_1A)}{P(A)} = \frac{\lambda_{21}e^{-\lambda_{12}T}}{\lambda_{21}e^{-\lambda_{12}T} + \lambda_{12}e^{-\lambda_{21}T}} \qquad T \geq 0$$

This answer checks out for $\lambda_{12} = \lambda_{21}$ and as $T \to 0$. Furthermore, if $\lambda_{21} \gg \lambda_{12}$, we would expect that, as $T \to \infty$, we would be increasingly likely to find the process in its slow transition (long-holding-time) state and our answer does exhibit this property.

example 2 Consider a service facility at which the arrival of customers is a Poisson process, with an average arrival rate of λ customers per hour. If customers find the facility fully occupied when they arrive, they enter a *queue* (a waiting line) and await their turns to be serviced on a first-come first-served basis. Once a customer leaves the queue and enters actual service, the time required by the facility to service him is an independent experimental value of an exponentially distributed random variable with an expected value of μ^{-1} hours $(\mu > \lambda)$.

Determine the limiting-state probabilities and the expected value for the *total* number of customers at the facility (in the queue and in service) if

(a) The facility can service only one customer at a time.
(b) The facility can service up to an infinite number of customers in parallel.

We shall study several other aspects of these situations as we answer the above questions. To begin our solution, we define the event S_i by

S_i: There are a total of i customers at the facility, where i includes both customers in the queue and those receiving service.

Since the customer arrival rate for either case is independent of the state of the system, the probability of an $S_i \to S_{i+1}$ transition in any incremental interval Δt is equal to $\lambda\ \Delta t$. The probability of completing a service and having an arrival in the same Δt is a second-order term.

a We are told that the service times are independent exponential random variables with the PDF $f_x(x_0) = \mu e^{-\mu x_0}$ for $x_0 \geq 0$. Parameter μ represents the maximum possible service rate at this facility. *If there were always at least one customer at the facility*, the completion of services would be a Poisson process with an average service completion rate of μ services per hour. Considering only first-order terms and not bothering with the *self-loops*, we have the state transition diagram

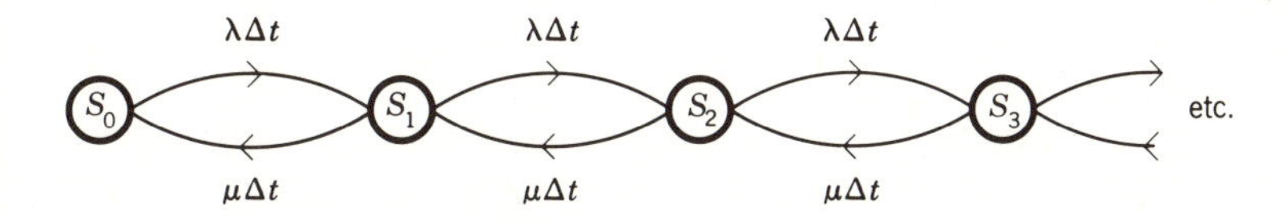

We are concerned with an infinite-state single-chain continuous-birth-and-death process, with

$$b_i = \lambda \qquad i = 0, 1, 2, \ldots \qquad d_i = \begin{cases} 0 & i = 0 \\ \mu & i = 1, 2, 3, \ldots \end{cases}$$

For the case of interest, $\mu > \lambda$, we shall assume that the limiting-state probabilities exist for this infinite-state process. If the maximum service rate were *less* than the average arrival rate, we would expect the length of the line to become infinite. We use the relations

$$P_{i+1} = \frac{b_i}{d_{i+1}} P_i \qquad i = 0, 1, 2, \ldots \qquad \text{and} \qquad \sum_i P_i = 1$$

and there follow

$$P_1 = \frac{\lambda}{\mu} P_0 \qquad P_2 = \frac{\lambda}{\mu} P_1 = \left(\frac{\lambda}{\mu}\right)^2 P_0 \qquad P_3 = \frac{\lambda}{\mu} P_2 = \left(\frac{\lambda}{\mu}\right)^3 P_0$$

$$P_i = \left(\frac{\lambda}{\mu}\right)^i P_0 \qquad \sum_i P_i = 1 = \sum_{i=0}^{\infty} \left(\frac{\lambda}{\mu}\right)^i P_0 = \left(1 - \frac{\lambda}{\mu}\right)^{-1} P_0$$

$$P_0 = 1 - \frac{\lambda}{\mu}$$

$$P_i = \left(1 - \frac{\lambda}{\mu}\right)\left(\frac{\lambda}{\mu}\right)^i \qquad i = 0, 1, 2, \ldots; \qquad \mu > \lambda$$

$$E(i) = \sum_i iP_i = \frac{\lambda}{\mu}\left(1 - \frac{\lambda}{\mu}\right)^{-1} \qquad \mu > \lambda$$

The expected value of the total number of customers at the facility when the process is in the steady state, $E(i)$, is obtained above by either using z transforms or noting the relation of the P_i's to the geometric PMF.

It is interesting to observe, for instance, that, if the average arrival rate is only 80% of the maximum average service rate, there will be, on the average, a total of four customers at the facility. When there are four customers present, three of them will be *waiting* to enter service even though the facility is empty 20% of all time. Such is the price of randomness.

Let's find the PDF for t, the total time (waiting for service and during service) spent by a randomly selected customer at this single-channel service facility. Customers arrive randomly, and the probability any customer will find exactly i *other* customers already at the facility is P_i. If he finds i other customers there already, a customer will leave after a total of $i + 1$ independent exponentially distributed service times are completed. Thus the conditional PDF for the waiting time of this customer is an Erlang PDF (Sec. 4-6) of order $i + 1$, and we have

$$f_t(t_0) = \sum_i f_{t|S_i}(t_0 \mid S_i)P(S_i) = \sum_i \frac{\mu^{i+1}t_0{}^i e^{-\mu t_0}}{i!} P_i$$

$$= \sum_{i=0}^{\infty} \frac{\mu^{i+1}t_0{}^i e^{-\mu t_0}}{i!} \left(\frac{\lambda}{\mu}\right)^i \left(1 - \frac{\lambda}{\mu}\right)$$

$$= (\mu - \lambda)e^{-(\mu-\lambda)t_0} \qquad t_0 \geq 0; \qquad \mu > \lambda$$

Thus, the total time any customer spends at the facility, for a process with independent exponentially distributed interarrival times and service times, is also an exponential random variable. The expected time spent at the facility is then

$$E(t) = (\mu - \lambda)^{-1} \qquad \mu > \lambda$$

and we can check this result by also using our result from Sec. 3-7, noticing that t is the sum of a random number $i + 1$ of independent service times,

$$E(t) = E(i + 1)E(x) = \left[\frac{\lambda}{\mu}\left(1 - \frac{\lambda}{\mu}\right)^{-1} + 1\right]\frac{1}{\mu} \stackrel{\checkmark}{=} (\mu - \lambda)^{-1}$$

b For this part, we consider the case where there are an infinite number of parallel service stations available at the facility and no customer has to wait before entering service. When there are exactly i customers in service at this facility, the probability that any particular customer will leave (complete service) in the next Δt is $\mu \Delta t$. *While there are i customers present,* we are concerned with departures representing services in any of i independent such processes and, to the first order, the probability of one departure in the next Δt is $i\mu \Delta t$. We have, again omitting the self-loops in the transition diagram,

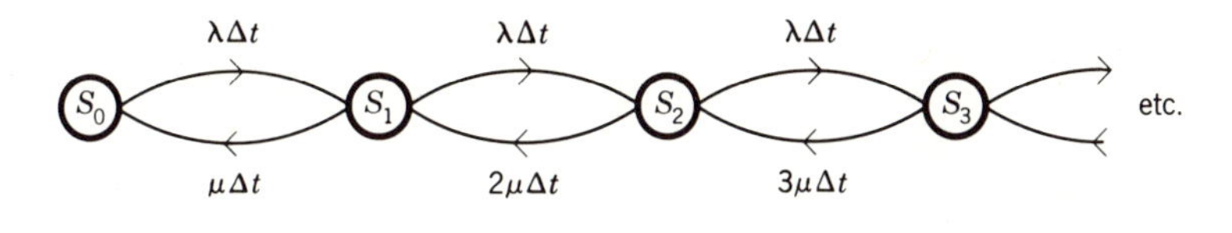

$b_i = \lambda \quad i = 0, 1, 2 \ldots \qquad\qquad d_i = i\mu \quad i = 0, 1, 2 \ldots$

Use of our simplified procedures for obtaining the limiting-state probabilities for birth-and-death processes proceeds:

$$P_{i+1} = \frac{b_i}{d_{i+1}} P_i \qquad i = 0, 1, 2, \ldots \qquad \sum_i P_i = 1$$

$$P_1 = \frac{\lambda}{\mu} P_0 \qquad P_2 = \frac{\lambda}{2\mu} P_1 = \frac{1}{2!}\left(\frac{\lambda}{\mu}\right)^2 P_0 \qquad P_3 = \frac{\lambda}{3\mu} P_2 = \frac{1}{3!}\left(\frac{\lambda}{\mu}\right)^3 P_0$$

$$P_i = \frac{1}{i!}\left(\frac{\lambda}{\mu}\right)^i P_0 \qquad \sum_i P_i = 1 = \sum_{i=0}^{\infty} \frac{1}{i!}\left(\frac{\lambda}{\mu}\right)^i P_0 = e^{\lambda/\mu} P_0 \qquad P_0 = e^{-\lambda/\mu}$$

$$P_i = \left(\frac{\lambda}{\mu}\right)^i \frac{e^{-\lambda/\mu}}{i!} \qquad i = 0, 1, 2, 3, \ldots \qquad E(i) = \frac{\lambda}{\mu} \qquad \mu > \lambda > 0$$

The limiting-state probabilities for this case form a Poisson PMF for the total number of customers at the facility (all of whom are in service) at a random time. As one would expect, P_0 is greater for this case than in part (a), and all other P_i's here are less than the corresponding quantities for that single-channel case. For instance, if $\lambda/\mu = 0.8$, this facility is completely idle a fraction

$$P_0 = e^{-(4/5)} = 0.45$$

of all time. This compares with $P_0 = 0.20$ for the same λ/μ ratio for the single-channel case in part (a).

example 3 A four-line switchboard services outgoing calls of four subscribers who never call each other. The durations of all phone calls are independent identically distributed exponential random variables with an expected value of μ^{-1}. For each subscriber, the interval between the end of any call and the time he places his next call is an independent exponential random variable with expected value λ^{-1}.

Four independent customers who never call each other and participate only in outgoing calls

Switchboard with capacity of four input lines

We shall assume that the system is in the steady state, neglect the possibility of busy signals, and use the state notation S_i: Exactly i of the input lines are active.

(a) Determine the limiting-state probabilities for the number of the input lines in use at any time.

(b) Given that there are at present three switchboard lines in use, determine the PDF for the waiting time until the next change of state.

(c) Determine the expected value of the number of busy input lines.

(d) What is the probability that there are exactly two lines busy at the switchboard the instants just before the arrivals of both members of a randomly selected pair of successive outgoing calls?

a Reasoning similar to that used in part (*b*) of the previous example leads us to the following observation. Given that there are exactly i input lines in use at the present time, the conditional probability of an $S_i \rightarrow S_{i+1}$ transition in the next Δt is $(4 - i)\lambda\,\Delta t$, and the conditional probability of an $S_i \rightarrow S_{i-1}$ transition in the next Δt is $i\mu\,\Delta t$. Neglecting the self-loops, we have the transition diagram

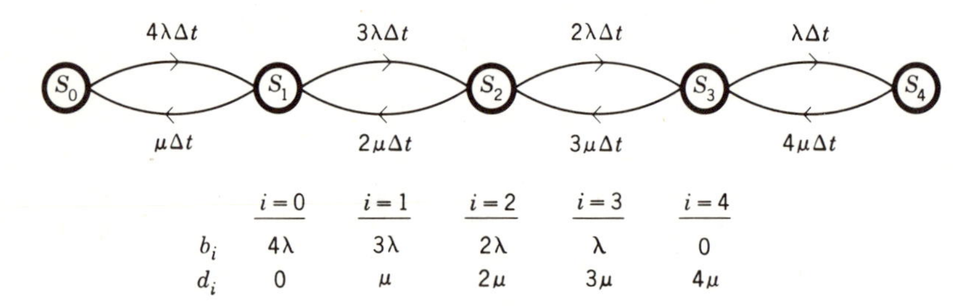

	$i=0$	$i=1$	$i=2$	$i=3$	$i=4$
b_i	4λ	3λ	2λ	λ	0
d_i	0	μ	2μ	3μ	4μ

This is a single-chain process with no periodicities; so the limiting-state probabilities will not depend on the initial conditions. Furthermore, it is a birth-and-death process; so we may use the special form of the simultaneous equations for the limiting-state probabilities,

$$P_{i+1} = \frac{b_i}{d_{i+1}} P_i \qquad i = 0, 1, 2, 3, 4 \qquad \sum_{i=0}^{4} P_i = 1$$

which result in

$$P_0 = \left(1 + \frac{\lambda}{\mu}\right)^{-4} \qquad P_1 = 4\frac{\lambda}{\mu}\left(1 + \frac{\lambda}{\mu}\right)^{-4} \qquad P_2 = 6\left(\frac{\lambda}{\mu}\right)^2\left(1 + \frac{\lambda}{\mu}\right)^{-4}$$

$$P_3 = 4\left(\frac{\lambda}{\mu}\right)^3\left(1 + \frac{\lambda}{\mu}\right)^{-4} \qquad P_4 = \left(\frac{\lambda}{\mu}\right)^4\left(1 + \frac{\lambda}{\mu}\right)^{-4}$$

As we would expect, if $\lambda \gg \mu$, P_4 is very close to unity and, if $\lambda \ll \mu$, P_0 is very close to unity.

b Given that the system is in state S_3, the probability that it will leave this state in the next Δt is $(3\mu + \lambda)\,\Delta t$, no matter how long it has been in S_3. Thus, the PDF for t, the exponential holding time in this state, is

$$f_t(t_0) = (3\mu + \lambda)e^{-(3\mu+\lambda)t_0} \qquad t_0 \geq 0$$

c Direct substitution of the P_i's obtained in part (*a*) into

$$E(i) = \sum_{i=0}^{4} iP_i$$

results in

$$E(i) = 4\frac{\lambda}{\mu}\left(1 + \frac{\lambda}{\mu}\right)^{-1}$$

and this answer agrees with our intuition for $\lambda \gg \mu$ and $\mu \gg \lambda$.

d This is as involved a problem as one is likely to encounter. Only the answer and a rough outline of how it may be obtained are given. The serious reader should be sure that he can supply the missing steps; the less serious reader may choose to skip this part.

$$\text{Answer} = \left(\frac{2\lambda P_2}{4\lambda P_0 + 3\lambda P_1 + 2\lambda P_2 + \lambda P_3}\right) \times \left(\frac{3\mu}{\lambda + 3\mu}\right) \times \left(\frac{\lambda}{\lambda + \mu}\right)$$

Probability first member of our pair of arrivals becomes 3d customer present at switchboard

Conditional probability that next change of state is due to completion of a call, given system is in state S_3

Conditional probability that next change of state is due to an arrival, given system is in state S_2

We have assumed that the "randomly selected" pair was chosen by selecting the first member of the pair by means of an equally likely choice among a large number of incoming calls. If the first member were to be chosen by selecting the first incoming call to follow a randomly selected instant, we would have a different situation.

PROBLEMS

5.01 For a series of dependent trials, the probability of success on any trial is given by $(k + 1)/(k + 3)$, where k is the number of successes in the previous three trials. Define a state description and set of transition probabilities which allow this process to be described as a Markov process. Draw the state transition diagram. Try to use the smallest possible number of states.

5.02 Consider the three-state discrete-transition Markov process

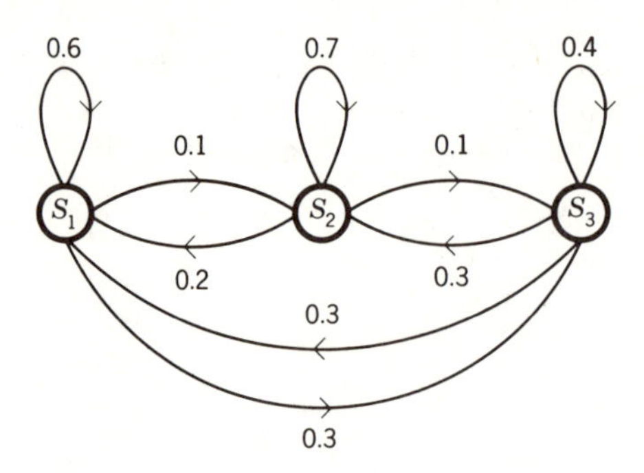

Determine the three-step transition probabilities $p_{11}(3)$, $p_{12}(3)$, and $p_{13}(3)$ both from a sequential sample space and by using the equation $p_{ij}(n+1) = \sum_k p_{ik}(n)p_{kj}$ in an effective manner.

5.03 We are observing and recording the outcomes of dependent flips of a coin at a distance on a foggy day. The probability that any flip will have the same outcome as the previous flip is equal to P. Our observations of the experimental outcomes are imperfect. In fact, the probability that we shall properly record the outcome of any trial is equal to F and is independent of all previous or future errors. We use the notation

h_n: We record the observation of the nth trial to be heads.
t_n: We record the observation of the nth trial to be tails.

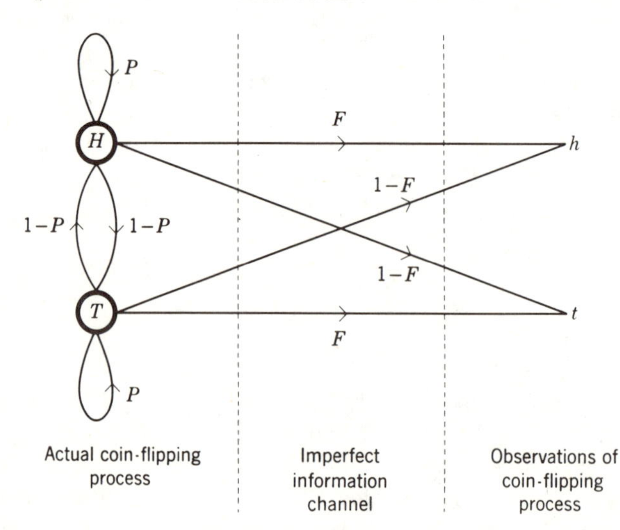

Can the possible sequences of our observations be modeled as the state history of a two-state Markov process?

5.04 **a** Identify the transient, recurrent, and periodic states of the discrete-state discrete-transition Markov process described by

$$[p] = \begin{bmatrix} 0.5 & 0 & 0 & 0 & 0.5 & 0 & 0 \\ 0.3 & 0.4 & 0 & 0 & 0.2 & 0.1 & 0 \\ 0 & 0 & 0.6 & 0.2 & 0 & 0.2 & 0 \\ 0 & 0 & 0 & 0.5 & 0 & 0 & 0.5 \\ 0.3 & 0.4 & 0 & 0 & 0.3 & 0 & 0 \\ 0 & 0 & 0.4 & 0.6 & 0 & 0 & 0 \\ 0 & 0 & 0 & 0.6 & 0 & 0 & 0.4 \end{bmatrix}$$

b How many chains are formed by the recurrent states of this process?

c Evaluate $\lim_{n\to\infty} p_{41}(n)$ and $\lim_{n\to\infty} p_{66}(n)$.

5.05 For the Markov process pictured here, the following questions may be answered by inspection:

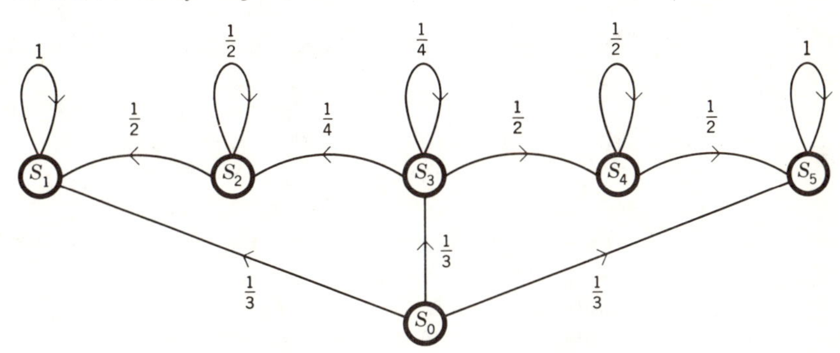

Given that this process is in state S_0 just before the first trial, determine the probability that:

a The process enters S_2 for the first time as the result of the Kth trial.

b The process never enters S_4.

c The process does enter S_2, but it also leaves S_2 on the trial after it entered S_2.

d The process enters S_1 for the first time on the third trial.

e The process is in state S_3 immediately after the Nth trial.

5.06 Days are either good (G), fair (F), or sad (S). Let F_n, for instance, be the event that the nth day is fair. Assume that the probability of having a good, fair, or sad day depends only on the condition of the previous day as dictated by the conditional probabilities

$$P(F_{n+1} \mid G_n) = 1/3 \qquad P(F_{n+1} \mid S_n) = 3/8 \qquad P(S_{n+1} \mid F_n) = 1/6$$
$$P(F_{n+1} \mid F_n) = 2/3 \qquad P(S_{n+1} \mid G_n) = 1/6 \qquad P(S_{n+1} \mid S_n) = 1/2$$

Assume that the process is in the steady state. A good day is worth \$1, a fair day is worth \$0, and a sad day is worth $-$\$1.

a Determine the expected value and the variance of the value of a randomly selected day.

b Determine the expected value and the variance of the value of a random two-day sequence. Compare with the above results, and comment.

c With the process in the steady-state at day zero, we are told that the sum value of days 13 and 14 was \$0. What is the probability that day 13 was a fair day?

5.07 The outcomes of successive flips of a particular coin are dependent and are found to be described fully by the conditional probabilities

$$\text{Prob}(H_{n+1} \mid H_n) = 3/4 \qquad \text{Prob}(T_{n+1} \mid T_n) = 2/3$$

where we have used the notation
Event H_k: Heads on kth toss   Event T_k: Tails on kth toss

We know that the first toss came up heads.

a Determine the probability that the *first* tail will occur on the kth toss ($k = 2, 3, 4, \ldots$).

b What is the probability that flip 5,000 will come up heads?

c What is the probability that flip 5,000 will come up heads and flip 5,002 will also come up heads?

d Given that flips 5,001, 5,002, . . . , $5{,}000 + m$ all have the same result, what is the probability that all of these m outcomes are heads? Simplify your answer as much as possible, and interpret your result for large values of m.

e We are told that the 375th head just occurred on the 500th toss. Determine the expected value of the number of additional flips required until we observe the 379th head.

5.08 In his new job for the city, Joe makes daily measurements of the level of the Sludge River. His predecessor established that these daily readings can be modeled by a Markov process and that there are only three possible river depths, zero, one, and two feet. It is also known that the river level never changes more than one foot per day. The city has compiled the following transition probabilities:

$$p_{01} = 1/4 \qquad p_{10} = 1/2 \qquad p_{12} = 1/4 \qquad p_{22} = 1/4$$

Let x_K represent Joe's reading of the river depth on his Kth day on the job. We are given that the reading on the day before he started was one foot.

Determine:

a The probability mass function for random variable x_1

b The probability that $x_{377} \neq x_{378}$

c The conditional probability mass function for x_{999}, given that $x_{1,000} = 1$

d The numerical values of

i $\lim_{n\to\infty} E(x_{n+1} - x_n)$ **ii** $\lim_{n\to\infty} E[(x_{n+1} - x_n)^2]$

iii $\lim_{n\to\infty} \frac{1}{n} \sum_{i=1}^{n} x_i$

e The z transform of the probability mass function $p_L(L_0)$, where random variable L is the smallest positive integer which satisfies $x_L \neq x_1$.

5.09 Mr. Mean Variance has the only key which locks or unlocks the door to Building 59, the Probability Building. He visits the door once each hour on the hour. When he arrives:
If the door is open, he locks it with probability 0.3.
If the door is locked, he unlocks it with probability 0.8.

a After he has been on the job several months, is he more likely to lock the door or to unlock it on a randomly selected visit?

b With the process in the steady state, Joe arrived at Building 59 two hours ahead of Harry. What is the probability that each of them found the door in the same condition?

c Given the door was open at the time Mr. Variance was hired, determine the z transform for the number of visits up to and including the one on which he unlocked the door himself for the first time.

5.10 Arrivals of potential customers on the street outside a one-pump gas station are noted to be a Poisson process with an average arrival rate of λ cars per hour. Potential customers will come in for gas if there are fewer than two cars already at the pump (including the one being attended to). If there are two cars already at the pump, the potential customers will go elsewhere.

It is noted that the amount of time required to service any car is an independent random variable with PDF

$$f_T(T_0) = \begin{cases} \mu e^{-\mu T_0} & T_0 \geq 0 \\ 0 & T_0 < 0 \end{cases}$$

a Give a physical interpretation of the constant μ.

b Write the differential equations relating the $P_n(t)$'s, where $P_n(t)$ is the probability of having n cars at the pump at time t. Do not solve the equations.

c Write and solve the equations for P_n, $n = 0, 1, 2, 3, \ldots$, where P_n is the steady-state probability of having a total of n cars at the pump.

d If the cars arrive at the average rate of 20 per hour and the average service time is two minutes per car, what is the probability that a potential customer will go elsewhere? What fraction of the attendant's time will be spent servicing cars?

e At the same salary, the owner can provide a more popular, but slower, attendant. This would raise the average service time to 2.5 minutes per car but would also increase λ from 20 to 28 cars per hour. Which attendant should he use to maximize his expected profit? Determine the percent change in the number of customers serviced per hour that would result from changing to the slower attendant.

5.11 At a single service facility, the interarrival times between successive customers are independent exponentially distributed random variables. The average customer arrival rate is 40 customers per hour.

When a total of two or fewer customers are present, a single attendant operates the facility and the service time for each customer is an exponentially distributed random variable with a mean value of two minutes.

Whenever there are three or more customers at the facility, the attendant is joined by an assistant. In this case, the service time is an exponentially distributed random variable with an expected value of one minute.

Assume the process is in the steady state.

a What fraction of the time are both of them free?

b What is the probability that both men are working at the facility the instant before a randomly selected customer arrives? The instant after he arrives?

c Each of the men is to receive a salary proportional to the expected value of the amount of time he is actually at work servicing customers. The constant of proportionality is the same for both men, and the sum of their salaries is to be \$100. Determine the salary of each man.

5.12 Only two taxicabs operate from a particular station. The total time it takes a taxi to service any customer and return to the station is an exponentially distributed random variable with a mean of $1/\mu$ hours. Arrivals of potential customers are modeled as a Poisson process with average rate of λ customers per hour. If any potential customer finds no taxi at the station at the instant he arrives, he walks to his destination and thus does not become an actual customer. The cabs always return directly to the station without picking up new customers. All parts of this problem are independent of statements in other parts.

a If $\lambda = \infty$, determine (in as simple and logical a manner as you can) the average number of customers served per hour.

b Using the notation $\rho = \mu/\lambda$, determine the steady-state probability that there is exactly one taxi at the station.

c If we survey a huge number of actual customers, what fraction of them will report that they arrived at an instant when there was exactly one taxi at the station?

d Taxi A has been destroyed in a collision with a Go-Kart, and we note that B is at the station at time $t = 0$. What are the expected value and variance of the time until B leaves the station with his fourth fare since $t = 0$?

5.13 All ships travel at the same velocity through a wide canal. Eastbound ship arrivals at the canal are a Poisson process with an average arrival rate λ_E ships per day. Westbound ships arrive as an independent Poisson process with average arrival rate λ_W per day. An indicator at a point in the canal is always pointing in the direction of travel of the most recent ship to pass it. Each ship takes T days to traverse the canal. Use the notation $\rho = \lambda_E/\lambda_W$ wherever possible.

a What is the probability that the next ship passing by the indicator causes it to change its indicated direction?

b What is the probability that an eastbound ship will see no westbound ships during its eastward journey through the canal?

c If we begin observing at an arbitrary time, determine the probability mass function $p_k(k_0)$, where k is the total number of ships we observe up to and including the seventh eastbound ship we see.

d If we begin observing at an arbitrary time, determine the probability density function $f_t(t_0)$, where t is the time until we see our seventh eastbound ship.

e Given that the pointer is pointing west:

i What is the probability that the next ship to pass it will be westbound?

ii What is the probability density function for the remaining time until the pointer changes direction?

5.14 A switchboard has two outgoing lines and is concerned only with servicing the outgoing calls of three customers who never call each other. When he is not on a line, each potential caller generates calls at a Poisson rate λ. Call lengths are exponentially distributed, with a mean call length of $1/\mu$. If a caller finds the switchboard blocked, he never tries to reinstitute that particular call.

a Determine the fraction of time that the switchboard is saturated.

b Determine the fraction of outgoing calls which encounter a saturated switchboard.

5.15 An illumination system contains $R + 3$ bulbs, each of which fails independently and has a life span described by the probability density function

$$f_t(t_0) = \lambda e^{-\lambda t_0} \qquad t_0 \geq 0$$

At the time of the third failure, the system is shut down, the dead bulbs are replaced, and the system is "restarted." The down time for the service operation is a random variable described by the probability density function

$$f_x(x_0) = \mu^2 x_0 e^{-\mu x_0} \qquad x_0 \geq 0$$

a Determine the mean and variance of y, the time from $t = 0$ until the end of the kth service operation.
b Determine the steady-state probability that all the lights are on.

5.16 Potential customers arrive at the input gate to a facility in a Poisson manner with average arrival rate λ. The facility will hold up to three customers including the one being serviced. Potential customers who arrive when the facility is full go elsewhere for service. Service time is an exponential random variable with mean $1/\mu$. Customers leave as soon as they are serviced. Service for actual customers is on a first-come first-served basis.
a If we select a random pair (see final comment in Sec. 5-8) of successive *potential* customers approaching the facility, what is the probability that they will eventually emerge as a pair of successive *actual* customers?
b If we select a random pair of successive *actual* customers leaving the facility, what is the probability that these customers arrived as successive *potential* customers at the facility input?
c Starting at a randomly selected time, what is the probability that *before the next actual customer arrives at the input gate* at least two customers would be observed leaving via the output gate?

5.17 Consider a K-state discrete-state Markov system with exponential holding times. The system is composed of a single chain.

$\lambda_{ij}\,\Delta t$ ($i \neq j$) = conditional probability that system will enter state S_j in the next Δt, given that present state is S_i

Other than in part (a) you may use the limiting-state probabilities as $P_1, P_2, \ldots, P_K$ in your answers.
a Write, in a simple form, a set of equations which determine the steady-state probabilities $P_1, P_2, \ldots, P_K$.
b Given that the process is at present in state S_i, what is the probability that two transitions from now it will again be in state S_i?
c What are the expected value and variance of the time the system spends during any visit to state S_i?
d Determine the average rate at which the system makes transitions.
e If we arrive at a random time with the process in the steady state, determine the probability density function for the time until the next transition after we arrive.

f Suppose that the process is a pure birth-and-death process and we arrive at a random time. What is the probability that the first transition we observe will be due to a birth? That both of the first two transitions we observe will be due to births?

5.18 An information source is always in one of m mutually exclusive, collectively exhaustive states $S_1, S_2, \ldots, S_m$. Whenever this source is in state S_i:

1 It produces printed messages in a Poisson manner at an average rate of μ_i messages per hour.

2 The conditional probability that the source will enter state S_j $(j \neq i)$ in the next incremental Δt is given by $\lambda_{ij}\,\Delta t$.

All messages are numbered consecutively and filed in a warehouse. The process is in the steady state and you may assume that the limiting state probabilities for the source, $P_1, P_2, \ldots, P_m$, are known quantities.

Each part of this problem is to be worked separately.

a Given that the process has been in state S_2 for the last three hours what is the probability that no messages were produced in the last 1.5 hours?

b Given that the process is *not* in state S_2, what is the probability that it will enter S_2 in the next incremental Δt?

c Determine the average rate at which messages are produced.

d What is the probability that the source will produce exactly two messages during any particular visit to state S_2?

e If we arrive at a random time to observe the process, what is the probability that we see at least one message generated before we observe the next state transition of the message source?

f If we select a random message from the file in the warehouse, what is the probability that it was produced when the source was in state S_2?

g If we select a pair of *consecutive* messages at random from the file, what is the probability that the source underwent exactly one change of state during the interval between the instants at which these two messages were produced?

h If we are told that, during the last 10 hours, the source process underwent exactly eight changes of state and spent

Exactly two hours in state S_3

Exactly five hours in state S_7

Exactly three hours in state S_8

determine the exact conditional PMF for the number of messages produced during the last 10 hours.

CHAPTER SIX

some fundamental limit theorems

Limit theorems characterize the mass behavior of experimental outcomes resulting from a large number of performances of an experiment. These theorems provide the connection between probability theory and the measurement of the parameters of probabilistic phenomena in the real world.

Early in this chapter, we discuss stochastic convergence, one important type of convergence for a sequence of random variables. This concept and an easily derived inequality allow us to establish one form of the law of large numbers. This law provides clarification of our earlier speculations (Sec. 3-6) regarding the relation, for large values of n, between the sum of n independent experimental values of random variable x and the quantity $nE(x)$.

We then discuss the Gaussian PDF. Subject to certain restrictions, we learn that the Gaussian PDF is often an excellent approximation to the actual PDF for the sum of many random variables, regardless of the forms of the PDF's for the individual random variables included in the sum.

This altogether remarkable result is known as the *central limit theorem.* A proof is presented for the case where the sum is composed of independent identically distributed random variables. Finally, we investigate several practical approximation procedures based on limit theorems.

6-1 The Chebyshev Inequality

The *Chebyshev inequality* states an upper bound on the probability that an experimental value of any random variable x will differ by at least any given positive quantity t from $E(x)$. In particular, the inequality will provide an upper bound on the quantity

$$\text{Prob}[|x - E(x)| \geq t]$$

in terms of t and σ_x. As long as the value of the standard deviation σ_x is known, other details of the PDF $f_x(x_0)$ are not relevant.

The derivation is simple. With $t > 0$, we have

$$\sigma_x{}^2 = \int_{x_0=-\infty}^{\infty} [x_0 - E(x)]^2 f_x(x_0)\, dx_0 \geq \int_{x_0=-\infty}^{E(x)-t} [x_0 - E(x)]^2 f_x(x_0)\, dx_0 + \int_{x_0=E(x)+t}^{\infty} [x_0 - E(x)]^2 f_x(x_0)\, dx_0$$

To obtain the above inequality, we note that the integrand in the leftmost integration is always positive. By removing an interval of length $2t$ from the range of that integral, we cannot increase the value of the integral. Inside the two integrals on the right-hand side of the above relation, it is always true that $|x - E(x)| \geq t$. We now replace $[x - E(x)]^2$ by t^2, which can never increase the value of the right-hand side, resulting in

$$\sigma_x{}^2 \geq \int_{x_0=-\infty}^{E(x)-t} t^2 f_x(x_0)\, dx_0 + \int_{x_0=E(x)+t}^{\infty} t^2 f_x(x_0)\, dx_0$$

After we divide both sides by t^2 and recognize the physical interpretation of the remaining quantity on the right-hand side, we have

$$\text{Prob}[|x - E(x)| \geq t] \leq \left(\frac{\sigma_x}{t}\right)^2$$

which is the Chebyshev inequality. It states, for instance, that the probability that an experimental value of any random variable x will be further than $K\sigma_x$ from $E(x)$ is always less than or equal to $1/K^2$

Since it is a rather weak bound, the Chebyshev inequality finds most of its applications in general theoretical work. For a random variable described by any particular PDF, better (though usually more complex) bounds may be established. We shall use the Chebyshev bound in Sec. 6-3 to investigate one form of the law of large numbers.

6-2 Stochastic Convergence

A *deterministic sequence* $\{x_n\} = x_1, x_2, \ldots$ is said to converge to the limit C if for every $\epsilon > 0$ we can find a finite n_0 such that

$$|x_n - C| < \epsilon \qquad \text{for all } n > n_0$$

If deterministic sequence $\{x_n\}$ does converge to the limit C, we write

$$\lim_{n\to\infty} x_n = C$$

Only for pathological cases would we expect to be able to make equally strong nonprobabilistic convergence statements for sequences of random variables. Several different types of convergence are defined for sequences of random variables. In this section we introduce and discuss one such definition, namely, that of *stochastic convergence*. We shall use this definition and the Chebyshev inequality to establish a form of the law of large numbers in the following section. (We defer any discussion of other forms of convergence for sequences of random variables until Sec. 6-9.)

A sequence of random variables, $\{y_n\} = y_1, y_2, y_3, \ldots$, is said to be *stochastically convergent* (or to *converge in probability*) to C if, for every $\epsilon > 0$, the condition $\lim_{n\to\infty} \text{Prob}(|y_n - C| > \epsilon) = 0$ is satisfied.

When a sequence of random variables, $\{y_n\}$, is known to be stochastically convergent to C, we must be careful to conclude *only* that the probability of the event $|y_n - C| > \epsilon$ vanishes as $n \to \infty$. We cannot conclude, for any value of n, that this event is impossible.

We may use the definition of a limit to restate the definition of stochastic convergence. Sequence $\{y_n\}$ is stochastically convergent to C if, for any $\epsilon > 0$ and any $\delta > 0$, it is possible to state a finite value of n_0 such that

$$\text{Prob}(|y_n - C| > \epsilon) < \delta \qquad \text{for all } n > n_0$$

Further discussion will accompany an application of the concept of stochastic convergence in the following section and a comparison with other forms of probabilistic convergence in Sec. 6-9.

6-3 The Weak Law of Large Numbers

A sequence of random variables, $\{y_n\}$, with finite expected values, is said to obey a *law of large numbers* if, in some sense, the sequence defined by

$$M_n = \frac{1}{n} \sum_{i=1}^{n} y_i$$

converges to its expected value. The type of convergence which applies determines whether the law is said to be *weak* or *strong*.

Let $y_1, y_2, \ldots$ form a sequence of independent identically distributed random variables with finite expected values $E(y)$ and finite variances σ_y^2. In this section we prove that the sequence

$$M_n = \frac{y_1 + y_2 + \cdots + y_n}{n}$$

is stochastically convergent to its expected value, and therefore the sequence $\{y_n\}$ obeys a (weak) law of large numbers.

For the conditions given above, random variable M_n is the average of n independent experimental values of random variable y. Quantity M_n is known as the *sample mean.*

From the definition of M_n and the property of expectations of sums, we have

$$E(M_n) = \frac{nE(y)}{n} = E(y)$$

and, because multiplying a random variable y by c defines a new random variable with a variance equal to $c^2\sigma_y^2$, we have

$$\sigma_{M_n}^2 = \frac{n\sigma_y^2}{n^2} = \frac{\sigma_y^2}{n} \qquad \sigma_{M_n} = \frac{\sigma_y}{\sqrt{n}}$$

To establish the weak law of large numbers for the case of interest, we simply apply the Chebyshev inequality to M_n to obtain

$$\text{Prob}[|M_n - E(M_n)| \geq \epsilon] \leq \left(\frac{\sigma_{M_n}}{\epsilon}\right)^2$$

and, substituting for M_n, $E(M_n)$, and σ_{M_n}, we find

$$\text{Prob}\left[\left|\frac{1}{n}\sum_{i=1}^{n} y_i - E(y)\right| \geq \epsilon\right] \leq \frac{\sigma_y^2}{n\epsilon^2}$$

(See Prob. 6.02 for a practical application of this relation.) Upon taking the limit as $n \to \infty$ for this equation there finally results

$$\lim_{n\to\infty} \text{Prob}\left[\left|\frac{1}{n}\sum_{i=1}^{n} y_i - E(y)\right| \geq \epsilon\right] = 0$$

which is known as the *weak law of large numbers* (in this case for random variable y). The law states that, as $n \to \infty$, the probability that the average of n independent experimental values of random variable y differs from $E(y)$ by more than any nonzero ϵ goes to zero. We have shown that as long as the variance of a random variable is finite, the random variable obeys the weak law of large numbers.

Neither the independence of the y_i's nor the finite variance σ_y^2 conditions are necessary for the $\{y_n\}$ sequence to obey a law of large numbers. Proof of these statements is beyond the scope of this book.

Let's apply our result to a situation where the y_i's are independent Bernoulli random variables with parameter P. Suppose that there are n trials and k is the number of successes. Using our result above, we have

$$\lim_{n\to\infty} \text{Prob}\left(\left|\frac{k}{n} - P\right| > \epsilon\right) = 0$$

which is known as the *Bernoulli law of large numbers.* This relation is one of the bases of the *relative-frequency* interpretation of probabilities. Often people read into it far more than it says.

For instance, let the trials be coin flips and the successes heads. If we flip the coin any number of times, it is still possible that all outcomes will be heads. If we know that P is a valid parameter of a coin-flipping process and we set out to estimate P by the experimental value of k/n, there is no value of n for which we could be *certain* that our experimental value was within an arbitrary $\pm\epsilon$ of the true value of parameter P.

The Bernoulli law of large numbers does *not* imply that k converges to the limiting value nP as $n \to \infty$. We know that the standard deviation of k, in fact, becomes infinite as $n \to \infty$ (see Sec. 3-6).

6-4 The Gaussian PDF

We consider a very important PDF which describes a vast number of probabilistic phenomena.

The *Gaussian* (or *normal*) *PDF* is defined to be

$$f_x(x_0) = \frac{1}{\sqrt{2\pi}\,\sigma} e^{-(x_0-m)^2/2\sigma^2} \qquad -\infty \leq x_0 \leq \infty$$

where we have written a PDF for random variable x with parameters m and σ. The s transform of this PDF is obtained from

$$f_x{}^T(s) = E(e^{-sx}) = \frac{1}{\sqrt{2\pi}\,\sigma} \int_{x_0=-\infty}^{\infty} e^{-sx_0} e^{-(x_0-m)^2/2\sigma^2}\,dx_0$$

$$= \frac{1}{\sqrt{2\pi}\,\sigma} \int_{x_0=-\infty}^{\infty} e^{-A(x_0)}\,dx_0$$

where the expression $A(x_0)$ is given by

$$A(x_0) = sx_0 + \frac{x_0{}^2}{2\sigma^2} + \frac{m^2}{2\sigma^2} - \frac{2mx_0}{2\sigma^2}$$

$$= \frac{1}{2\sigma^2}\{[x_0 + (s\sigma^2 - m)]^2 - s^2\sigma^4 + 2ms\sigma^2\}$$

In the above equation, we have carried out the algebraic operation known as "completing the square." We may substitute this result into the expression for $f_x{}^T(s)$,

$$f_x{}^T(s) = e^{(s^2\sigma^2/2)-sm} \int_{x_0=-\infty}^{\infty} \frac{1}{\sqrt{2\pi}\,\sigma} e^{-(x_0+s\sigma^2-m)^2/2\sigma^2}\,dx_0$$

Note that the integral is equal to the total area under a Gaussian PDF for which m has been replaced by the expression $m - s\sigma^2$. This change displaces the PDF with regard to the x_0 axis but does not change the *total* area under the curve. Since the total area under any PDF must be unity, we have found the transform of a Gaussian PDF with parameters m and σ to be

$$f_x{}^T(s) = e^{(s^2\sigma^2/2)-sm}$$

Using the familiar relations

$$E(x) = -\left[\frac{d}{ds} f_x{}^T(s)\right]_{s=0}$$

$$\sigma_x{}^2 = \left\{\frac{d^2}{ds^2} f_x{}^T(s) - \left[\frac{d}{ds} f_x{}^T(s)\right]^2\right\}_{s=0}$$

we evaluate the expected value and the variance of the Gaussian PDF to be

$$E(x) = m \qquad \sigma_x{}^2 = \sigma^2$$

We have learned that a normal PDF is specified by two parameters, the expected value $E(x)$ and the standard deviation σ_x. The normal PDF and its s transform are

$$f_x(x_0) = \frac{1}{\sqrt{2\pi}\,\sigma_x} e^{-[x_0-E(x)]^2/2\sigma_x{}^2} \qquad -\infty \le x_0 \le \infty$$

$$f_x{}^T(s) = e^{(s^2\sigma_x{}^2/2)-sE(x)}$$

A sketch of the normal PDF, in terms of $E(x)$ and σ_x, follows:

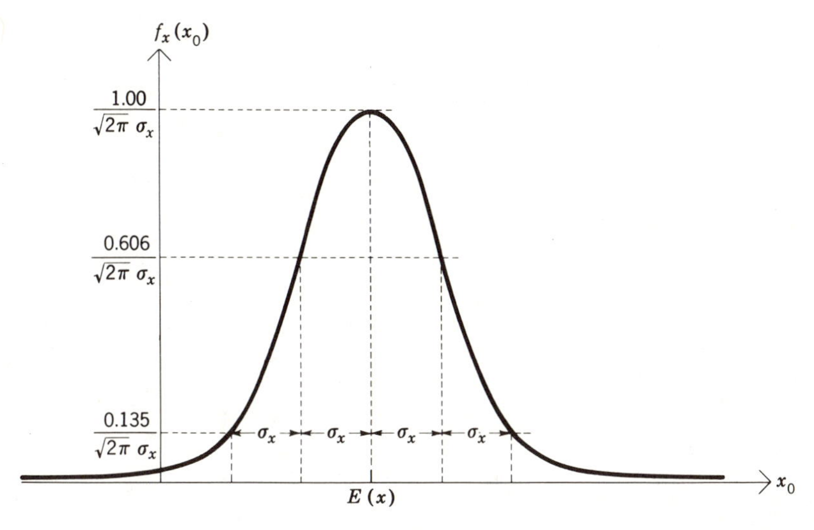

Suppose that we are interested in the PDF for random variable w, defined to be the sum of two *independent* Gaussian random variables x and y. We use the s transforms

$$f_x{}^T(s) = e^{(s^2\sigma_x{}^2/2)-sE(x)} \qquad \text{and} \qquad f_y{}^T(s) = e^{(s^2\sigma_y{}^2/2)-sE(y)}$$

and the relation for the transform of a PDF for the sum of independent random variables (Sec. 3-5) to obtain

$$f_w{}^T(s) = f_x{}^T(s)f_y{}^T(s) = e^{(s^2/2)(\sigma_x{}^2+\sigma_y{}^2)-s[E(x)+E(y)]}$$

We recognize this to be the transform of another Gaussian PDF. Thus we have found that

The PDF for a sum of independent Gaussian random variables is itself a Gaussian PDF.

A similar property was obtained for Poisson random variables in Sec. 4-7.

When one becomes involved in appreciable numerical work with PDF's of the same form but with different parameters, it is often desirable to suppress the parameters $E(x)$ and $\sigma_x{}^2$ of any particular situation and work in terms of a *unit* (or *normalized*, or *standardized*) random variable. A unit random variable has an expected value of zero, and its standard deviation is equal to unity. The unit random variable y for random variable x is obtained by subtracting $E(x)$ from x

and dividing this difference by σ_x

$$y = \frac{x - E(x)}{\sigma_x}$$

and we may demonstrate the properties,

$$E(y) = E\left[\frac{x - E(x)}{\sigma_x}\right] = \frac{E(x)}{\sigma_x} - \frac{E(x)}{\sigma_x} = 0$$

$$\sigma_y^2 = E\{[y - E(y)]^2\} = E\left\{\left[\frac{x - E(x)}{\sigma_x} - 0\right]^2\right\} = \frac{E\{[x - E(x)]^2\}}{\sigma_x^2} = 1$$

It happens that the cumulative distribution function, $p_{x\leq}(x_0)$, for a Gaussian random variable cannot be found in closed form. We shall be working with tables of the CDF for a Gaussian random variable, and it is for this reason that we are interested in the discussion in the previous paragraph. We shall be concerned with the unit normal PDF, which is given by

$$f_y(y_0) = \frac{1}{\sqrt{2\pi}} e^{-y_0^2/2} \qquad -\infty \leq y_0 \leq \infty, \quad E(y) = 0, \quad \sigma_y = 1$$

We define the function $\Phi(y_0)$ to be the CDF for the unit normal PDF,

$$\Phi(y_0) \equiv p_{y\leq}(y_0) = \int_{\alpha_0=-\infty}^{y_0} f_y(\alpha_0)\, d\alpha_0 = \frac{1}{\sqrt{2\pi}} \int_{\alpha_0=-\infty}^{y_0} e^{-\alpha_0^2/2}\, d\alpha_0$$

and extensive tables of $\Phi(y_0)$ exist.

To make use of tables of $\Phi(y_0)$ for a Gaussian (but unstandardized) random variable x, we need only recall the relation

$$y = \frac{x - E(x)}{\sigma_x}$$

and thus, for the CDF of Gaussian random variable x, we have

$$p_{x\leq}(x_0) = p_{y\leq}\left[\frac{x_0 - E(x)}{\sigma_x}\right] = \Phi\left[\frac{x_0 - E(x)}{\sigma_x}\right]$$

The argument on the right side of the above equation is equal to the number of standard deviations σ_x by which x_0 exceeds $E(x)$. If values of x_0 are measured in units of standard deviations from $E(x)$, tables of the CDF for the unit normal PDF may be used directly to obtain values of the CDF $p_{x\leq}(x_0)$.

Since a Gaussian PDF is symmetrical about its expected value, the CDF may be fully described by tabulating it only for values above (or below) its expected value. The following is a brief four-place table of $\Phi(y_0)$, the CDF for a unit normal random variable:

y_0	$\Phi(y_0)$	y_0	$\Phi(y_0)$	y_0	$\Phi(y_0)$	y_0	$\Phi(y_0)$
0.00	0.5000	1.00	0.8413	2.00	0.9772	3.00	0.9987
0.10	0.5398	1.10	0.8643	2.10	0.9821	3.10	0.9990
0.20	0.5793	1.20	0.8849	2.20	0.9861	3.20	0.9993
0.30	0.6179	1.30	0.9032	2.30	0.9893	3.30	0.9995
0.40	0.6554	1.40	0.9192	2.40	0.9918	3.40	0.9997
0.50	0.6915	1.50	0.9332	2.50	0.9938	3.60	0.9998
0.60	0.7257	1.60	0.9452	2.60	0.9953		
0.70	0.7580	1.70	0.9554	2.70	0.9965		
0.80	0.7881	1.80	0.9641	2.80	0.9974		
0.90	0.8159	1.90	0.9713	2.90	0.9981		

To obtain $\Phi(y_0)$ for $y_0 < 0$, we note that the area under $f_y(y_0)$ is equal to unity and that the shaded areas in the following sketches are equal

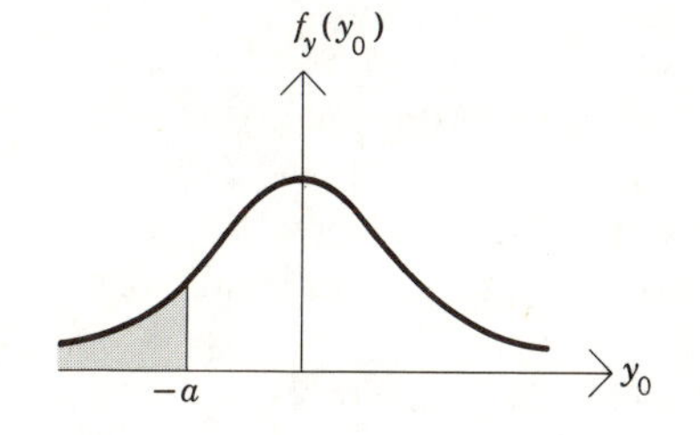

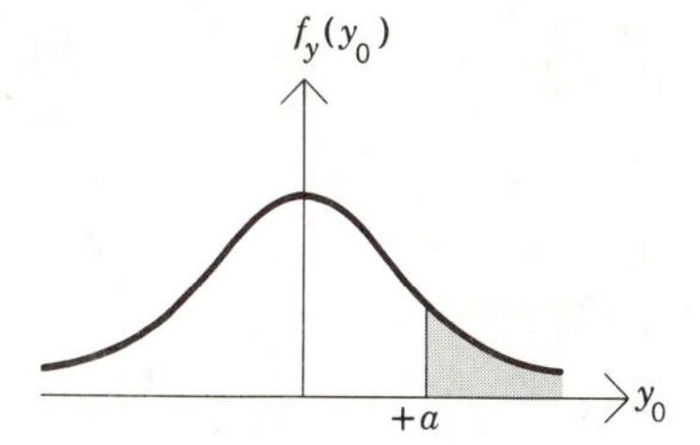

and we may use the relation

$$\Phi(-y_0) = 1 - \Phi(y_0)$$

We present a brief example. Suppose that we wish to determine the probability that an experimental value of a Gaussian random variable x falls within $\pm k\sigma_x$ of its expected value, for $k = 1$, 2, and 3. Thus, we wish to evaluate

$$\text{Prob}[|x - E(x)| \leq k\sigma_x] \qquad \text{for } k = 1, 2, \text{ and } 3$$

Since k is already in units of standard deviations, we may use the $\Phi(y_0)$ table directly according to the relation

$$\text{Prob}[|x - E(x)| \leq k\sigma_x] = \Phi(k) - \Phi(-k) = \Phi(k) - [1 - \Phi(k)]$$

$$= 2\Phi(k) - 1 = \begin{cases} 0.682 & k = 1 \\ 0.954 & k = 2 \\ 0.997 & k = 3 \end{cases}$$

Our result states, for instance, that the probability that an experimental value of any Gaussian random variable x falls within an interval of total length $4\sigma_x$ which is centered on $E(x)$ is equal to 0.954.

6-5 Central Limit Theorems

Let random variable r be defined to be the sum of n independent identically distributed random variables, each of which has a finite expected value and a finite variance. It is an altogether remarkable fact that, *as $n \to \infty$, the CDF $p_{r\leq}(r_0)$ approaches the CDF of a Gaussian random variable, regardless of the form of the PDF for the individual random variables in the sum.* This is one simple case of a *central limit theorem.*

Every central limit theorem states some particular set of conditions for which the CDF for a sum of n random variables will approach the CDF of a Gaussian random variable as $n \to \infty$.

As long as it is not always true that a particular few of the member random variables dominate the sum, the *identically distributed* (same PDF) condition is not essential for a central limit theorem to apply to the sum of a large number of independent random variables. The *independence* condition may also be relaxed, subject to certain other restrictions which are outside the scope of this presentation.

Since many phenomena may be considered to be the result of a large number of factors, central limit theorems are of great practical significance, especially when the effects of these factors are either purely additive ($r = x_1 + x_2 + \cdots$) or purely multiplicative ($\log r = \log x_1 + \log x_2 + \cdots$).

We first undertake a brief digression to indicate why we are stating central limit theorems in terms of the CDF $p_{r\leq}(r_0)$ rather than the PDF $f_r(r_0)$.

There are several ways in which the members of a sequence of deterministic functions $\{g_n(x_0)\} = g_1(x_0), g_2(x_0), \ldots$ can "approach" the corresponding members of a sequence of functions $\{h_n(x_0)\}$ in the limit as $n \to \infty$. However, the simplest and most easily visualized manner is *point-by-point* convergence; namely, if for any particular x_0 and for any $\epsilon > 0$ we can always find an n_0 for which

$$|g_n(x_0) - h_n(x_0)| < \epsilon \qquad \text{for all } n > n_0$$

This is the type of convergence we wish to use in our statement of the central limit theorem.

Consider a case where random variable r is defined to be the sum of n independent experimental values of a Bernoulli random variable with parameter P. For any value of n we know that r will be a binomial random variable with PMF

$$p_r(r_0) = \binom{n}{r_0} P^{r_0}(1 - P)^{n-r_0} \qquad r_0 = 0, 1, 2, \ldots, n$$

This PMF, written as a PDF, will always include a set of $n + 1$ impulses and be equal to zero between the impulses. Thus, $f_r(r_0)$ can never approach a Gaussian PDF on a point-by-point basis. However, it is possible for the CDF of r to approach the CDF for a Gaussian random variable on a point-by-point basis as $n \to \infty$, and the central limit theorem given in the first paragraph of this section states that this is indeed the case.

We now present a proof of the form of the central limit theorem stated in the opening paragraph of this section.

1 Let $x_1, x_2, \ldots, x_n$ be independent identically distributed random variables, each with finite expected value $E(x)$ and finite variance σ_x^2. We define random variable r to be $r = x_1 + x_2 + \cdots + x_n$, and we wish to show that the CDF $p_{r\leq}(r_0)$ approaches the CDF of a Gaussian random variable as $n \to \infty$.

2 From the independence of the x_i's and the definition of r, we have

$$f_r^T(s) = [f_x^T(s)]^n$$

3 Note that for any random variable y, defined by $y = ar + b$, we may obtain $f_y^T(s)$ in terms of $f_r^T(s)$ from the definition of the s transform as follows:

$$f_y^T(s) = E(e^{-sy}) = E(e^{-ars}e^{-bs}) = e^{-sb} \int_{-\infty}^{\infty} e^{-asr_0} f_r(r_0)\, dr_0$$

We may recognize the integral in the above equation to obtain

$$f_y^T(s) = e^{-sb} f_r^T(as)$$

We shall apply this relation to the case where y is the standardized random variable for r,

$$y = \frac{r - E(r)}{\sigma_r} = \frac{r - nE(x)}{\sqrt{n}\,\sigma_x} \qquad a = \frac{1}{\sqrt{n}\,\sigma_x} \qquad b = -\frac{\sqrt{n}\,E(x)}{\sigma_x}$$

to obtain $f_y^T(s)$ from the expression for $f_r^T(s)$ of step 2.

$$f_y^T(s) = \left[e^{sE(x)/\sigma_x\sqrt{n}}\; f_x^T\left(\frac{s}{\sqrt{n}\,\sigma_x}\right)\right]^n$$

So far, we have found the s transform for y, the standardized sum of n independent identically distributed random variables,

$$y = \frac{x_1 + x_2 + \cdots + x_n - nE(x)}{\sqrt{n}\,\sigma_x}$$

4 The above expression for $f_y^T(s)$ may be written with $e^{sE(x)/\sigma_x\sqrt{n}}$ and $f_x^T\left(\frac{s}{\sqrt{n}\,\sigma_x}\right)$ each approximated suitably near to $s = 0$. These approximations are found to be

$$e^{sE(x)/\sigma_x\sqrt{n}} \approx 1 + \frac{E(x)}{\sigma_x}\left(\frac{s}{\sqrt{n}}\right) + \frac{[E(x)]^2}{2\sigma_x^2}\left(\frac{s}{\sqrt{n}}\right)^2$$

$$f_x^T\left(\frac{s}{\sqrt{n}\,\sigma_x}\right) \approx 1 - \frac{E(x)}{\sigma_x}\left(\frac{s}{\sqrt{n}}\right) + \frac{E(x^2)}{2\sigma_x^2}\left(\frac{s}{\sqrt{n}}\right)^2$$

When we multiply and collect terms, for suitably small s (or, equivalently, for suitably large n) we have

$$f_y^T(s) \approx \left[1 + \frac{1}{2}\frac{s^2}{n}\right]^n$$

5 We use the relation

$$\lim_{n\to\infty}\left(1 + \frac{a}{n}\right)^n = e^a$$

to take the limit as $n \to \infty$ of the approximation for $f_y^T(s)$ obtained in step 4. This results in

$$\lim_{n\to\infty} f_y^T(s) = e^{s^2/2}$$

and we have shown that the s transform of the PDF for random variable y approaches the transform of a unit normal PDF. This does not tell us *how* (or if) the PDF $f_y(y_0)$ approaches a Gaussian PDF on a point-by-point basis. But a relation known as the *continuity theorem* of transform theory may be invoked to assure us that

$$\lim_{n\to\infty} f_{y\leq}(y_0) = \Phi(y_0)$$

[This theorem assures us that, if $\lim_{n\to\infty} f_{y_n}^T(s) = f_w^T(s)$ *and* if $f_w^T(s)$ is a continuous function, then the CDF for random variable y_n converges (on a point-by-point basis) to the CDF of random variable w. This convergence need not be defined at discontinuities of the limiting CDF.]

6 Since y is the standardized form of r, we simply substitute into the above result for $f_{y\leq}(y_0)$ and conclude the following.

If $r = x_1 + x_2 + \cdots + x_n$ and $x_1, x_2, \ldots, x_n$ are independent identically distributed random variables each with finite expected value $E(x)$ and finite standard deviation σ_x, we have

$$\lim_{n\to\infty} p_{r\leq}(r_0) = \Phi\left[\frac{r_0 - E(r)}{\sigma_r}\right] \qquad E(r) = nE(x) \qquad \sigma_r = \sqrt{n}\,\sigma_x$$

This completes our proof of one central limit theorem.

6-6 Approximations Based on the Central Limit Theorem

We continue with the notation $r = x_1 + x_2 + \cdots + x_n$, where the random variables $x_1, x_2, \ldots, x_n$ are mutually independent and identically distributed, each with finite expected value $E(x)$ and finite variance σ_x^2. If every member of the sum happens to be a Gaussian random variable, we know (from Sec. 6-4) that the PDF $f_r(r_0)$ will also be Gaussian for any value of n. *Whatever* the PDF for the individual members of the sum, one central limit theorem states that, as $n \to \infty$, we have

$$p_{r\leq}(r_0) \to \Phi\left[\frac{r_0 - E(r)}{\sigma_r}\right]$$

As $n \to \infty$, the CDF for r approaches the CDF for that Gaussian random variable which has the same mean and variance as r.

If we wish to use the approximation

$$p_{r\leq}(r_0) \approx \Phi\left[\frac{r_0 - E(r)}{\sigma_r}\right]$$

for "large" but finite values of n, and the individual x_i's are not Gaussian random variables, there are no simple general results regarding the precision of the approximation.

If the individual terms in the sum are described by any of the more common PDF's (with finite mean and variance), $f_r(r_0)$ rapidly ($n \approx 5$ or 10) approaches a Gaussian curve in the vicinity of $E(r)$. Depending on the value of n and the degree of symmetry expected in $f_r(r_0)$, we generally expect $f_r(r_0)$ to be poorly approximated by a Gaussian curve in ranges of r_0 more than some number of standard deviations distant from $E(r)$. For instance, even if the x_i's can take on only positive experimental values, the use of an approximation based on the central limit theorem will always result in some nonzero probability that the experimental value of $x_1 + x_2 + \cdots + x_n$ will be negative.

The discussion of the previous paragraph, however crude it may be, should serve to emphasize the *central* property of approximations based on the central limit theorem.

As one example of the use of an approximation based on the central limit theorem, let random variable r be defined to be the sum of 48 independent experimental values of random variable x, where the PDF for x is given by

$$f_x(x_0) = \begin{cases} 1 & \text{if } 0 < x_0 \leq 1 \\ 0 & \text{otherwise} \end{cases}$$

We wish to determine the probability that an experimental value of r falls in the range $22.0 < r \leq 25.0$. By direct calculation we easily obtain

$$E(x) = 0.5 \qquad \sigma_x^2 = 1/12$$
$$E(r) = 48E(x) = 24.0 \qquad \sigma_r^2 = 48\sigma_x^2 = 4.0$$

In using the central limit theorem to approximate

$$\text{Prob}(22.0 < r \leq 25.0)$$

we are approximating the true PDF $f_r(r_0)$ in the range $22 < r \leq 25$ by the Gaussian PDF

$$f_r(r_0) \approx \frac{1}{\sqrt{2\pi}\,\sigma_r} e^{-[r_0 - E(r)]^2/2\sigma_r^2} = \frac{1}{2\sqrt{2\pi}} e^{-(r_0-24)^2/8}$$

If we wish to evaluate $\text{Prob}(22.0 < r \leq 25.0)$ directly from the table for the CDF of the unit normal PDF, the range of interest for random variable r should be measured in units of σ_r from $E(r)$. We have

$$\begin{aligned}\text{Prob}(22.0 < r \leq 25.0) &= \text{Prob}\{-1.0\sigma_r < [r - E(r)] \leq 0.5\sigma_r\} \\ &= \Phi(0.5) - \Phi(-1.0) \\ &= \Phi(0.5) - [1 - \Phi(1.0)] \\ &= 0.6915 - 1.0000 + 0.8413 \\ &= 0.5328\end{aligned}$$

It happens that this is a very precise approximation. In fact, by simple convolution (or by our method for obtaining derived distributions in sample space) one can show, for the given $f_x(x_0)$, that even for $n = 3$ or $n = 4$, $f_r(r_0)$ becomes very close to a Gaussian PDF over most of the possible range of random variable r (see Prob. 6.10). However, a similar result for such very small n may not exist for several other common PDF's (see Prob. 6.11).

6-7 Using the Central Limit Theorem for the Binomial PMF

We wish to use an approximation based on the central limit theorem to approximate the PMF for a discrete random variable. Assume we are interested in events defined in terms of k, the number of successes in n trials of a Bernoulli process with parameter P. From earlier work (Sec. 4-1) we know that

$$p_k(k_0) = \binom{n}{k_0} P^{k_0}(1 - P)^{n-k_0} \qquad k_0 = 0, 1, 2, \ldots, n$$

and if a and b are integers with $b > a$, there follows

$$\text{Prob}(a \leq k \leq b) = \sum_{k_0=a}^{b} \binom{n}{k_0} P^{k_0}(1 - P)^{n-k_0}$$

Should this quantity be of interest, it would generally require a very unpleasant calculation. So we might, for large n, turn to the central limit theorem, noting that

$$k = x_1 + x_2 + \cdots + x_n$$

where each x_i is an independent Bernoulli random variable.

If we applied the central limit theorem, subject to no additional considerations, we would have

$$\text{Prob}(a \leq k \leq b) \approx \Phi\left[\frac{b - E(k)}{\sigma_k}\right] - \Phi\left[\frac{a - E(k)}{\sigma_k}\right] \qquad E(k) = nP \quad \sigma_k = \sqrt{nP(1 - P)}$$

We have approximated the probability that a binomial random variable k falls in the range $a \leq k \leq b$ by the area under a normal curve over this range. In many cases this procedure will yield excellent results. By looking at a picture of this situation, we shall suggest one simple improvement of the approximation.

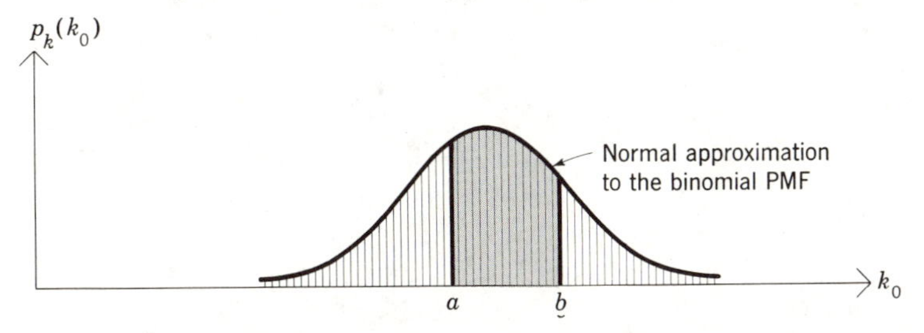

The bars of $p_k(k_0)$ are shown to be about the same height as the approximating normal curve. This must be the case if n is large enough for the CDF's of $p_k(k_0)$ and the approximating normal curve to increase by about the same amount for each unit distance along the k_0 axis (as a result of the central limit theorem). The shaded area in this figure represents the approximation to Prob$(a \leq k \leq b)$ which results from direct substitution, where we use the CDF for a normal curve whose expected value and variance are the same as those of $p_k(k_0)$.

By considering the above sketch, we might expect that a more reasonable procedure could be suggested to take account of the discrete nature of k. In particular, it appears more accurate to associate the area under the normal curve between $k_0 - 0.5$ and $k_0 + 0.5$ with the probability of the event that random variable k takes on experimental value k_0. This not only seems better on a term-by-term basis than direct use of the central-limit-theorem approximation, but we can also show one extreme case of what may happen when this suggested improvement is not used. Notice (from the above sketch) that, if we have $b = a + 1$ [with a and b in the vicinity of $E(x)$], direct use of the CDF for the normal approximating curve will produce an approximation which is about 50% of the correct probability,

$$\text{Prob}(a \leq k \leq a + 1) = p_k(a) + p_k(a + 1)$$

When using the central limit theorem to approximate the binomial PMF, the adoption of our suggested improvement leads us to write,

$$\text{Prob}(a \leq k \leq b) \approx \Phi\left[\frac{b + \frac{1}{2} - nP}{\sqrt{nP(1 - P)}}\right] - \Phi\left[\frac{a - \frac{1}{2} - nP}{\sqrt{nP(1 - P)}}\right]$$

This result, a special case of the central limit theorem, is known as the *DeMoivre-Laplace limit theorem.* It can be shown to yield an improvement over the case in which the $\pm\frac{1}{2}$ terms are not used. These corrections may be significant when a and b are close $[(b - a)/\sigma_k < 1]$ or when a or b is near the peak of the approximating Gaussian PDF.

For example, suppose that we flipped a fair coin 100 times, and let k equal the number of heads. If we wished to approximate $\text{Prob}(48 \leq k \leq 51)$, the $\pm\frac{1}{2}$ corrections at the end of the range of k would clearly make a significant contribution to the accuracy of the approximation. On the other hand, for a quantity such as $\text{Prob}(23 \leq k \leq 65)$, the effect of the $\pm\frac{1}{2}$ is negligible.

One must always question the validity of approximations; yet it is surprising how well the DeMoivre-Laplace limit theorem applies for even a narrow range of k [near $E(k)$] when n is not very large. We shall do one such problem three ways. After obtaining these solutions, we shall comment on some limitations of this approximation technique.

Consider a set of 16 Bernoulli trials with $P = 0.5$. We wish to determine the probability that the number of successes, k, takes on an experimental value equal to 6, 7, or 8. First we do the exact calculation,

$$\text{Prob}(6 \leq k \leq 8) = \sum_{k_0=6}^{8} \binom{16}{k_0} P^{k_0}(1 - P)^{16-k_0} = 0.49313$$

If we carelessly make direct use of the normal approximation, we have

$$\text{Prob}(6 \leq k \leq 8) \approx \Phi\left(\frac{8 - 8}{2}\right) - \Phi\left(\frac{6 - 8}{2}\right) = 0.34135$$

which, for reasons we have discussed, is poor indeed. Finally, if we use the $\pm\frac{1}{2}$ correction of the DeMoivre-Laplace theorem, we find

$$\text{Prob}(6 \leq k \leq 8) \approx \Phi\left(\frac{8 + \frac{1}{2} - 8}{2}\right) - \Phi\left(\frac{6 - \frac{1}{2} - 8}{2}\right) = 0.49306$$

which is within 0.02% of the correct value.

If P is too close either to zero or to unity for a given n, the resulting binomial PMF will be very asymmetric, with its peak very close to $k_0 = 0$ or to $k_0 = n$ and any Gaussian approximation will be poor. A reasonable (but arbitrary) rule of thumb for determining whether the DeMoivre-Laplace approximation to the binomial PMF may be employed [in the vicinity of $E(k)$] is to require that

$$nP > 3\sigma_k \qquad n(1 - P) > 3\sigma_k \qquad \text{with } \sigma_k = \sqrt{nP(1 - P)}$$

The better the margin by which these constraints are satisfied, the larger the range about $E(k)$ for which the normal approximation will yield satisfactory results.

6-8 The Poisson Approximation to the Binomial PMF

We have noted that the DeMoivre-Laplace limit theorem will not provide a useful approximation to the binomial PMF if either P or $1 - P$ is very small. When either of these quantities is too small for a given value of n, any Gaussian approximation to the binomial will be unsatisfactory. The Gaussian curve will remain symmetrical about $E(k)$, even though that value may be only a fraction of a standard deviation from the lowest or highest possible experimental value of the binomial random variable.

If n is large and P is small such that the DeMoivre-Laplace theorem may not be applied, we may take the limit of

$$p_k(k_0) = \binom{n}{k_0} P^{k_0}(1 - P)^{n-k_0}$$

by letting $n \to \infty$ and $P \to 0$ while always requiring $nP = \mu$. Our result will provide a very good term-by-term approximation for the significant members [k_0 nonnegative and within a few σ_k of $E(k)$] of the binomial PMF for large n and small P.

First, we use the relation $nP = \mu$ to write

$$p_k(k_0) = \frac{n(n-1) \cdots (n - k_0 + 1)}{k_0!} \left(\frac{\mu}{n}\right)^{k_0} \left(1 - \frac{\mu}{n}\right)^{n-k_0}$$

and, rearranging the terms, we have

$$p_k(k_0) = \frac{n(n-1) \cdots (n - k_0 + 1)}{n^{k_0}} \frac{\mu^{k_0}}{k_0!} \left(1 - \frac{\mu}{n}\right)^{n-k_0}$$

Finally, we take the limit as $n \to \infty$ to obtain

$$\lim_{n \to \infty} p_k(k_0) = \frac{\mu^{k_0} e^{-\mu}}{k_0!} = \frac{(nP)^{k_0} e^{-(nP)}}{k_0!}$$

The above result is known, for obvious reasons, as the *Poisson approximation to the binomial PMF.*

For a binomial random variable k, we may note that, as $P \to 0$, the $E(k)/\sigma_k$ ratio is very nearly equal to $\sqrt{nP}$. For example, if $n = 100$ and $P = 0.01$, the expected value $E(k) = nP = 1.0$ is only one standard deviation from the minimum possible experimental value of k. Under these circumstances, the normal approximation (DeMoivre-Laplace) is poor, but the Poisson approximation is quite accurate for the small values of k_0 at which we find most of the probability mass of $p_k(k_0)$.

As an example, for the case $n = 100$ and $P = 0.01$, we find

	$k_0 = 0$	$k_0 = 1$	$k_0 = 3$	$k_0 = 10$
Exact value of $p_k(k_0)$	0.3660	0.3697	0.0610	$7 \cdot 10^{-8}$
Poisson approximation	0.3679	0.3679	0.0613	$10 \cdot 10^{-8}$
DeMoivre-Laplace approximation	0.2420	0.3850	0.0040	$<2 \cdot 10^{-18}$

6-9 A Note on Other Types of Convergence

In Sec. 6-2, we defined any sequence $\{y_n\}$ of random variables to be stochastically convergent (or to converge in probability) to C if, for every $\epsilon > 0$, the condition

$$\lim_{n \to \infty} \text{Prob}(|y_n - C| > \epsilon) = 0$$

is satisfied.

Let A_n denote the event $|y_n - C| < \epsilon$. By using the definition of a limit, an equivalent statement of the condition for stochastic convergence is that, for any $\epsilon > 0$ and any $\delta > 0$, we can find an n_0 such that

$$\text{Prob}(A_n) > 1 - \delta \qquad \text{for all } n > n_0$$

However, it does *not* follow from stochastic convergence that for any $\epsilon > 0$ and any $\delta > 0$ we can necessarily find an n_0 such that

$$\text{Prob}(A_{n_0+1} A_{n_0+2} A_{n_0+3} \cdots) > 1 - \delta$$

For a stochastically convergent sequence $\{y_n\}$, we would conclude that, *for all $n > n_0$, with n_0 suitably large:*

1 It is very probable that *any particular* y_n is within $\pm\epsilon$ of C.
2 It is not necessarily very probable that *every* y_n is within $\pm\epsilon$ of C.

A stronger form of convergence than stochastic convergence is known as *convergence with probability* 1 (or *convergence almost everywhere*). The sequence $\{y_n\}$ of random variables is defined to converge with probability 1 to C if the relation

$$\text{Prob}(\lim_{n\to\infty} y_n = C) = 1$$

is satisfied. Convergence with probability 1 implies stochastic convergence, but the converse is not true. Furthermore it can be shown (by using measure theory) that convergence with probability 1 does require that, for any $\epsilon > 0$ and any $\delta > 0$, we can find an n_0 such that

$$\text{Prob}(A_{n_0+1}A_{n_0+2}A_{n_0+3} \cdots) > 1 - \delta$$

For a sequence $\{y_n\}$ convergent to C with probability 1, we would conclude that, for one thing, the sequence is also stochastically convergent to C and also that, *for all* $n > n_0$, *with* n_0 *suitably large:*
1 It is very probable that *any particular* y_n is within $\pm\epsilon$ of C.
2 It is also very probable that *every* y_n is within $\pm\epsilon$ of C.

A third form of convergence, *mean-square convergence* (or *convergence in the mean*) of a sequence $\{y_n\}$ of random variables is defined by the relation

$$\lim_{n\to\infty} E[(y_n - C)^2] = 0$$

It is simple to show that mean-square convergence implies (but is not implied by) stochastic convergence (see Prob. 6.20). Mean-square convergence does not imply and is not implied by convergence with probability 1.

Determination of the necessary and sufficient conditions for sequences of random variables to display various forms of convergence, obey various laws of large numbers, and obey central limit theorems is well beyond the scope of our discussion. We do remark, however, that, because of the limited tools at our disposal, the law of large numbers obtained in Sec. 6-3 is unnecessarily weak.

PROBLEMS

6.01 Let x be a random variable with PDF $f_x(x_0) = \lambda e^{-\lambda x_0}$ for $x_0 > 0$. Use the Chebyshev inequality to find an upper bound on the quantity

$$\text{Prob}[|x - E(x)| \geq d]$$

as a function of d. Determine also the true value of this probability as a function of d.

6.02 In Sec. 6-3, we obtained the relation

$$\text{Prob}\left[\left|\frac{1}{n}\sum_{i=1}^{n} y_i - E(y)\right| \geq \epsilon\right] \leq \frac{\sigma_y^2}{n\epsilon^2}$$

where $y_1, y_2, \ldots$ were independent identically distributed random variables. If y is known to have a Bernoulli PMF with parameter P, use this relation to find how large n must be if we require that $(1/n) \sum_{i=1}^{n} y_i$ be within ± 0.01 of P with a probability of at least 0.95. (Remember that this is a loose bound and the resulting value of n may be unnecessarily large. See Prob. 6.09.)

6.03 Let $x_1, x_2, \ldots$ be independent experimental values of a random variable with PDF $f_x(x_0) = \mu_{-1}(x_0 - 0) - \mu_{-1}(x_0 - 1)$. Consider the sequence defined by

$$y_n = \max(x_1, x_2, \ldots, x_n)$$

Determine whether or not the sequence $\{y_n\}$ is stochastically convergent.

6.04 Consider a Gaussian random variable x, with expected value $E(x) = m$ and variance $\sigma_x^2 = (m/2)^2$.

a Determine the probability that an experimental value of x is negative.

b Determine the probability that the sum of four independent experimental values of x is negative.

c Two independent experimental values of x (x_1, x_2) are obtained. Determine the PDF and the mean and variance for:

i $ax_1 + bx_2$ **ii** $ax_1 - bx_2$ **iii** $|x_1 - x_2|$ $a, b > 0$

d Determine the probability that the product of four independent experimental values of x is negative.

6.05 A useful bound for the area under the *tails* of a unit normal PDF is obtained from the following inequality for $a \geq 0$:

$$p_{x\leq}(-a) = \int_a^\infty \frac{1}{\sqrt{2\pi}} e^{-x_0^2/2}\, dx_0 \leq \int_a^\infty \frac{1}{\sqrt{2\pi}} \frac{x_0}{a} e^{-x_0^2/2}\, dx_0$$

Use this result to obtain lower bounds on the probability that an experimental value of a Gaussian random variable x is within

a $\pm\sigma_x$ **b** $\pm 2\sigma_x$ **c** $\pm 4\sigma_x$

of its expected value. Compare these bounds with the true values of these probabilities and with the corresponding bounds obtained from the Chebyshev inequality.

6.06 A noise signal x may be considered to be a Gaussian random variable with an expected value of zero and a variance of σ_x^2. Assume that any experimental value of x will cause an error in a digital communication system if it is larger than $+A$.

a Determine the probability that any particular experimental value of x will cause an error if:

i $\sigma_x^2 = 10^{-3}A^2$ **ii** $\sigma_x^2 = 10^{-1}A^2$
iii $\sigma_x^2 = A^2$ **iv** $\sigma_x^2 = 4A^2$

b For a given value of A, what is the largest σ_x^2 may be to obtain an error probability for any experimental value of x less than 10^{-3}? 10^{-6}?

6.07 The weight of a Pernotti Parabolic Pretzel, w, is a continuous random variable described by the probability density function

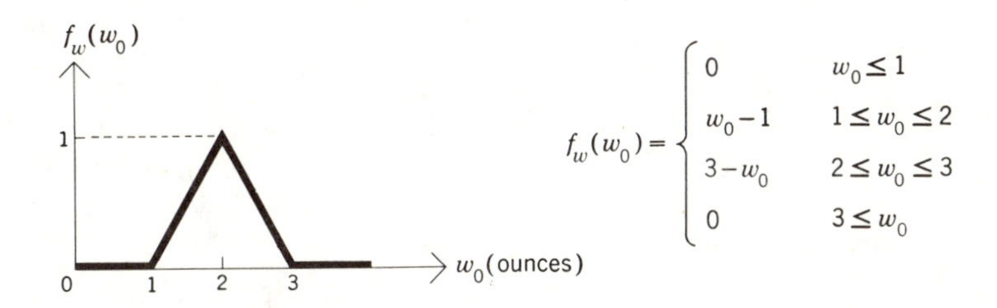

a What is the probability that 102 pretzels weigh more than 200 ounces?

b If we select 4 pretzels independently, what is the probability that exactly 2 of the 4 will each have the property of weighing more than 2 ounces?

c What is the smallest integer (the pretzels are not only inedible, they are also unbreakable) N for which the total weight of N pretzels will exceed 200 ounces with probability 0.990?

6.08 The energy of any individual particle in a certain system is an independent random variable with probability density function

$$f_E(E_0) = \begin{cases} 2e^{-2E_0} & E_0 \geq 0 \\ 0 & E_0 < 0 \end{cases}$$

The total system energy is the sum of the energies of the individual particles.

Numerical answers are required for parts (a), (b), (c), and (e).

a If there are 1,600 particles in the system, determine the probability that there are between 780 and 840 energy units in the system.

b What is the largest number of particles the system may contain if the probability that its total energy is less than 440 units must be at least 0.9725?

c Each particle will escape from the system if its energy exceeds (ln 3)/2 units. If the system originally contained 4,800 particles, what is the probability that at least 1,700 particles will escape?

d If there are 10 particles in the system, determine an *exact* expression for the PDF for the total energy in the system.

e Compare the second and fourth moments of the answer to (d) with those resulting from a central-limit-theorem approximation.

6.09 Redo Prob. 6.02, using an approximation based on the central limit theorem rather than the Chebyshev inequality.

6.10 Determine and plot the precise PDF for r, the sum of four independent experimental values of random variable x, where the PDF for x is $f_x(x_0) = \mu_{-1}(x_0 - 0) - \mu_{-1}(x_0 - 1)$. Compare the numerical values of $f_r(r_0)$ and its Gaussian approximation at $r_0 = E(r) \pm K\sigma_r$, for $K = 0$, 1, and 2.

6.11 Let r be the sum of four independent experimental values of an exponential random variable. Compare the numerical values of $f_r(r_0)$ and its central limit theorem approximation at $r_0 = E(r) \pm K\sigma_r$, for $K = 0$, 1, and 2.

6.12 A certain town has a Saturday night movie audience of 600 who must choose between two comparable movie theaters. Assume that the movie-going public is composed of 300 couples, each of which independently flips a fair coin to decide which theater to patronize.

a Using a central limit theorem approximation, determine how many seats each theater must have so that the probability of exactly one theater running out of seats is less than 0.1.

b Repeat, assuming that each of the 600 customers makes an independent decision (instead of acting in pairs).

6.13 For 3,600 independent tosses of a fair coin:

a Determine a number n such that the probability is 0.5 that the number of heads resulting will be between 1,780 and n.

b Determine the probability that the number of heads is within $\pm 1\%$ of its expected value.

6.14 Reyab aspirin has exactly 10^7 users, all of them fitfully loyal. The number of tablets consumed by any particular customer on any one day is a random variable k, described by the probability mass function

$$p_k(k_0) = \frac{4 - k_0}{10} \qquad k_0 = 0, 1, 2, 3$$

Each customer's Reyab consumption is independent of that of all other customers. Reyab is sold only in 100-tablet bottles. On a day when a customer consumes exactly k_0 tablets, he purchases a new 100-tablet bottle of Reyab with probability $k_0/100$.

a Determine the mean and variance of k, the random variable describing the Reyab consumption of any particular customer in one day.

b Determine the mean and variance of t, the random variable describing the *total* number of Reyab tablets consumed in one day.

c Determine the probability that the *total* number of tablets consumed on any day will differ by more than $\pm 5{,}000$ tablets from the average daily total consumption.

d Determine the probability that a particular customer buys a new bottle of Reyab on a given day.

e What is the probability that a randomly selected tablet of Reyab (it was marked at the factory) gets consumed on a day when its owner consumes exactly two Reyab tablets?

f The Clip Pharmacy supplies exactly 30 Reyab customers with their entire requirements. What is the probability that this store sells exactly four bottles of aspirin on a particular day?

6.15 A population is sampled randomly (with replacement) to estimate S, the fraction of smokers in that population. Determine the sample size n such that the probability that the estimate is within ± 0.02 of the true value is at least 0.95. In other words, determine the smallest value of n such that

$$\text{Prob}\left(\left|\frac{\text{number of smokers counted}}{n} - S\right| < 0.02\right) \geq 0.95$$

6.16 Consider the following model for the weight gain of a prehistoric neopalenantioctipus. His (or her) weight gain in pounds on any particular day was an independent discrete random variable k, with the PMF

$$p_k(k_0) = \begin{cases} 0.74 & k_0 = 0.50 \\ 0.25 & k_0 = 4.00 \\ 0.01 & k_0 = 200.0 \end{cases}$$

Using this crude model, determine an approximation to the PDF for the weight of one such animal when it expired at the ripe age of 100,000 days.

6.17 Consider the number of 3s which result from 600 tosses of a fair six-sided die.

a Determine the probability that there are exactly 100 3s, using a form of Stirling's approximation for $n!$ which is very accurate for these values,

$$n! \approx \sqrt{2\pi}\, e^{-n} n^{n+0.5}$$

b Use the Poisson approximation to the binomial PMF to obtain the probability that there are exactly 100 3s.

c Repeat part (b), using the central limit theorem intelligently.

d Use the Chebyshev inequality to find a lower bound on the probability that the number of 3s is between 97 and 103 inclusive, between 90 and 110 inclusive, and between 60 and 140 inclusive.

e Repeat part (d), using the central limit theorem and employing the DeMoivre-Laplace result when it appears relevant. Compare your answers with those obtained above, and comment.

6.18 A coin is tossed n times. Each toss is an independent Bernoulli trial with probability of heads P. Random variable x is defined to be the number of heads observed. For each of the following expressions, either find the value of K which makes the statement true for *all* $n \geq 1$, or state that no such value of K exists.

Example: $E(x) = An^k$ Answer: $k = 1$

a $E\left(\frac{x}{n}\right) = Bn^k$ **b** $E\{[x - E(x)]^2\} = Cn^k$ **c** $E(x^2) = Dn^k$

In the following part, consider only the case for large n:

d $\text{Prob}\left(\left|P - \frac{x}{n}\right| \leq Fn^k\right) = 0.15$

6.19 Each performance of a particular experiment is said to generate one "experimental value" of random variable x described by the probability density function

$$f_x(x_0) = \begin{cases} 1 & \text{if } 0 < x_0 \leq 1 \\ 0 & \text{otherwise} \end{cases}$$

The experiment is performed K times, and the resulting successive (and independent) experimental values are labeled $x_1, x_2, \ldots, x_K$.

a Determine the probability that x_2 and x_4 are the two largest of the K experimental values.

b Given that $x_1 + x_2 > 1.00$, determine the conditional probability that the smaller of these first two experimental values is smaller than 0.75.

c Determine the numerical value of

$$\lim_{K\to\infty} \text{Prob}\left(\left|\frac{K}{2} - \sum_{i=1}^{K} x_i\right| < \frac{K^{1/3}}{2\sqrt{3}}\right)$$

A formal proof is not required, but your reasoning should be fully explained.

d If $r = \min(x_1,x_2, \ldots ,x_K)$ and $s = \max(x_1,x_2, \ldots ,x_K)$, determine the joint probability density function $f_{r,s}(r_0,s_0)$ for all values of r_0 and s_0.

6.20 Use the Chebyshev inequality to prove that stochastic convergence is assured for any sequence of random variables which converges in the mean-square sense.

6.21 A simple form of the Cauchy PDF is given by

$$f_x(x_0) = [\pi(1 + x_0^2)]^{-1} \qquad -\infty \leq x_0 \leq \infty$$

a Determine $f_x^T(s)$. (You may require a table of integrals.)
b Let y be the sum of K independent samples of x. Determine $f_y^T(s)$.
c Would you expect $\lim_{K\to\infty} f_y(y_0)$ to become Gaussian? Explain.

6.22 The number of rabbits in generation i is n_i. Variable n_{i+1}, the number of rabbits in generation $i + 1$, depends on random effects of light, heat, water, food, and predator population, as well as on the number n_i. The relation is

$$p_{n_{i+1}}(jn_i) = \begin{cases} 0.2 & j = 1 \\ 0.3 & j = 2 \\ 0.5 & j = 3 \end{cases}$$

This states, for instance, that with probability 0.5 there will be three times as many rabbits in generation $i + 1$ as there were in generation i.

The rabbit population in generation 1 was 2. Find an approximation to the PMF for the number of rabbits in generation 12. (*Hint:* To use the central limit theorem, you must find an expression involving the *sum* of random variables.)

CHAPTER SEVEN

an introduction to statistics

The *statistician* suggests probabilistic models of reality and investigates their validity. He does so in an attempt to gain insight into the behavior of physical systems and to facilitate better predictions and decisions regarding these systems. A primary concern of statistics is *statistical inference,* the drawing of inferences from data.

The discussions in this chapter are brief, based on simple examples, somewhat incomplete, and always at an introductory level. Our major objectives are (1) to introduce some of the fundamental issues and methods of statistics and (2) to indicate the nature of the transition required as one moves from probability theory to its applications for *statistical reasoning.*

We begin with a few comments on the relation between statistics and probability theory. After identifying some prime issues of concern in statistical investigations, we consider common methods for the study of these issues. These methods generally represent the viewpoint of *classical* statistics. Our concluding sections serve as a brief introduction to the developing field of *Bayesian* (or *modern*) statistics.

7-1 Statistics Is Different

Probability theory is axiomatic. Fully defined probability problems have unique and precise solutions. So far we have dealt with problems which are wholly abstract, although they have often been based on probabilistic *models* of reality.

The field of statistics is different. Statistics is concerned with the relation of such models to actual physical systems. The methods employed by the statistician are arbitrary ways of *being reasonable* in the application of probability theory to physical situations. His primary tools are probability theory, a mathematical sophistication, and common sense.

To use an extreme example, there simply is no unique *best* or *correct* way to extrapolate the gross national product five years hence from three days of rainfall data. In fact, there is no best way to predict the rainfall for the fourth day. But there are many ways to try.

7-2 Statistical Models and Some Related Issues

In contrast to our work in previous chapters, we are now concerned both with models of reality and reality itself. It is important that we keep in mind the differences between the statistician's model (and its implications) and the actual physical situation that is being modeled.

In the real world, we may design and perform experiments. We may observe certain *characteristics of interest* of the experimental outcomes. If we are studying the behavior of a coin of suspicious origin, a characteristic of interest might be the number of heads observed in a certain number of tosses. If we are testing a vaccine, one characteristic of interest could be the observed immunity rates in a control group and in a vaccinated group.

What is the nature of the statistician's model? From whatever knowledge he has of the physical mechanisms involved and from his past experience, the statistician postulates a probabilistic model for the system of interest. He anticipates that this model will exhibit a probabilistic behavior *in the characteristics of interest* similar to that of the physical system. The details of the model might or might not be closely related to the actual nature of the physical system.

If the statistician is concerned with the coin of suspicious origin.

he might suggest a model which is a Bernoulli process with probability P for a head on any toss. For the study of the vaccine, he might suggest a model which assigns a probability of immunity P_1 to each member of the control group and assigns a probability of immunity P_2 to each member of the vaccinated group.

We shall consider some of the questions which the statistician asks about his models and learn how he employs experimental data to explore these questions.

1 Based on some experimental data, does a certain model seem reasonable or at least not particularly unreasonable? This is the domain of *significance testing.* In a significance test, the statistician speculates on the likelihood that data similar to that actually observed would be generated by *hypothetical* experiments with the model.

2 Based on some experimental data, how do we express a preference among several postulated models? (These models might be similar models differing only in the values of their parameters.) When one deals with a selection among several hypothesized models, he is involved in a matter of *hypothesis testing.* We shall learn that hypothesis testing and significance testing are very closely related.

3 Given the form of a postulated model of the physical system and some experimental data, how may the data be employed to establish the most desirable values of the parameters of the model? This question would arise, for example, if we considered the Bernoulli model for flips of the suspicious coin and wished to adjust parameter P to make the model as compatible as possible with the experimental data. This is the domain of *estimation.*

4 We may be uncertain of the appropriate parameters for our model. However, from previous experience with the physical system and from other information, we may have convictions about a reasonable PDF for these parameters (which are, to us, random variables). The field of *Bayesian analysis* develops an efficient framework for combining such "prior knowledge" with experimental data. Bayesian analysis is particularly suitable for investigations which must result in *decisions* among several possible future courses of action.

The remainder of this book is concerned with the four issues introduced above. The results we shall obtain are based on *subjective* applications of concepts of probability theory.

7-3 Statistics: Sample Values and Experimental Values

In previous chapters, the phrase "experimental value" always applied to what we might now consider to be *the outcome of a hypothetical experiment with a model of a physical system.* Since it is important that we be able to distinguish between consequences of a model and consequences of reality, we establish two definitions.

EXPERIMENTAL VALUE: Refers to actual data which must, of course, be obtained by the performance of (real) experiments with a physical system

SAMPLE VALUE: Refers to the outcome resulting from the performance of (hypothetical) experiments with a model of a physical system

These particular definitions are not universal in the literature, but they will provide us with an explicit language.

Suppose that we perform a hypothetical experiment with our model n times. Let random variable x be the characteristic of interest defined on the possible experimental outcomes. We use the notation x_i to denote the random variable defined on the ith performance of this hypothetical experiment. The set of random variables $(x_1,x_2, \ldots ,x_n)$ is defined to be a *sample of size* n of random variable x. A sample of size n is a collection of random variables whose probabilistic behavior is specified by our model. Hypothesizing a model is equivalent to specifying a compound PDF for the members of the sample.

We shall use the word *statistic* to describe any function of some random variables, $q(u,v,w, \ldots)$. We may use for the argument of a statistic either the members of a sample or actual experimental values of the random variables. The former case results in what is known as a *sample value* of the statistic. When experimental values are used for $u,v,w, \ldots$, we obtain an *experimental value* of the statistic. Given a specific model for consideration, we may, in principle, derive the PDF for the sample value of any statistic from the compound PDF for the members of the sample. If our model happens to be correct, this PDF would also describe the experimental value of the statistic.

Much of the field of statistics hinges on the following three steps:

1 Postulate a model for the physical system of interest.

2 Based on this model, select a desirable statistic for which:

The PDF for the sample value of the statistic may be calculated in a useful form.

Experimental values of the statistic may be obtained from reality.

3 Obtain an experimental value of the statistic, and comment on the likelihood that a similar value would result from the use of the proposed model instead of reality.

The operation of deriving the PDF's and their means and variances for useful statistics is often very complicated, but there are a few cases of frequent interest for which some of these calculations are not too involved. *Assuming that the x_i's in our sample are always independent and identically distributed,* we present some examples.

One fundamental statistic of the sample $(x_1, x_2, \ldots, x_n)$ is the *sample mean* M_n, whose definition, expected value, and variance were introduced in Sec. 6-3

$$M_n \equiv \frac{1}{n} \sum_{i=1}^{n} x_i \qquad E(M_n) = E(x) \qquad \sigma_{M_n}^{\ 2} = \frac{\sigma_x^{\ 2}}{n}$$

and our proof, for the case $\sigma_x^2 < \infty$, showed that M_n obeyed (at least) the weak law of large numbers. If characteristic x is in fact described by *any* PDF with a finite variance, we can with high probability use M_n as a good estimate of $E(x)$ by using a large value of n, since we know that M_n converges stochastically to $E(x)$.

It is often difficult to determine the exact expression for $f_{M_n}(M)$, the PDF for the sample mean. Quite often we turn to the central limit theorem for an approximation to this PDF. Our interests in the PDF's for particular statistics will become clear in later sections.

Another important statistic is S_n^2, the *sample variance*. The definition of this particular random variable is given by

$$S_n^2 \equiv \frac{1}{n} \sum_{i=1}^{n} (x_i - M_n)^2$$

where M_n is the sample mean as defined earlier. We may expand the above expression,

$$S_n^2 = \frac{1}{n} \sum_{i=1}^{n} x_i^2 - \frac{2}{n} M_n \sum_{i=1}^{n} x_i + M_n^2 = \frac{1}{n} \sum_{i=1}^{n} x_i^2 - M_n^2$$

This is a more useful form of S_n^2 for the calculation of its expectation

$$E(S_n^2) = \frac{1}{n} E\left(\sum_{i=1}^{n} x_i^2\right) - E(M_n^2)$$

The expectation in the first term is the expected value of a sum and may be simplified by

$$E\left(\sum_{i=1}^{n} x_i^2\right) = E(x_1^2 + x_2^2 + \cdots + x_n^2) = nE(x^2)$$

The calculation of $E(M_n^2)$ requires a few intermediate steps,

$$E(M_n^2) = E\left[\left(\frac{1}{n}\sum_{i=1}^{n} x_i\right)^2\right] = E\left(\frac{1}{n^2}\sum_{i=1}^{n} x_i^2\right) + E\left(\frac{1}{n^2}\sum_{\substack{l=1 \\ l\neq j}}^{n}\sum_{j=1}^{n} x_l x_j\right)$$

$$= \left(\frac{1}{n}\right)^2 \{nE(x^2) + n(n-1)[E(x)]^2\}$$

In the last term of the above expression, we have used the fact that, for $l \neq j$, x_l and x_j are independent random variables. Returning to our expression for $E(S_n^2)$, the expected value of the sample variance, we have

$$E(S_n^2) = \frac{1}{n} nE(x^2) - \frac{1}{n} E(x^2) - \left(1 - \frac{1}{n}\right)[E(x)]^2$$

$$= \left(1 - \frac{1}{n}\right)\{E(x^2) - [E(x)]^2\} = \frac{n-1}{n}\sigma_x^2$$

Thus, we see that for samples of a large size, the *expected value* of the sample variance is very close to the variance of random variable x. The poor agreement between $E(S_n^2)$ and σ_x^2 for small n is most reasonable when one considers the definition of the sample variance for a sample of size 1.

We shall not investigate the variance of the sample variance. However, the reader should realize that a result obtainable from the previous equation, namely,

$$\lim_{n\to\infty} E(S_n^2) = \sigma_x^2$$

does not necessarily mean, in itself, that an experimental value of S_n^2 for large n is with high probability a good estimate of σ_x^2. We would need to establish that S_n^2 at least obeys a weak law of large numbers before we could have confidence in an experimental value of S_n^2 (for large n) as a *good* estimator of σ_x^2. For instance, $E(S_n^2) \approx \sigma_x^2$ for large n does not even require that the variance of S_n^2 be finite.

7-4 Significance Testing

Assume that, as a result of preliminary modeling efforts, we have proposed a model for a physical system and we are able to determine the PDF for the sample value of q, the statistic we have selected. In *significance testing*, we work in the event space for statistic q, using this PDF, which would also hold for experimental values of q *if* our model were correct. We wish to evaluate the hypothesis that our model is correct.

In the event space for q we define an event W, known as the *improbable* event. We may select for our improbable event any particular event of probability α, where α is known as the *level of significance* of the test. *After* event W has been selected, we obtain an experimental value of statistic q. Depending on whether or not the experimental value of q falls within the improbable event W, we reach one of two conclusions as a result of the significance test. These conclusions are

1 *Rejection of the hypothesis.* The experimental value q fell within the improbable event W. If our hypothesized model were correct, our observed experimental value of the statistic would be an improbable result. Since we did in fact obtain such an experimental value, we believe it to be unlikely that our hypothesis is correct.

2 *Acceptance of the hypothesis.* The experimental value of q fell in W'. If our hypothesis were true, the observed experimental value of the statistic would not be an improbable event. Since we did in fact obtain such an experimental value, the significance test has not provided us with any particular reason to doubt the hypothesis.

We discuss some examples and further details, deferring general comments until we are more familiar with significance testing.

Suppose that we are studying a coin-flipping process to test the hypothesis that the process is a Bernoulli process composed of fair $(P = \frac{1}{2})$ trials. Eventually, we shall observe 10,000 flips, and we have selected as our statistic k the number of heads in 10,000 flips. Using the central limit theorem, we may, for our purposes, approximate the sample value of k as a continuous random variable with a Gaussian PDF as shown below:

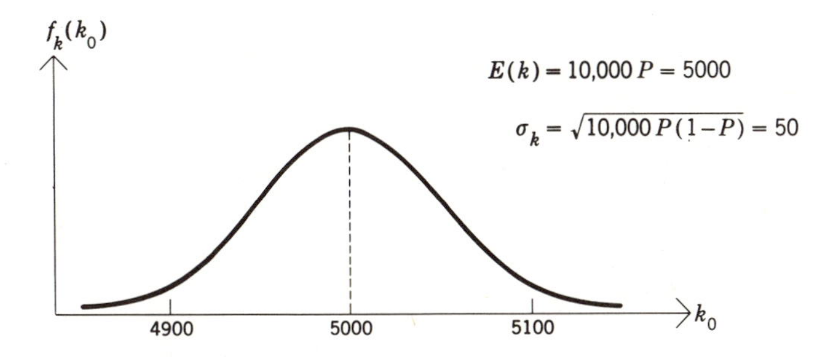

Thus we have the conditional PDF for statistic k, given our hypothesis is correct. If we set α, the probability of the "improbable" event at 0.05, many events could serve as the improbable event W. Several such choices for W are shown below in an event space for k, with $P(W)$ indicated by the area under the PDF $f_k(k_0)$.

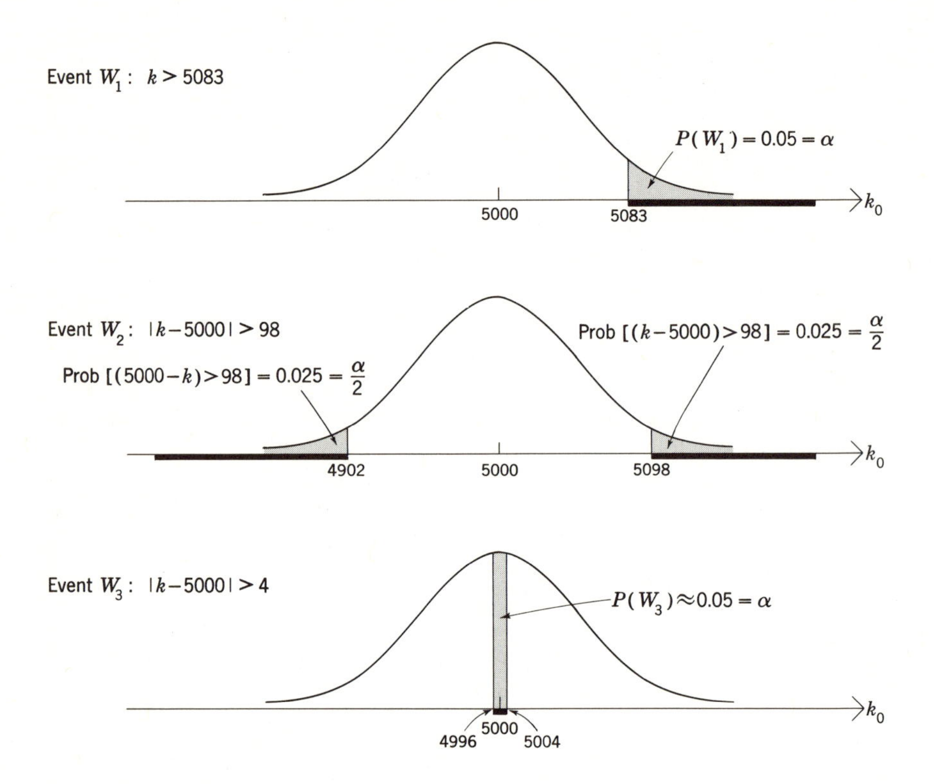

The heavy line appearing on the k_0 axis in each of the above sketches represents one possible selection of an improbable event at the 0.05 level of significance.

That part of the event space for a statistic which is included in the improbable event W is called the *critical range* of the statistic. If the experimental value of the statistic falls into the critical region, the hypothesis is "rejected"; otherwise it is "accepted." Note that the level of significance of the test is actually equal to the conditional probability that a hypothesis will be rejected, given that it is correct.

A reasonable choice of the improbable event must depend on the actual problem at hand. *In a significance test, one is, in effect, testing his hypothesis against all other hypotheses, with no particular alternatives in mind.* If the hypothesis being tested is not correct, some other hypothesis (stated or unstated) is correct. The critical region is placed where we believe other hypotheses are more likely to place the experimental value of the statistic than is the particular hypothesis under test. This may be viewed as "setting a trap" for outcomes due to other hypotheses, and it often results in a decision to make the acceptance region W' as small as possible. There can be no escape from the fact that this type of statistical reasoning is necessarily an arbitrary and subjective procedure, but it is a procedure that most people would consider superior to guessing.

Let's return to the example of the coin-tossing process and assume that we have agreed to set α, the level of significance (or the conditional probability of the improbable event, given that the model is correct) equal to 0.05. If we have no general feelings about possible alternative hypotheses, we would expect to trap most other hypotheses most often by making our acceptance region, the complement of the critical region, as small as possible. For this purpose, we would select W_2 in the above sketch as our choice for the improbable event W.

If we had suspicions that the most likely alternative hypotheses were of the form "P greater than 0.5," we would want the critical region to cover values of our statistic most favored by the alternative hypotheses. We would therefore select W_1 of the three improbable events shown above. Most often, however, significance testing refers to testing one hypothesis with no others in mind, and the acceptance region is generally made as small as possible for a given level of significance.

The choice of the level of significance is rather arbitrary. There are a few popular conventional values of α, and these include 0.05, 0.02, and 0.01. The smaller the level of significance, the less likely we are to reject our hypothesis if it is true and the more likely we are to accept our hypothesis if it is false. In most cases, one would expect the choice of the level of significance to depend on the relative costs of the two possible types of errors which may result from the test, *false acceptance* of the hypothesis and *false rejection* of the hypothesis.

Consider one additional example of the specification of a significance test. Suppose that our model for characteristic x of a certain process is that x is a random variable described by a Gaussian PDF with $\sigma_x = 1$. We have

$$f_x(x_0) = \frac{1}{\sqrt{2\pi}} e^{-(x_0-r)^2/2} \qquad -\infty \leq x_0 \leq \infty$$

and we do not know the value of r, the expected value of x. Ten independent experimental values of characteristic x have been obtained, and we wish to test the hypothesis that parameter r is equal to zero.

Using x_i as the notation for the ith experimental value of x, we arbitrarily elect to use the statistic

$$y = x_1 + x_2 + \cdots + x_{10}$$

since we know (from the properties of sums of independent Gaussian random variables) that the PDF for the sample value of y is

$$f_y(y_0) = \frac{1}{\sqrt{2\pi}\sqrt{10}} e^{-(y_0-10r)^2/(2\cdot 10)} \qquad -\infty \leq y_0 \leq \infty$$

In a significance test we work with the conditional PDF for our statistic, given that our hypothesis is true. For this example, we have

$$f_y(y_0) = \frac{1}{\sqrt{2\pi}\,\sqrt{10}} e^{-y_0^2/20} \qquad -\infty \le y_0 \le \infty$$

Assume that we have decided to test at the 0.05 level of significance and that, with no particular properties of the possible alternative hypotheses in mind, we choose to make the acceptance region for the significance test as small as possible. This leads to a rejection region of the form $|y| > A$. The following sketch applies,

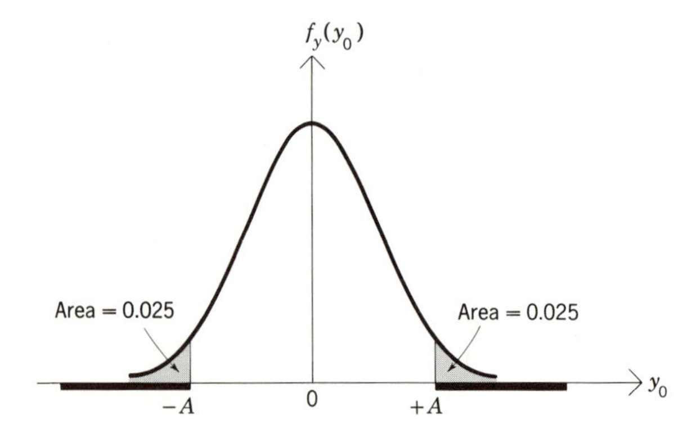

and A is determined by

$$\text{Prob}(y \ge A) = 0.025 = 1 - \Phi\left(\frac{A - 0}{\sqrt{10}}\right)$$

$$\Phi\left(\frac{A}{\sqrt{10}}\right) = 0.975 \qquad A \approx 6.2 \qquad \text{from table in Sec. 6-4}$$

Thus, at the 0.05 level, we shall reject our hypothesis that $E(x) = 0$ if it happens that the magnitude of the sum of the 10 experimental values of x is greater than 6.2.

We conclude this section with several brief comments on significance testing:

1 The use of different statistics, based on samples of the same size and the same experimental values, may result in different conclusions from the significance test, even if the acceptance regions for both statistics are made as small as possible (see Prob. 7.06).

2 In our examples, it happened that the only parameter in the PDF's for the statistics was the one whose value was specified by the hypothesis. In the above example, if σ_x^2 were not specified and we wished to make no assumptions about it, we would have had to *try* to find a statistic whose PDF depended on $E(x)$ but not on σ_x^2.

3 Even if the outcome of a significance test results in acceptance of the hypothesis, there are probably many other more accurate (and less accurate) hypotheses which would also be accepted as the result of similar significance tests upon them.

4 Because of the imprecise statement of the alternative hypotheses for a significance test, there is little we can say in general about the relative desirability of several possible statistics based on samples of the same size. One desires a statistic which, in its event space, *discriminates* as sharply as possible between his hypothesis and other hypotheses. In almost all situations, increasing the size of the sample will contribute to this discrimination.

5 The formulation of a significance test does not allow us to determine the a priori probability that a significance test will result in an incorrect conclusion. Even if we can agree to accept an a priori probability $P(H_0)$ that the hypothesis H_0 is true (before we undertake the test), we are still unable to evaluate the probability of an incorrect outcome of the significance test. Consider the following sequential event space picture for any significance test:

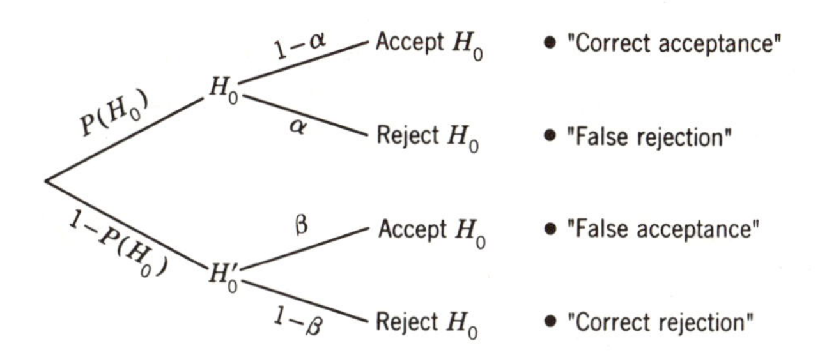

The lack of specific alternatives to H_0 prevents us from calculating a priori probabilities for the bottom two event points, even if we accept a value (or range of values) for $P(H_0)$. We have no way to estimate β, the conditional probability of acceptance of H_0 given H_0 is incorrect.

6 One value of significance testing is that it often leads one to discard particularly poor hypotheses. In most cases, statistics based on large enough samples are excellent for this purpose, and this is achieved with a rather small number of assumptions about the situation under study.

7-5 Parametric and Nonparametric Hypotheses

Two examples of significance tests were considered in the previous section. In both cases, the PDF for the statistic resulting from the model contained a parameter. In the first example, the parameter was P, the probability of success for a Bernoulli process. In the second example, the parameter of the PDF for the statistic was r, the expected value of a Gaussian random variable. The significance tests were performed on hypotheses which specified values for these parameters.

If, in effect, we assume the given form of a model and test hypotheses which specify values for parameters of the model, we say that we are testing *parametric* hypotheses. The hypotheses in both examples were parametric hypotheses.

Nonparametric hypotheses are of a broader nature, often with regard to the general form of a model or the form of the resulting PDF for the characteristic of interest. The following are some typical nonparametric hypotheses:

1 Characteristic x is normally distributed.
2 Random variables x and y have identical marginal PDF's, that is, $f_x(u) = f_y(u)$ for all values of u.
3 Random variables x and y have unequal expected values.
4 The variance of random variable x is greater than the variance of random variable y.

In principle, significance testing for parametric and nonparametric hypotheses follows exactly the same procedure. In practice, the determination of useful statistics for nonparametric tests is often a very difficult task. To be useful, the PDF's for such statistics must not depend on unknown quantities. Furthermore, one strives to make as few additional assumptions as possible before testing nonparametric hypotheses. Several nonparametric methods of great practical value, however, may be found in most elementary statistics texts.

7-6 Hypothesis Testing

The term *significance test* normally refers to the evaluation of a hypothesis H_0 in the absence of any useful information about alternative hypotheses. An evaluation of H_0 in a situation where the alternative hypotheses $H_1, H_2, \ldots$ are specified is known as a *hypothesis test.*

In this section we discuss the situation where it is known that there are only two possible parametric hypotheses $H_0(Q = Q_0)$ and $H_1(Q = Q_1)$. We are using Q to denote the parameter of interest.

To perform a hypothesis test, we select one of the hypotheses, H_0 (called the *null* hypothesis), and subject it to a significance test based on some statistic q. If the experimental value of statistic q falls into the critical (or rejection) region W, defined (as in Sec. 7-4) by

$$\text{Prob}(q \text{ in } W \mid H_0) = \text{Prob}(q \text{ in } W \mid Q = Q_0) = \alpha$$

we shall "reject" H_0 and "accept" H_1. Otherwise we shall accept H_0 and reject H_1. In order to discuss the choice of the "best" possible critical region W for a given statistic in the presence of a specific alternative hypothesis H_1, consider the two possible errors which may result from the outcome of a hypothesis test.

Suppose that H_0 were true. If this were so, the only possible error would be to reject H_0 in favor of H_1. The conditional probability of this type of error (called an *error of type* I, or *false rejection*) given H_0 is true is

$$\text{Prob}(\text{reject } H_0 \mid Q = Q_0) = \text{Prob}(q \text{ in } W \mid Q = Q_0) = \alpha$$

Suppose that H_0 is false and H_1 is true. Then the only type of error we could make would be to accept H_0 and reject H_1. The conditional probability of this type of error (called an *error of type* II, or *false acceptance*) given H_1 is true is

$$\text{Prob}(\text{accept } H_0 \mid Q = Q_1) = \text{Prob}(q \text{ not in } W \mid Q = Q_1) = \beta$$

It is important to realize that α and β are conditional probabilities which apply in different conditional event spaces. Furthermore, for *significance* testing (in Sec. 7-4) we did not know enough about the alternative hypotheses to be able to evaluate β. When we are concerned with a hypothesis test, this is no longer the case.

Let's return to the example of 10,000 coin tosses and a Bernoulli model of the process. Assume that we consider only the two alternative hypotheses $H_0(P = 0.5)$ and $H_1(P = 0.6)$. These hypotheses lead to two alternative conditional PDF's for k, the number of heads. We have

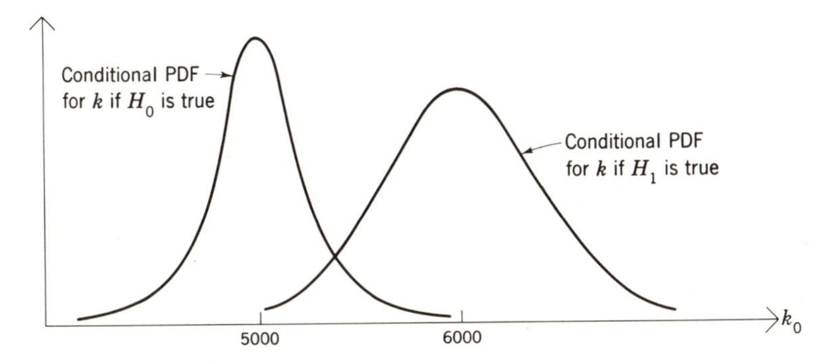

In this case, for any given α (the conditional probability of false rejection) we desire to select a critical region which will minimize β (the conditional probability of false acceptance). It should be clear that, for this example, the most desirable critical region W for a given α will be a continuous range of k on the right. For a given value of α, we may now identify α and β as areas under the conditional PDF's for k, as shown below:

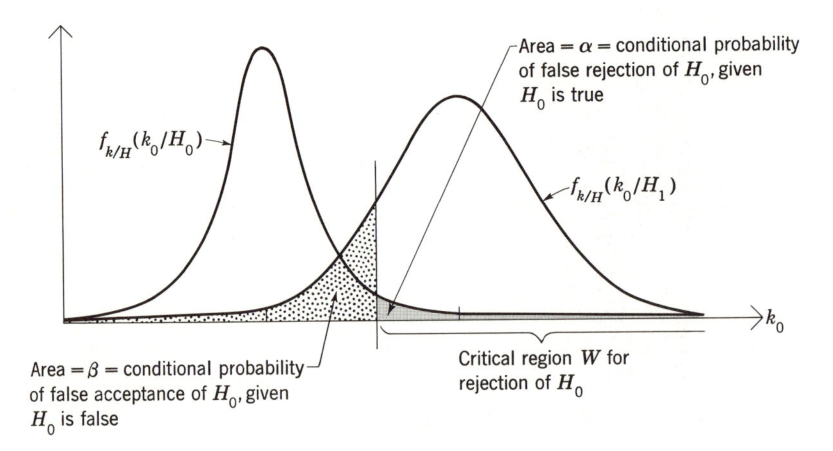

In practice, the selection of a pair of values α and β would usually depend on the relative costs of the two possible types of errors and some a priori estimate of the probability that H_0 is true (see Prob. 7.10).

Consider a sequential event space for the performance of a hypothesis test upon H_0 with one specific alternative hypothesis H_1:

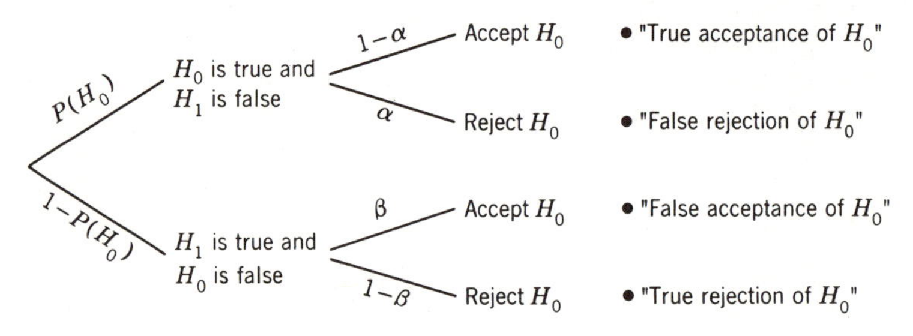

If we are willing to assign an a priori probability $P(H_0)$ to the validity of H_0, we may then state that the probability (to us) that this hypothesis test will result in an incorrect conclusion is equal to

$$\alpha P(H_0) + \beta[1 - P(H_0)]$$

Even if we are uncomfortable with any step which involves the assumption of $P(H_0)$, we may still use the fact that

$$0 \leq P(H_0) \leq 1$$

and the previous expression to obtain the bounds

$$\min(\alpha,\beta) \leq \text{Prob(incorrect conclusion)} \leq \max(\alpha,\beta)$$

We now comment on the selection of the statistic q. For any hypothesis test, a desirable statistic would be one which provides good discrimination between H_0 and H_1. For one thing, we would like the ratio

$$\frac{f_{q|H_0}(q_0 \mid H_0)}{f_{q|H_1}(q_0 \mid H_1)}$$

to be as large as possible in the acceptance region W' and to be as small as possible in the rejection region W. This would mean that, for any experimental value of statistic q, we would be relatively unlikely to accept the wrong hypothesis.

We might decide that the best statistic, q, is one which (for a given sample size of a given observable characteristic) provides the minimum β for any given α. Even when such a best statistic does exist, however, the derivation of the form of this best statistic and its conditional PDF's may be very difficult.

7-7 Estimation

Assume that we have developed the form of a model for a physical process and that we wish to determine the most desirable values for some parameters of this model. The general theory of using experimental data to estimate such parameters is known as the *theory of estimation.*

When we perform a hypothesis test with a rich set of alternatives, the validity of several suggested forms of a model may be under question. For our discussion of estimation, we shall take the viewpoint that the general form of our model is not to be questioned. We wish here only to estimate certain parameters of the process, given that the form of the model is correct. Since stating the form of the model is equivalent to stating the form of the PDF for characteristic x of the process, determining the parameters of the model is similar to adjusting the parameters of the PDF to best accommodate the experimental data.

Let $Q_n(x_1,x_2, \ldots ,x_n)$ be a statistic whose sample values are a function of a sample of size n and whose experimental values are a function of n independent experimental values of random variable x. Let Q be a parameter of our model or of its resulting PDF for random variable x. We shall be interested in those statistics Q_n whose experimental values happen to be good estimates of parameter Q. Such statistics are known as *estimators.*

Some examples of useful estimators follow. We might use the average value of n experimental values of x, given by

$$Q_n = \frac{1}{n}(x_1 + x_2 + \cdots + x_n) = M_n$$

as an estimate of the parameter $E(x)$. We have already encountered this statistic several times. [Although it is, alas, known as the *sample mean* (M_n), we must realize that, like any other statistic, it has both sample and experimental values. A similar comment applies to our next example of an estimator.] Another example of a statistic which may serve as an estimator is that of the use of the sample variance, given by

$$Q_n = \frac{1}{n}\sum_{i=1}^{n}(x_i - M_n)^2 = S_n^2$$

to estimate the variance of the PDF for random variable x. For a final example, we might use the maximum of n experimental values of x, given by

$$Q_n = \max(x_1,x_2, \ldots ,x_n)$$

to estimate the largest possible experimental value of random variable x.

Often we are able to suggest many reasonable estimators for a particular parameter Q. Suppose, for instance, that it is known that $f_x(x_0)$ is symmetric about $E(x)$, that is,

$$f_x[E(x) + a] = f_x[E(x) - a] \qquad \text{for all } a$$

and we wish to estimate $E(x)$ using some estimator $Q_n(x_1,x_2, \ldots ,x_n)$. We might use the estimator

$$Q_{n1} = \frac{1}{n}\sum_{i=1}^{n} x_i$$

or the estimator

$$Q_{n2} = \frac{\max(x_1,x_2, \ldots ,x_n) - \min(x_1,x_2, \ldots ,x_n)}{2}$$

or we could list the x_i in increasing order by defining

$$y_i = i\text{th smallest member of } (x_1,x_2, \ldots ,x_n)$$

and use for our estimator of $E(x)$ the statistic

$$Q_{n3} = \begin{cases} y_{(n+1)/2} & n \text{ odd} \\ \frac{1}{2}(y_{n/2} + y_{(n+2)/2}) & n \text{ even} \end{cases}$$

Any of these three estimators might turn out to be the most desirable, depending on what else is known about the form of $f_x(x_0)$ and also depending, of course, on our criterion for desirability.

In the following section, we introduce some of the properties relevant to the selection and evaluation of useful estimators.

7-8 Some Properties of Desirable Estimators

A sequence of estimates $Q_1, Q_2, \ldots$ of parameter Q is called *consistent* if it converges stochastically to Q as $n \to \infty$. That is, Q_n is a consistent estimator of Q if

$$\lim_{n\to\infty} \text{Prob}(|Q_n - Q| > \epsilon) = 0 \qquad \text{for any } \epsilon > 0$$

In Chap. 6, we proved that, given that σ_x^2 is finite, the sample mean M_n is stochastically convergent to $E(x)$. Thus, the sample mean is a consistent estimator of $E(x)$. If an estimator is known to be consistent, we would become confident of the accuracy of estimates based on very large samples. However, consistency is a limit property and may not be relevant for small samples.

A sequence of estimates $Q_1, Q_2, \ldots$ of parameter Q is called *unbiased* if the expected value of Q_n is equal to Q for all values

$$n = 1, 2, \ldots$$

That is, Q_n is an unbiased estimate for Q if

$$E(Q_n) = Q \qquad \text{for } n = 1, 2, \ldots$$

We noted (Sec. 7-3) that the sample mean M_n is an unbiased estimator for $E(x)$. We also noted that, for the expected value of the sample variance, we have

$$E(S_n^2) = \frac{n-1}{n}\sigma_x^2$$

and thus the sample variance is not an unbiased estimator of σ_x^2. However, it is true that

$$\lim_{n\to\infty} E(S_n^2) = \sigma_x^2$$

Any such estimator Q_n, which obeys

$$\lim_{n\to\infty} E(Q_n) = Q$$

is said to be an *asymptotically unbiased* estimator of Q. If Q_n is an unbiased (or asymptotically unbiased) estimator of Q, this property alone does not assure us of a good estimate when n is very large. We should also need some evidence that, as n grows, the PDF for Q_n becomes adequately concentrated near parameter Q.

The *relative efficiency* of two unbiased estimators is simply the ratio of their variances. We would expect that, the smaller the variance of an unbiased estimator Q_n, the more likely it is that an experimental value of Q_n will give an accurate estimate of parameter Q. We would say the *most efficient* unbiased estimator for Q is the unbiased estimator with the minimum variance.

We now discuss the concept of a *sufficient* estimator. Consider the n-dimensional sample space for the values $x_1, x_2, \ldots, x_n$. In general, when we go from a point in this space to the corresponding value of the estimator $Q_n(x_1,x_2, \ldots ,x_n)$, one of two things must happen. Given that our model is correct, either Q_n contains all the information in the experimental outcome $(x_1,x_2, \ldots ,x_n)$ relevant to the estimation of parameter Q, or it does not. For example, it is true for some estimation problems (and not for some others) that

$$Q_n = \sum_{i=1}^{n} x_i$$

contains all the information relevant to the estimation of Q which may be found in $(x_1,x_2, \ldots ,x_n)$. The reason we are interested in this matter is that we would expect to make the best use of experimental data by using estimators which take advantage of *all* relevant information in the data. Such estimators are known as sufficient estimators. The formal definition of sufficiency does not follow in a simple form from this intuitive discussion.

To state the mathematical definition of a sufficient estimator, we shall use the notation

$\lfloor x \rfloor = \lfloor x_1 \quad x_2 \quad \cdots \quad x_n \rfloor$ representing an n-dimensional random variable

$\lfloor x_0 \rfloor = \lfloor x_{10} \quad x_{20} \quad \cdots \quad x_{n0} \rfloor$ representing any particular value of $\lfloor x \rfloor$

Our model provides us with a PDF for $\lfloor x \rfloor$ in terms of a parameter Q which we wish to estimate. This PDF for $\lfloor x \rfloor$ may be written as

$$f_{\lfloor x \rfloor}(\lfloor x_0 \rfloor) = g(\lfloor x_0 \rfloor, Q) \qquad \text{where } g \text{ is a function only of } \lfloor x_0 \rfloor \text{ and } Q$$

If we are given the experimental value of estimator Q_n, this is at least partial information about $\lfloor x \rfloor$ and we could hope to use it to calculate the resulting conditional PDF for $\lfloor x \rfloor$,

$$f_{\lfloor x \rfloor | Q_n}(\lfloor x_0 \rfloor \mid Q_n) = h(\lfloor x_0 \rfloor, Q, Q_n)$$

where h is a function only of $\lfloor x_0 \rfloor$, Q, and Q_n. If and only if the PDF h does not depend on parameter Q after the value of Q_n is given, we *define* Q_n to be a sufficient estimator for parameter Q.

A few comments may help to explain the apparent distance between our simple intuitive notion of a sufficient statistic and the formal definition in the above paragraph. We are estimating Q because we do not know its value. Let us accept for a moment the notion that Q is (to us) a random variable and that our knowledge about it is given by some a priori PDF. When we say that a sufficient estimator Q_n will contain all the information about Q which is to be found in $(x_1,x_2, \ldots ,x_n)$, the implication is that the conditional PDF for Q, given Q_n, will be identical to the conditional PDF for Q, given the values $(x_1,x_2, \ldots ,x_n)$. Because classical statistics does not provide a framework for viewing our uncertainties about unknown constants in terms of such PDF's, the above definition has to be worked around to be in terms of other PDF's. Instead of stating that Q_n tells us everything about Q which might be found in $(x_1,x_2, \ldots ,x_n)$, our formal definition states that Q_n tells us everything about $(x_1,x_2, \ldots ,x_n)$ that we could find out by knowing Q.

In this section we have discussed the concepts of consistency, bias, relative efficiency, and sufficiency of estimators. We should also note that actual estimates are normally accompanied by *confidence limits*. The statistician specifies a quantity δ for which, given that his model is correct, the probability that the "random interval" $Q_n \pm \delta$ will fall such that it happens to include the true value of parameter Q is equal to some value such as 0.95 or 0.98. Note that it is the location of the interval centered about the experimental value of the estimator, and not the true value of parameter Q, which is considered to be the random phenomenon when one states confidence limits. We shall not explore the actual calculation of confidence limits in this text. Although there are a few special (simple) cases, the general problem is of an advanced nature.

7-9 Maximum-likelihood Estimation

There are several ways to obtain a desirable estimate for Q, an unknown parameter of a proposed statistical model. One method of estimation will be introduced in this section. A rather different approach will be indicated in our discussion of Bayesian analysis.

To use the method of *maximum-likelihood estimation*, we first obtain an experimental value for some sample $(x_1,x_2, \ldots ,x_n)$. We then determine which of all possible values of parameter Q maximizes the *a priori* probability of the observed experimental value of the sample (or of some statistic of the sample). Quantity Q^*, that possible value of Q which maximizes this a priori probability, is known as the maximum-likelihood estimator for parameter Q.

The a priori probability of the observed experimental outcome is calculated under the assumption that the model is correct. Before expanding on the above definition (which is somewhat incomplete) and commenting upon the method, we consider a simple example.

Suppose that we are considering a Bernoulli process as the model for a series of coin flips and that we wish to estimate parameter P, the probability of heads (or success), by the method of maximum-likelihood. Our experiment will be the performance of n flips of the coin and our sample $(x_1,x_2, \ldots ,x_n)$ represents the exact sequence of resulting Bernoulli random variables.

The a priori probability of any particular sequence of experimental outcomes which contains exactly k heads out of a total of n flips is given by

$$P^k(1 - P)^{n-k} \qquad k = 0, 1, \ldots , n$$

To find P^*, the maximum-likelihood estimator for P, we use elementary calculus to determine which value of P, in the range $0 \leq P \leq 1$, maximizes the above a priori probability for any experimental value of k. Differentiating with respect to P, setting the derivative equal to zero, and checking that we are in fact maximizing the above expression, we finally obtain

$$P^* = \frac{k}{n}$$

which is the maximum-likelihood estimator for parameter P if we observe exactly k heads during the n trials.

In our earlier discussion of the Bernoulli law of large numbers (Sec. 6-3) we established that this particular maximum-likelihood estimator satisfies the definition of a consistent estimator. By performing the calculation

$$E(P^*) = E\left(\frac{k}{n}\right) = \frac{1}{n}E(k) = \frac{nP}{n} = P$$

we find that this estimator is also unbiased.

Note also that, for this example, maximum-likelihood estimation based on either of two different statistics will result in the same expression for P^*. We may use an n-dimensional statistic (the sample itself) which is a finest-grain description of the experimental outcome or we may use the alternative statistic k, the number of heads observed. (It happens that k/n is a sufficient estimator for parameter P of a Bernoulli process.)

We now make a necessary expansion of our original definition of maximum-likelihood estimation. If the model under consideration results in a continuous PDF for the statistic of interest, the probability associated with any particular experimental value of the statistic is zero. For this case, let us, for an n-dimensional statistic, view the problem in an n-dimensional event space whose coordinates represent the n components of the statistic. Our procedure will be to determine that possible value of Q which maximizes the a priori probability of the event represented by an n-dimensional incremental cube, centered about the point in the event space which represents the observed experimental value of the statistic.

The procedure in the preceding paragraph is entirely similar to the procedure used earlier for maximum-likelihood estimation when the statistic is described by a PMF. For the continuous case, we work with incremental events centered about the event point representing the observed experimental outcome. The result can be restated in a simple manner. If the statistic employed is described by a continuous PDF, we maximize the appropriate *PDF evaluated at,* rather than the *probability of,* the observed experimental outcome.

As an example, suppose that our model for an interarrival process is that the process is Poisson. This assumption models the first-order interarrival times as independent random variables, each with PDF

$$f_x(x_0) = \lambda e^{-\lambda x_0} \qquad x_0 > 0$$

In order to estimate λ, we shall consider a sample (r,s,t,u,v) composed of five independent values of random variable x. Our statistic is the sample itself. The compound PDF for this statistic is given by

$$f_{r,s,t,u,v}(r_0,s_0,t_0,u_0,v_0)$$
$$= \begin{cases} \lambda e^{-\lambda r_0}\lambda e^{-\lambda s_0}\lambda e^{-\lambda t_0}\lambda e^{-\lambda u_0}\lambda e^{-\lambda v_0} & \text{if } r_0,\ s_0,\ t_0,\ u_0,\ v_0 \geq 0 \\ 0 & \text{otherwise} \end{cases}$$
$$= \begin{cases} \lambda^5 e^{-\lambda(r_0+s_0+t_0+u_0+v_0)} & \text{if } r_0,\ s_0,\ t_0,\ u_0,\ v_0 \geq 0 \\ 0 & \text{otherwise} \end{cases}$$

Maximization of this PDF with respect to λ for any particular experimental outcome (r_0,s_0,t_0,u_0,v_0) leads to the maximum-likelihood estimator

$$\lambda^* = \frac{5}{r_0 + s_0 + t_0 + u_0 + v_0}$$

which seems reasonable, since this result states that the maximum-likelihood estimator of the average arrival rate happens to be equal to the experimental value of the average arrival rate. [We used the (r,s,t,u,v) notation instead of $(x_1,x_2, \ldots ,x_n)$ to enable us to write out the compound PDF for the sample in our more usual notation.]

Problem 7.15 assists the reader to show that λ^* is a consistent estimator which is biased but asymptotically unbiased. It also happens that λ^* (the number of interarrival times divided by their sum) is a sufficient estimator for parameter λ.

In general, maximum-likelihood estimators can be shown to have a surprising number of useful properties, both with regard to theoretical matters and with regard to the simplicity of practical application of the method. For situations involving very large samples, there are few people who disagree with the reasoning which gives rise to this arbitrary but most useful estimation technique.

However, serious problems do arise if one attempts to use this estimation technique for decision problems involving small samples or if one attempts to establish that maximum likelihood is a truly *fundamental* technique involving fewer assumptions than other methods of estimation.

Suppose that we have to make a large wager based on the true value of P in the above coin example. There is time to flip the coin only five times, and we observe four heads. Very few people would be willing to use the maximum-likelihood estimate for P, $\frac{4}{5}$, as their estimator for parameter P if there were large stakes involved in the accuracy of their estimate. Since maximum likelihood depends on a simple maximization of an *unweighted* PDF, there seems to be an uncomfortable implication that *all* possible values of parameter P were equally likely *before* the experiment was performed. We shall return to this matter in our discussion of Bayesian analysis.

7-10 Bayesian Analysis

A *Bayesian* believes that any quantity whose value he does not know is (to him) a random variable. He believes that it is possible, at any time, to express his state of knowledge about such a random variable in the form of a PDF. As additional experimental evidence becomes available, Bayes' theorem is used to combine this evidence with the previous PDF in order to obtain a new a posteriori PDF representing his updated state of knowledge. The PDF expressing the analyst's state of knowledge serves as the quantitative basis for any *decisions* he is required to make.

Consider the Bayesian analysis of Q, an unknown parameter of a postulated probabilistic model of a physical system. We assume that the outcomes of experiments with the system may be described by the resulting experimental values of continuous random variable x, the characteristic of interest.

Based on past experience and all other available information, the Bayesian approach begins with the specification of a PDF $f_Q(Q_0)$, the analyst's a priori PDF for the value of parameter Q. As before, the model specifies the PDF for the sample value of characteristic x, given the value of parameter Q. Since we are now regarding Q as another random variable, the PDF for the sample value of x with parameter Q is to be written as the conditional PDF,

$f_{x|Q}(x_0 \mid Q_0)$ = conditional PDF for the sample value of characteristic x, given that the value of parameter Q is equal to Q_0

Each time an experimental value of characteristic x is obtained, the continuous form of Bayes' theorem

$$f_{Q|x}(Q_0 \mid x_0) = \frac{f_{x,Q}(x_0,Q_0)}{f_x(x_0)} = \frac{f_{x|Q}(x_0 \mid Q_0) f_Q(Q_0)}{\int_{Q_0} f_{x|Q}(x_0 \mid Q_0) f_Q(Q_0)\, dQ_0} \equiv f'_Q(Q_0)$$

is used to obtain the a posteriori PDF $f'_Q(Q_0)$, describing the analyst's new state of knowledge about the value of parameter Q. This PDF $f'_Q(Q_0)$ serves as the basis for any present decisions and also as the *a priori* PDF for any future experimentation with the physical system.

The Bayesian analyst utilizes his state-of-knowledge PDF to resolve issues such as:

1 Given a function $C(Q^t,Q^*)$, which represents the penalty associated with estimating Q^t, the true value of parameter Q, by an estimate Q^*, determine that estimator Q^* which minimizes the expected value of $C(Q^t,Q^*)$. (For example, see Prob. 2.05.)

2 Given the function $C(Q^t,Q^*)$, which represents the cost of imperfect estimation, and given another function which represents, as a function of n, the cost of n repeated experiments on the physical system, specify the experimental test program which will minimize the expected value of the total cost of experimentation and estimation.

As one example of Bayesian analysis, assume that a Bernoulli model has been accepted for a coin-flipping process and that we wish to investigate parameter P, the probability of success (heads) for this model. We shall discuss only a few aspects of this problem. One should keep in mind that there is probably a cost associated with each flip of the coin and that our general objective is to combine our prior convictions about the value of P with some experimental evidence to obtain a suitably accurate and economical estimate P^*.

The Bayesian analyst begins by stating his entire assumptive structure in the form of his a priori PDF $f_P(P_0)$. Although this is necessarily an inexact and somewhat arbitrary specification, no estimation procedure, classical or Bayesian, can avoid this (or an equivalent) step. We continue with the example, deferring a more general discussion to Sec. 7-12.

Four of many possible choices for $f_P(P_0)$ are shown below:

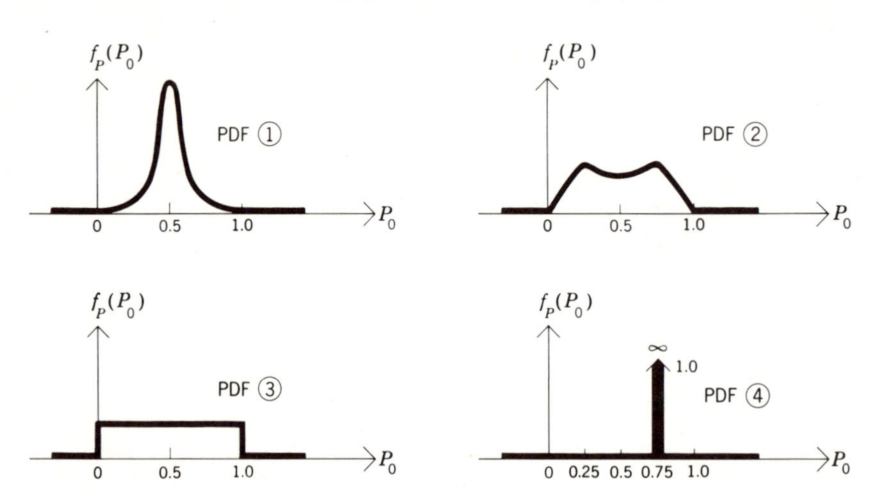

A priori PDF ① could represent the prior convictions of one who believes, "Almost all coins are fair or very nearly fair, and I don't see anything special about this coin." If it is believed that the coin is probably biased, but the direction of the bias is unknown, PDF ② might serve as $f_P(P_0)$. There might be a person who claims, "I don't know anything about parameter P, and the least biased approach is represented by PDF ③." Finally, PDF ④ is the a priori state of knowledge for a person who is certain that the value of P is equal to 0.75. In fact, since PDF ④ allocates all its probability to this single possible value of P, there is nothing to be learned from experimentation. For PDF ④, the a posteriori PDF will be identical to the a priori PDF, no matter what experimental outcomes may be obtained.

Because the design of complete test programs is too involved for our introductory discussion, assume that some external considerations have dictated that the coin is to be flipped exactly N_0 times. We wish to see how the experimental results (exactly K_0 heads in N_0 tosses) are used to update the original a priori PDF $f_P(P_0)$.

The Bernoulli model of the process leads us to the relation

$$p_{K|P}(K_0 \mid P_0) = \binom{N_0}{K_0} P_0{}^{K_0}(1 - P_0)^{N_0 - K_0} \qquad K_0 = 0, 1, 2, \ldots, N_0$$

where we are using a PMF because of the discrete nature of K, the characteristic of interest. The equation for using the experimental outcome to update the a priori PDF $f_P(P_0)$ to obtain the a posteriori PDF $f'_P(P_0)$ is thus, from substitution into the continuous form of Bayes' theorem, found to be,

$$f'_P(P_0) = f_{P|K}(P_0 \mid K_0) = \frac{\binom{N_0}{K_0} P_0^{K_0}(1 - P_0)^{N_0-K_0} f_P(P_0)}{\int_{P_0=0}^{1} \binom{N_0}{K_0} P_0^{K_0}(1 - P_0)^{N_0-K_0} f_P(P_0)\, dP_0}$$

and we shall continue our consideration of this relation in the following section.

In general, we would expect that, the narrower the a priori PDF $f_p(P_0)$, the more the experimental evidence required to obtain an a posteriori PDF which is appreciably different from the a priori PDF. For very large amounts of experimental data, we would expect the effect of this evidence to dominate all but the most unreasonable a priori PDF's, with the a posteriori PDF $f'_P(P_0)$ becoming heavily concentrated near the true value of parameter P.

7-11 Complementary PDF's for Bayesian Analysis

For the Bayesian analysis of certain parameters of common probabilistic processes, such as the situation in the example of Sec. 7-10, some convenient and efficient procedures have been developed.

The general calculation for an a posteriori PDF is unpleasant, and although it may be performed for any a priori PDF, it is unlikely to yield results in a useful form. To simplify his computational burden, the Bayesian often takes advantage of the obviously imprecise specification of his a priori state of knowledge. In particular, he elects that, whenever it is possible, he will select the a priori PDF from a family of PDF's which has the following three properties:

1. The family should be rich enough to allow him to come reasonably close to a statement of his subjective state of knowledge.
2. Individual members of the family should be determined by specifying the value of a few parameters. It would not be realistic to pretend that the a priori PDF represents very precise information.
3. The family should make the above updating calculation as simple as possible. In particular, if one member of the family is used as the a priori PDF, then, for any possible experimental outcome, the resulting a posteriori PDF should simply be another member of the family. One should be able to carry out the updating calculation by merely using the experimental results to modify the parameters of the a priori PDF to obtain the a posteriori PDF.

The third item in the above list is clearly a big order. We shall not investigate the existence and derivation of such families here. However, when families of PDF's with this property do exist for the estimation of parameters of probabilistic processes, such PDF's are said to be *complementary* (or *conjugate*) PDF's for the process being studied. A demonstration will be presented for the example of the previous section.

Consider the *beta* PDF for random variable P with parameters k_0 and n_0. It is convenient to write this PDF as $\mathcal{B}_P(P_0 \mid k_0, n_0)$, defined by

$$\mathcal{B}_P(P_0 \mid k_0, n_0) = C(k_0,n_0)P_0^{k_0-1}(1 - P_0)^{n_0-k_0-1} \qquad \begin{cases} 0 \le P_0 \le 1 \\ k_0 \ge 0 \\ n_0 \ge k_0 \end{cases}$$

where $C(k_0,n_0)$ is simply the normalization constant

$$C(k_0,n_0) = \left[\int_{P_0=0}^{1} P_0^{k_0-1}(1 - P_0)^{n_0-k_0-1}\, dP_0\right]^{-1}$$

and several members of this family of PDF's are shown below:

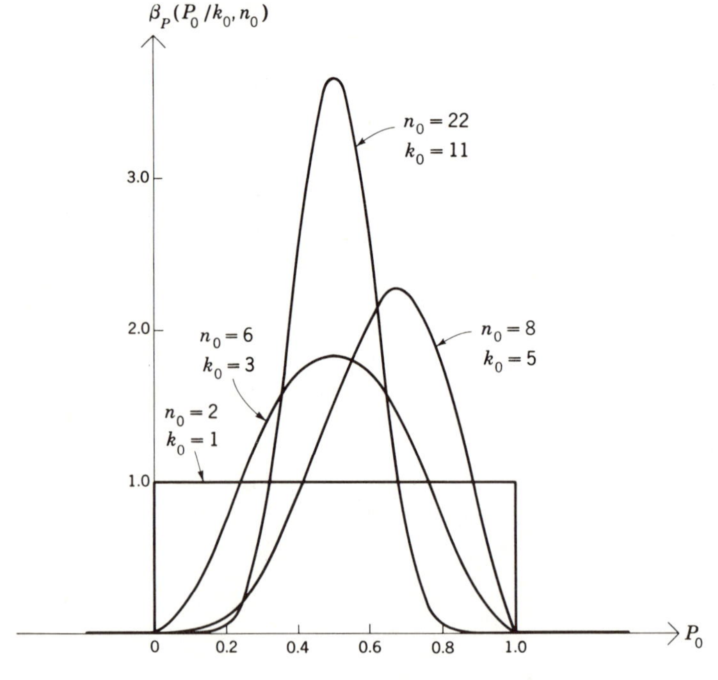

An individual member of this family may be specified by selecting values for its mean and variance rather than by selecting constants k_0 and n_0 directly. Although techniques have been developed to allow far more structured PDF's, the Bayesian often finds that these two parameters $E(P)$ and σ_P^2 allow for an adequate expression of his prior beliefs about the unknown parameter P of a Bernoulli model.

Direct substitution into the relation for $f'_P(P_0)$, the a posteriori

PDF for our example, establishes that if the Bayesian starts out with the a priori PDF

$$f_P(P_0) = \mathcal{B}_P(P_0 \mid k_0, n_0)$$

and then observes exactly K_0 successes in N_0 Bernoulli trials, the resulting a posteriori PDF is

$$f'_P(P_0) = f_{P|K}(P_0 \mid K_0) = \mathcal{B}_P(P_0 \mid k_0 + K_0, n_0 + N_0)$$

Thus, for the estimation of parameter P for a Bernoulli model, use of a beta PDF for $f_P(P_0)$ allows the a posteriori PDF to be determined by merely using the experimental values K_0 and N_0 to modify the parameters of the a priori PDF. Using the above sketch, we see, for instance, that, if $f_P(P_0)$ were the beta PDF with $k_0 = 3$ and $n_0 = 6$, an experimental outcome of two successes in two trials would lead to the a posteriori beta PDF with $k_0 = 5$ and $n_0 = 8$.

It is often the case, as it is for our example, that the determination of parameters of the a priori PDF can be interpreted as assuming a certain "equivalent past experience." For instance, if the cost structure is such that we shall choose our best estimate of parameter P to be the expectation of the a posteriori PDF, the resulting estimate of parameter P, which we call P^*, turns out to be

$$P^* = \frac{K_0 + k_0}{N_0 + n_0}$$

This same result could have been obtained by the method of maximum-likelihood estimation, had we agreed to combine a bias of k_0 successes in n_0 hypothetical trials with the actual experimental data.

Finally, we remark that the use of the beta family for estimating parameter P of a Bernoulli process has several other advantages. It renders quite simple the otherwise most awkward calculations for what is known as *preposterior analysis.* This term refers to an exploration of the nature of the a posteriori PDF and its consequences *before* the tests are performed. It is this feature which allows one to optimize a test program and design effective experiments without becoming bogged down in hopelessly involved detailed calculations.

7-12 Some Comments on Bayesian Analysis and Classical Statistics

There is a large literature, both mathematical and philosophical, dealing with the relationship between classical statistics and Bayesian analysis. In order to indicate some of the considerations in a relatively brief manner, some imprecise generalizations necessarily appear in the following discussion.

The Bayesian approach represents a significant departure from the more conservative classical techniques of statistical analysis. Classical techniques are often particularly appropriate for purely scientific investigations and for matters involving large samples. Classical procedures attempt to require the least severe possible assumptive structure on the part of the analyst. Bayesian analysis involves a more specific assumptive structure and is often described as being *decision-oriented.* Some of the most productive applications of the Bayesian approach are found in situations where prior convictions and a relatively small amount of experimentation must be combined in a rational manner to make decisions among alternative future courses of action.

There is appreciable controversy about the degree of the difference between classical and Bayesian statistics. The Bayesian states his entire assumptive structure in his a priori PDF; his methods require no further arbitrary steps once this PDF is specified. It is true that he is often willing to state a rather sharp a priori PDF which heavily weights his prior convictions. But the Bayesian also points out that all statistical procedures of any type involve similar (although possibly weaker) statements of prior convictions. The assumptive structures of classical statistics are less visible, being somewhat submerged in established statistical tests and the choice of statistics.

Any two Bayesians would begin their analyses of the same problem with somewhat different a priori PDF's. If their work led to conflicting terminal decisions, their different assumptions are apparent in their a priori PDF's and they have a clear common ground for further discussions. The common ground between two different classical procedures which result in conflicting advice tends to be less apparent.

Objection is frequently made to the arbitrary nature of the a priori PDF used by the Bayesian. One frequently hears that this provides an arbitrary bias to what might otherwise be a scientific investigation. The Bayesian replies that all tests involve a form of bias and that he prefers that *his* bias be rational. For instance, in considering the method of maximum likelihood for the estimation of parameter P of a Bernoulli process, we noted the implication that all possible values of P were equally likely before the experiments. Otherwise, the method of maximum likelihood would maximize a *weighted* form of that function of P which represents the a priori probability of the observed experimental outcome.

Continuing this line of thought, the Bayesian contends that, for anybody who has ever seen a coin, how could any bias be less rational than that of a priori PDF ③ in the example of Sec. 7-10? Finally, he would note that there is nothing fundamental in starting out with a uniform PDF over the possible values of P as a manifestation of "minimum bias." Parameter P is but one arbitrary way to characterize the process; other parameters might be, for example,

$$U = \frac{1}{P} \qquad V = P^2 + P \ln (P + 1)$$

and professing that all possible values of one of these parameters be equally likely would lead to different results from those obtained by assuming the uniform PDF over all possible values of parameter P. The Bayesian believes that, since it is impossible to avoid bias, one can do no better than to assume a rational rather than naïve form of bias.

We should remark in closing that, because we considered a particularly simple estimation problem, we had at our disposal highly developed Bayesian procedures. For multivariate problems or for tests of nonparametric hypotheses, useful Bayesian formulations do not necessarily exist.

PROBLEMS

7.01 Random variable M_n, the sample mean, is defined to be the average value of n independent experimental values of random variable x. Determine the *exact* PDF (or PMF) for M_n and its expected value and variance if:

a $f_x(x_0) = \dfrac{\lambda^k x_0{}^{k-1} e^{-\lambda x_0}}{(k-1)!} \qquad k = 1, 2, 3, \ldots ; \quad x_0 \geq 0$

b $f_x(x_0) = \dfrac{1}{4\sqrt{2\pi}} e^{-(x_0-8)^2/32} \qquad -\infty \leq x_0 \leq \infty$

c $p_x(x_0) = \dfrac{\mu^{x_0} e^{-\mu}}{x_0!} \qquad x_0 = 0, 1, 2, \ldots$

d $p_x(x_0) = P(1-P)^{x_0-1} \qquad x_0 = 1, 2, 3, \ldots$

7.02 Our model for a process states that x is a random variable described by the PDF

$$f_x(x_0) = \begin{cases} 1 & \text{if } r < x_0 \leq r + 1 \\ 0 & \text{otherwise} \end{cases}$$

and we do not know the value of r. For the following questions, assume that the form of our model is correct.

a We may use the average value of 48 independent experimental values of random variable x to estimate the value of r from the relation

$$M_n \approx E(x) = \int_r^{r+1} x_0 f_x(x_0)\, dx_0 = r + \tfrac{1}{2}$$

What is the probability that our estimate of r obtained in this way will be within ± 0.01 of the true value? Within ± 0.05 of the true value?

b We may use the largest of our 48 experimental values as our estimate of the quantity $r + 1$, thus obtaining another estimate of the value of parameter r. What is the probability that our estimate of r obtained this way is within $(+0, -0.02)$ of the true value? Within $(+0, -0.10)$ of the true value?

7.03 **a** Use methods similar to those of Sec. 7-3 to derive a reasonably simple expression for the variance of the sample variance.

b Does the sequence of sample variances $(S_1^2, S_2^2, \ldots)$ for a Gaussian random variable obey the weak law of large numbers? Explain.

7.04 There are 240 students in a literature class ("Proust, Joyce, Kafka, and Mickey Spillane"). Our model states that x, the numerical grade for any individual student, is an independent Gaussian random variable with a standard deviation equal to $10\sqrt{2}$. Assuming that our model is correct, we wish to perform a significance test on the hypothesis that $E(x)$ is equal to 60.

Determine the highest and lowest class averages which will result in the acceptance of this hypothesis:

a At the 0.02 level of significance

b At the 0.50 level of significance

7.05 We have accepted a Bernoulli model for a certain physical process involving a series of discrete trials. We wish to perform a significance test on the hypothesis that P, the probability of success on any trial, is equal to 0.50. Determine the rejection region for tests at the 0.05 level of significance if we select as our statistic

a Random variable r, the number of trials up to and including the 900th success

b Random variable s, the number of successes achieved in a total of 1,800 trials

The expected number of coin flips for each of these significance tests is equal. Discuss the relative merits of these tests. Consider the two ratios $\sigma_r/E(r)$ and $\sigma_s/E(s)$. Is the statistic with the smaller standard-deviation to expected-value ratio necessarily the better statistic?

7.06 Random variable x is known to be described by the PDF

$$f_x(x_0) = \begin{cases} \frac{1}{A} & \text{if } 0 < x_0 \leq A \\ 0 & \text{otherwise} \end{cases}$$

but we do not know the value of parameter A. Consider the following statistics, each of which is based on a set of five independent experimental values $(x_1, x_2, \ldots, x_5)$ of random variable x:

$$r = 0.2(x_1 + x_2 + \cdots + x_5)$$
$$s = \max(x_1, x_2, \ldots, x_5)$$
$$t = 0.5(x_1 + x_2)$$

We wish to test the hypothesis $A = 2.0$ at the 0.5 level of significance. (A significance test using statistic r, for example, is referred to as T_r.)

Without doing too much work, can you suggest possible values of the data $(x_1, x_2, \ldots, x_5)$ which would result in:

a Acceptance only on T_r (and rejection on T_s and T_t)? Acceptance only on T_s? Acceptance only on T_t?

b Rejection only on T_r (and acceptance on T_s and T_t)? Rejection only on T_s? Rejection only on T_t?

c Acceptance on all three tests? Rejection on all three tests?

If the hypothesis is accepted on all three tests, does that mean it has passed an equivalent single significance test at the $1 - (0.5)^3$ level of significance?

7.07 Al, the bookie, plans to place a bet on the number of the round in which Bo might knock out Ci in their coming (second) fight. Al assumes only the following details for his model of the fight:

1 Ci can survive exactly 50 solid hits. The 51st solid hit (if there is one) finishes Ci.

2 The times between solid hits by Bo are independent random variables with the PDF

$$f_t(t_0) = \lambda e^{-\lambda t_0} \qquad t_0 \geq 0$$

3 Each round is three minutes.

Al hypothesizes that $\lambda = \frac{1}{22}$ (hits per second). Given the result of the previous fight (Ci won), at what significance level can Al accept his hypothesis $H_0(\lambda = \frac{1}{22})$? In the first fight Bo failed to come out for round 7—Ci lasted at least six rounds. Discuss any additional assumptions you make.

7.08 We are sure that the individual grades in a class are normally distributed about a mean of 60.0 and have standard deviation σ equal to either 5.0 or 8.0. Consider a hypothesis test of the null hypothesis $H_0(\sigma = 5.0)$ with a statistic which is the experimental value of a single grade.

a Determine the acceptance region for H_0 if we wish to set the conditional probability of false rejection (the level of significance) at 0.10.

b For the above level of significance and critical region, determine the conditional probability of acceptance of H_0, given $\sigma = 8.0$.

c How does increasing the number of experimental values averaged in the statistic contribute to your confidence in the outcome of this hypothesis test?

d Suggest some appropriate statistics for a hypothesis test which is intended to discriminate between $H_0(\sigma = 5.0)$ and $H_1(\sigma = 8.0)$.

e If we use H_1 as a model for the grades, what probability does it allot to grades less than 0 or greater than 100?

7.09 A random variable x is known to be characterized by either a Gaussian PDF with $E(x) = 20$ and $\sigma_x = 4$ or by a Gaussian PDF with $E(x) = 25$ and $\sigma_x = 5$. Consider the null hypothesis $H_0[E(x) = 20, \sigma_x = 4]$. We wish to test H_0 at the 0.05 level of significance. Our statistic is to be the sum of three experimental values of random variable x.

a Determine the conditional probability of false acceptance of H_0.

b Determine the conditional probability of false rejection of H_0.

c Determine an upper bound on the probability that we shall arrive at an incorrect conclusion from this hypothesis test.

d If we agree that one may assign an a priori probability of 0.6 to the event that H_0 is true, determine the probabilities that this hypothesis test will result in:

i False acceptance of H_0

ii False rejection of H_0

iii An incorrect conclusion

7.10 A random variable x is known to be the sum of k independent identically distributed exponential random variables, each with an expected value equal to $(k\lambda)^{-1}$. We have only two hypotheses for the value of parameter k; these are $H_0(k = 64)$ and $H_1(k = 400)$. Before we obtain any experimental data, our a priori guess is that these two hypotheses are equally likely.

The statistic for our hypothesis test is to be the sum of four independent experimental values of x. We estimate that false acceptance of H_0 will cost us \$100, false rejection of H_0 will cost us \$200, and any correct outcome of the test is worth \$500 to us.

Determine approximately the rejection region for H_0 which maximizes the expected value of the outcome of this hypothesis test.

7.11 A Bernoulli process satisfies either $H_0(P = 0.5)$ or $H_1(P = 0.6)$. Using the number of successes observed in n trials as our statistic, we wish to perform a hypothesis test in which α, the conditional probability of false rejection of H_0, is equal to 0.05. What is the smallest value of n for which this is the case if β, the conditional probability of false acceptance of H_0, must also be no greater than 0.05?

7.12 A hypothesis test based on the statistic

$$M_n = \frac{x_1 + x_2 + \cdots + x_n}{n}$$

is to be used to choose between two hypotheses

$$H_0[E(x) = 0, \sigma_x = 2] \qquad H_1[E(x) = 1, \sigma_x = 4]$$

for the PDF of random variable x which is known to be Gaussian.

a Make a sketch of the possible points (α,β) in an α,β plane for the cases $n = 1$ and $n = 4$. (α and β are, respectively, the conditional probabilities of false rejection and false acceptance.)

b Sketch the ratio of the two conditional PDF's for random variable M_n (given H_0, given H_1) as a function of M_n for the cases $n = 1$ and $n = 4$. Discuss the properties of a desirable statistic that might be exhibited on such a plot.

7.13 Expanding on the statement of Prob. 7.06, consider the statistic

$$s_n = \max(x_1, x_2, \ldots, x_n)$$

as an estimator of parameter A.

a Is this estimator biased? Is it asymptotically biased?

b Is this estimator consistent?

c Carefully determine the maximum-likelihood estimator for A, based only on the experimental value of the statistic s_n.

7.14 Suppose that we flip a coin until we observe the lth head. Let n be the number of trials up to and including the lth head. Determine the maximum-likelihood estimator for P, the probability of heads. Another experiment would involve flipping the coin n (a predetermined number) times and letting the random variable be l, the number of heads in the n trials. Determine the maximum-likelihood estimator for P for the latter experiment. Discuss your results.

7.15 We wish to estimate λ for a Poisson process. If we let $(x_1,x_2, \ldots ,x_n)$ be independent experimental values of n first-order interarrival times, we find (Sec. 7-9) that λ_n^*, the maximum-likelihood estimator for λ, is given by

$$\lambda_n^* = n \left(\sum_{i=1}^{n} x_i \right)^{-1}$$

a Show that $E(\lambda_n^*) = n\lambda/(n - 1)$.

b Determine the exact value of the variance of random variable λ_n^* as a function of n and λ.

c Is λ_n^* a biased estimator for λ? Is it asymptotically biased?

d Is λ_n^* a consistent estimator for λ?

e Based on what we know about λ_n^*, can you suggest a desirable unbiased consistent estimator for λ?

Another type of maximum-likelihood estimation for the parameter λ of a Poisson process appears in the following problem.

7.16 Assume that it is known that occurrences of a particular event constitute a Poisson process in time. We wish to investigate the parameter λ, the average number of arrivals per minute.

a In a predetermined period of T minutes, exactly n arrivals are observed. Derive the maximum-likelihood estimator λ^* for λ based on this data.

b In 10,000 minutes 40,400 arrivals are observed. At what significance level would the hypothesis $\lambda = 4$ be accepted?

c Prove that the maximum-likelihood estimator derived in (a) is an unbiased estimator for λ.

d Determine the variance of λ^*.

e Is λ^* a consistent estimator for λ?

7.17 The volumes of gasoline sold in a month at each of nine gasoline stations may be considered independent random variables with the PDF

$$f_v(v_0) = \frac{1}{\sqrt{2\pi}\,\sigma_v} e^{-[v_0 - E(v)]^2/2\sigma_v^2} \qquad -\infty \leq v_0 \leq +\infty$$

a Assuming that $\sigma_v = 1$, find E^*, the maximum-likelihood estimator for $E(v)$ when we are given only V, the total gasoline sales for all nine stations, for a particular month.

b Without making any assumptions about σ_v, determine σ_v^* and E^*, the maximum-likelihood estimators for σ_v and $E(v)$.

c Is the value of E^* in (b) an unbiased estimator for $E(v)$?

7.18 Consider the problem of estimating the parameter P (the probability of heads) for a particular coin. To begin, we agree to assume the following a priori probability mass function for P:

$$p_P(P_0) = \begin{cases} 0.1 & P_0 = 0.4 \\ 0.8 & P_0 = 0.5 \\ 0.1 & P_0 = 0.6 \end{cases}$$

We are now told that the coin was flipped n times. The first flip resulted in heads, and the remaining $n - 1$ flips resulted in tails.

Determine the a posteriori PMF for P as a function of n for $n \geq 2$. Prepare neat sketches of this function for $n = 2$ and for $n = 5$.

7.19 Given a coin from a particular source, we decide that parameter P (the probability of heads) for a toss of this coin is (to us) a random variable with probability density function

$$f_P(P_0) = \begin{cases} K(1 - P_0)^8 P_0{}^8 & \text{if } 0 \leq P_0 \leq 1 \\ 0 & \text{otherwise} \end{cases}$$

We proceed to flip the coin 10 times and note an experimental outcome of six heads and four tails. Determine, within a normalizing constant, the resulting a posteriori PDF for random variable P.

7.20 Consider a Bayesian estimation of λ, the unknown average arrival rate for a Poisson process. Our state of knowledge about λ leads us to describe it as a random variable with the PDF

$$f_\lambda(\lambda_0) = \frac{a^k \lambda_0{}^{k-1} e^{-a\lambda_0}}{(k-1)!} \qquad \lambda_0 \geq 0$$

where k is a positive integer.

a If we observe the process for a predetermined interval of T units of time and observe exactly N arrivals, determine the a posteriori PDF for random variable λ. Speculate on the general behavior of this PDF for very large values of T.

b Determine the expected value of the a priori and a posteriori PDF's for λ. Comment on your results.

c Before the experiment is performed, we are required to give an estimate λ_G for the true value of λ. We shall be paid $100 - 500(\lambda_G - \lambda)^2$ dollars as a result of our guess. Determine the value of λ_G which maximizes the expected value of the guess.

APPENDIX ONE

further reading

This appendix presents several suggestions for further reading, including a few detailed references. Only a few works, all of relatively general interest, are listed. Unless stated otherwise, the books below are at a level which should be accessible to the reader of this text. No attempt has been made to indicate the extensive literature pertaining to particular fields of application.

References are listed by author and date, followed by a brief description. Complete titles are tabulated at the end of this appendix. For brevity, the present volume is referred to as FAPT in the annotations.

1 *Some Relevant Philosophy and the History of Probability Theory*

DAVID (1962) Engaging history of some of the earliest developments in probability theory. Attention is also given to the colorful personalities involved.

KYBURG AND SMOKLER (1963) Essays by Borel, de Finetti, Koopman, Ramsey, Savage, and Venn on a matter of significance in applied probability theory, the topic of *subjective* probability.

LAPLACE (1825) Interesting discussions of philosophical issues related to probability theory and its applications to real world issues.

TODHUNTER (1865) The classic reference for the early history of probability theory.

2 *Introductory Probability Theory and Its Applications*

FELLER (1957) Thorough development of the discrete case with a vast supply of interesting topics and applications. Contains a large body of fundamental material on combinatorial analysis and the use of transforms in the study of discrete renewal processes which is not included in FAPT.

FISZ (1963) Large, scholarly, and relatively complete text treating probability theory and classical mathematical statistics. Tightly written. A very desirable reference work.

GNEDENKO (1962) Respected text with much coverage common to FAPT, but at a more advanced mathematical level. Includes a brief introduction to mathematical statistics.

KOLMOGOROV (1933) A short, original, and definitive work which established the axiomatic foundation of modern mathematical probability theory. Every student of applied probability theory will profit from spending at least several hours with this exceptional document. Although many sections are presented at an advanced level, the reader will rapidly achieve some understanding of the nature of those topics which are required for a rigorous theoretical foundation but neglected in a volume such as FAPT. As one significant example, he will learn that our third axiom of probability theory (known formally as the axiom of *finite additivity*) must be replaced by another axiom (specifying *countable additivity*) in order to deal properly with probability in continuous sample spaces.

LOÈVE (1955) Significantly more advanced than Gnedenko, this is a mathematical exposition of probability theory. Limited concern with applications and physical interpretation.

PAPOULIS (1965) An effective, compact presentation of applied probability theory, followed by a detailed investigation of random processes with emphasis on communication theory. Especially recommended for electrical engineers.

PARZEN (1960) More formal, appreciably more detailed presentation at a mathematical level slightly above FAPT. More concern with mathematical rather than physical interpretation. A lucid, valuable reference work.

PFEIFFER (1965) A more formal development at about the same level as FAPT. With care, patience, and illustration the author introduces matters of integration, measure, etc., not mentioned in FAPT. A recommended complement to FAPT for readers without training in theoretical mathematics who desire a somewhat more rigorous foundation. Contains an annotated bibliography at the end of each chapter.

PITT (1963) A concise statement of introductory mathematical probability theory, for readers who are up to it. Essentially self-contained, but the information density is very great.

3 *Random Processes*

COX (1962) Compact, readable monograph on the theory of renewal processes with applications.

COX and MILLER (1965) General text on the theory of random processes with applications.

COX and SMITH (1961) Compact, readable monograph which introduces some aspects of elementary queuing problems.

DAVENPORT and ROOT (1958) Modern classic on the application of random process theory to communication problems.

DOOB (1953) Very advanced text on the theory of random processes for readers with adequate mathematical prerequisites. (Such people are unlikely to encounter FAPT.)

FISZ (1963) Cited above. Contains a proof of the ergodic theorem for discrete-state discrete-transition Markov processes stated in Chap. 5 of FAPT.

HOWARD (1960) An entirely clear, brief introduction to the use of Markov models for decision making in practical situations with economic consequences.

HOWARD (in preparation) Detailed investigation of Markov models and their applications in systems theory.

LEE (1960) Lucid introductory text on communication applications of random process theory.

MORSE (1958) Clear exposition of Markov model applications in queuing theory aspects of a variety of practical operational situations.

PAPOULIS (1965) Cited above.

PARZEN (1962) Relatively gentle introduction to random process theory with a wide range of representative examples.

4 *Classical and Modern Statistics*

CHERNOFF and MOSES (1959) An elementary, vivid introduction to decision theory.

CRAMER (1946) A thorough, mathematically advanced text on probability and mathematical statistics.

FISZ (1963) Cited above.

FRASER (1958) Clear presentation of elementary classical statistical theory and its applications.

FREEMAN (1963) The last half of this book is a particularly logical, readable presentation of statistical theory at a level somewhat more advanced than Fraser. Contains many references and an annotated bibliography of texts in related fields.

MOOD and GRAYBILL (1963) One of the most popular and successful basic treatments of the concepts and methods of classical statistics.

PRATT, RAIFFA, and SCHLAIFER (1965) From elementary probability theory through some frontiers of modern statistical decision theory with emphasis on problems with economic consequences.

RAIFFA and SCHLAIFER (1961) An advanced, somewhat terse text on modern Bayesian analysis. Lacks the interpretative material and detailed explanatory examples found in the preceding reference.

SAVAGE (1954) An inquiry into the underlying concepts of statistical theory. Does not require advanced mathematics.

WILLIAMS (1954) A gentle, animated introduction to game theory—the study of decision making in competitive, probabilistic situations.

5 *Complete Titles of Above References*

CHERNOFF, H., and L. MOSES: "*Elementary Decision Theory,*" John Wiley & Sons, Inc., New York, 1959.

COX, D. R.: "*Renewal Theory,*" John Wiley & Sons, Inc., New York, 1962.

———, and H. D. MILLER: "*The Theory of Stochastic Processes,*" John Wiley & Sons, Inc., New York, 1965.

———, and W. L. SMITH: "*Queues,*" John Wiley & Sons, Inc., New York, 1961.

CRAMER, H.: "*Mathematical Methods of Statistics,*" Princeton University Press, Princeton, N.J., 1946.

DAVENPORT, W. B., JR., and W. L. ROOT: "*Random Signals and Noise,*" McGraw-Hill Book Company, New York, 1958.

DAVID, F. N.: "*Games, Gods, and Gambling,*" Hafner Publishing Company, Inc., New York, 1962.

DOOB, J. L.: "*Stochastic Processes,*" John Wiley & Sons, Inc., New York, 1953.

FELLER, W.: "*An Introduction to Probability Theory and Its Applications,*" vol. 1, 2d ed., John Wiley & Sons, Inc., New York, 1957.

FISZ, M.: "*Probability Theory and Mathematical Statistics,*" John Wiley & Sons, Inc., New York, 1963.

FRASER, D. A. S.: "*Statistics: An Introduction,*" John Wiley & Sons, Inc., New York, 1958.

FREEMAN, H.: "*Introduction to Statistical Inference,*" Addison-Wesley Publishing Company, Inc., Reading, Mass., 1963.

GNEDENKO, B. V.: "*Theory of Probability,*" Chelsea, New York, 1962.

HOWARD, R. A.: "*Dynamic Programming and Markov Processes,*" The M.I.T. Press, Cambridge, Mass., 1960.

———: (in preparation), John Wiley & Sons, Inc., New York.

KOLMOGOROV, A. N.: "*Foundations of the Theory of Probability*" (Second English Edition), Chelsea, New York, 1956.

KYBURG, H. E., JR., and H. E. SMOKLER: "*Studies in Subjective Probability,*" John Wiley & Sons, Inc., New York, 1964.

LAPLACE, *A Philosophical Essay on Probabilities*, 1825 (English translation), Dover Publications, Inc., New York, 1951.

LEE, Y. W.: "*Statistical Theory of Communication,*" John Wiley & Sons, Inc., New York, 1960.

LOÈVE, M.: "*Probability Theory,*" D. Van Nostrand Company, Inc., Princeton, N.J., 1963.

MOOD, A. M., and F. A. GRAYBILL: "*Introduction to the Theory of Statistics,*" McGraw-Hill Book Company, New York, 1963.

MORSE, P. M.: "*Queues, Inventories, and Maintenance,*" John Wiley & Sons, Inc., New York, 1958.

PAPOULIS, A.: "*Probability, Random Variables, and Stochastic Processes,*" McGraw-Hill Book Company, New York, 1965.

PARZEN, E.: "*Modern Probability Theory and Its Applications,*" John Wiley & Sons, Inc., New York, 1960.

———: "*Stochastic Processes,*" Holden-Day, San Francisco, 1962.

PFEIFFER, P. E.: "*Concepts of Probability Theory,*" McGraw-Hill Book Company, New York, 1965.

PITT, H. R.: "*Integration, Measure, and Probability,*" Hafner Publishing Company, Inc., New York, 1963.

PRATT, J. W., H. RAIFFA, and R. SCHLAIFER: "*Introduction to Statistical Decision Theory*" (preliminary edition), McGraw-Hill Book Company, New York, 1965.

RAIFFA, H., and R. SCHLAIFER: "*Applied Statistical Decision Theory,*" Harvard Business School, Division of Research, Boston, 1961.

SAVAGE, L. J.: "*The Foundations of Statistics,*" John Wiley & Sons, Inc., New York, 1954.

TODHUNTER, I.: "*A History of a Mathematical Theory of Probability from the Time of Pascal to that of Laplace*" 1865, Chelsea, New York, 1949.

WILLIAMS, J. D.: "*The Compleat Strategyst, Being a Primer on the Theory of Games of Strategy,*" McGraw-Hill Book Company, New York, 1954.

common PDF's, PMF's, and their means, variances, and transforms

The most common elementary PDF's and PMF's are listed below. Only those transforms which may be expressed in relatively simple forms are included.

In each entry involving a single random variable, the random variable is denoted by x. For compound PDF's and PMF's, the random variables are denoted by two or more of the symbols u, v, w, x, and y. Symbols k, m, and n are used exclusively for nonnegative integer parameters. Unless there is a special reason for using other symbols, all parameters are denoted by the symbols a, b, and c.

The nomenclature used for PDF's and PMF's is not universal. Several of the entries below are, of course, special cases or generalizations of other entries.

Bernoulli PMF

$$p_x(x_0) = \begin{cases} 1 - P & x_0 = 0 \\ P & x_0 = 1 \\ 0 & \text{otherwise} \end{cases}$$

$$0 < P < 1$$

$$E(x) = P \qquad \sigma_x^2 = P(1 - P)$$

$$p_x^T(z) = 1 - P + zP$$

Beta PDF

$$f_x(x_0) = \begin{cases} c(a,b)x_0^{a-1}(1 - x_0)^{b-1} & 0 < x_0 < 1 \\ 0 & \text{otherwise} \end{cases}$$

$$a > 0 \qquad b > 0$$

$$c(a,b) = \frac{(a + b - 1)!}{(a - 1)!(b - 1)!}$$

$$E(x) = \frac{a}{a + b} \qquad \sigma_x^2 = \frac{ab}{(a + b)^2(a + b + 1)}$$

Binomial PMF

$$p_x(x_0) = \begin{cases} \binom{n}{x_0} P^{x_0}(1 - P)^{n-x_0} & x_0 = 0,1,2, \ldots ,n \\ 0 & \text{otherwise} \end{cases}$$

$$0 < P < 1 \qquad n = 1, 2, 3, \ldots$$

$$E(x) = nP \qquad \sigma_x^2 = nP(1 - P)$$

$$p_x^T(z) = (1 - P + zP)^n$$

Bivariate-normal PDF

$$f_{x,y}(x_0,y_0) = \frac{\exp\left(-\frac{1}{2(1 - \rho^2)}\left\{\left[\frac{x - E(x)}{\sigma_x}\right]^2 + \left[\frac{y - E(y)}{\sigma_y}\right]^2 - 2\rho\,\frac{x - E(x)}{\sigma_x}\,\frac{y - E(y)}{\sigma_y}\right\}\right)}{2\pi\sigma_x\sigma_y\sqrt{1 - \rho^2}}$$

$$-\infty < x_0 < \infty \qquad -\infty < y_0 < \infty$$

$$\sigma_x > 0 \qquad \sigma_y > 0 \qquad -1 < \rho < 1$$

$$f_{x,y}^T(s_1,s_2) = E(e^{-s_1x}e^{-s_2y}) = \exp\left[-s_1E(x) - s_2E(y) + \tfrac{1}{2}(s_1^2\sigma_x^2 + 2\rho s_1s_2\sigma_x\sigma_y + s_2^2\sigma_y^2)\right]$$

Cauchy PDF

$$f_x(x_0) = \frac{1}{\pi} \frac{a}{a^2 + (x_0 - b)^2} \qquad -\infty < x_0 < \infty$$

$a > 0 \qquad -\infty < b < \infty$

$E(x) \equiv b \qquad \sigma_x^2 = \infty$

[The above value of $E(x)$ is a common definition. Although $E(x) = b$ seems intuitive from the symmetry of $f_x(x_0)$, note that

$$\int_{-\infty}^{\infty} x_0 f_x(x_0)\, dx_0$$

has no unique value for the Cauchy PDF.]

$f_x^T(s) = e^{-bs-a|s|}$

Chi-square PDF

$$f_x(x_0) = \begin{cases} \left[\left(\frac{n}{2} - 1\right)!\right]^{-1} 2^{-n/2} x_0^{(n/2)-1} e^{-x_0/2} & x_0 > 0 \\ 0 & \text{otherwise} \end{cases}$$

$n = 1, 2, 3, \ldots$

$E(x) = n \qquad \sigma_x^2 = 2n$

$f_x^T(s) = (1 + 2s)^{-n/2}$

Erlang PDF

$$f_x(x_0) = \begin{cases} \dfrac{a^n x_0^{n-1} e^{-ax_0}}{(n-1)!} & x_0 > 0 \\ 0 & \text{otherwise} \end{cases}$$

$a > 0 \qquad n = 1, 2, 3, \ldots$

$E(x) = na^{-1} \qquad \sigma_x^2 = na^{-2}$

$f_x^T(s) = a^n(s + a)^{-n}$

Exponential PDF

$$f_x(x_0) = \begin{cases} ae^{-ax_0} & x_0 > 0 \\ 0 & \text{otherwise} \end{cases}$$

$a > 0$

$E(x) = a^{-1} \qquad \sigma_x^2 = a^{-2}$

$f_x^T(s) = a(s + a)^{-1}$

Gamma PDF

$$f_x(x_0) = \begin{cases} \dfrac{x_0{}^a e^{-x_0/b}}{a!b^{a+1}} & x_0 > 0 \\ 0 & \text{otherwise} \end{cases}$$

$a > -1 \qquad b > 0$

$E(x) = (a+1)b \qquad \sigma_x{}^2 = (a+1)b^2$

$f_x{}^T(s) = (1 + bs)^{-a-1}$

Geometric PMF

$$p_x(x_0) = \begin{cases} P(1-P)^{x_0-1} & x_0 = 1, 2, 3, \ldots \\ 0 & \text{otherwise} \end{cases}$$

$0 < P < 1$

$E(x) = P^{-1} \qquad \sigma_x{}^2 = (1-P)P^{-2}$

$p_x{}^T(z) = zP[1 - z(1-P)]^{-1}$

Hypergeometric PMF

$$p_x(x_0) = \begin{cases} \dbinom{m}{x_0}\dbinom{n}{k-x_0}\Big/\dbinom{m+n}{k} & x_0 = 0, 1\ 2, \ldots, k \\ 0 & \text{otherwise} \end{cases}$$

$m = 1, 2, 3, \ldots \qquad n = 1, 2, 3, \ldots \qquad k = 1, 2, 3, \ldots, (m+n)$

$$E(x) = \frac{mk}{m+n} \qquad \sigma_x{}^2 = \frac{mnk(m+n-k)}{(m+n)^2(m+n-1)}$$

Laplace PDF

$$f_x(x_0) = \frac{a}{2} e^{-a|x_0-b|} \qquad -\infty < x_0 < \infty$$

$a > 0 \qquad -\infty < b < \infty$

$E(x) = b \qquad \sigma_x{}^2 = 2a^{-2}$

$f_x{}^T(s) = a^2 e^{-bs}(a^2 - s^2)^{-1}$

Log-normal PDF

$$f_x(x_0) = \begin{cases} \dfrac{\exp\{-[\ln(x_0 - a) - b]^2/2\sigma^2\}}{\sqrt{2\pi}\,\sigma(x_0 - a)} & x_0 > a \\ 0 & \text{otherwise} \end{cases}$$

$$\sigma > 0 \qquad -\infty < a < \infty \qquad -\infty < b < \infty$$

$$E(x) = a + e^{b+0.5\sigma^2} \qquad \sigma_x^2 = e^{2b+\sigma^2}(e^{\sigma^2} - 1)$$

Maxwell PDF

$$f_x(x_0) = \begin{cases} \sqrt{2/\pi}\, a^3 x_0^2 e^{-a^2x_0^2/2} & x_0 > 0 \\ 0 & \text{otherwise} \end{cases}$$

$$a > 0$$

$$E(x) = \sqrt{8/\pi}\, a^{-1} \qquad \sigma_x^2 = (3 - 8/\pi)a^{-2}$$

Multinomial PMF

$$p_{u,v,\ldots,y}(u_0,v_0, \ldots ,y_0) = \frac{n!p_u^{u_0}p_v^{v_0} \cdots p_y^{y_0}}{u_0!v_0! \cdots y_0!}$$

$$u_0 = 0, 1, \ldots, n \qquad v_0 = 0, 1, \ldots, n \quad \cdots \quad y_0 = 0, 1, \ldots, n$$

$$u_0 + v_0 + \cdots + y_0 = n$$

$$p_u + p_v + \cdots + p_y = 1 \qquad 0 < p_u, p_v, \ldots, p_y < 1$$

$$E(u) = np_u \qquad E(v) = np_v \quad \cdots \quad E(y) = np_y$$

$$\sigma_u^2 = np_u(1 - p_u) \qquad \sigma_v^2 = np_v(1 - p_v) \quad \cdots \quad \sigma_y^2 = np_y(1 - p_y)$$

Normal PDF

$$f_x(x_0) = \frac{e^{-[x_0-E(x)]^2/2\sigma_x^2}}{\sqrt{2\pi}\,\sigma_x} \qquad -\infty < x_0 < \infty$$

$$\sigma_x > 0 \qquad -\infty < E(x) < \infty$$

$$f_x^T(s) = e^{-sE(x)+(s^2\sigma_x^2/2)}$$

Pascal PMF

$$p_x(x_0) = \begin{cases} \binom{x_0 - 1}{n - 1} P^n(1 - P)^{x_0-n} & x_0 = n, n + 1, n + 2, \ldots \\ 0 & \text{otherwise} \end{cases}$$

$$0 < P < 1 \qquad n = 1, 2, 3, \ldots$$

$$E(x) = nP^{-1} \qquad \sigma_x^2 = n(1 - P)P^{-2}$$

$$p_x^T(z) = (zP)^n[1 - z(1 - P)]^{-n}$$

Poisson PMF

$$p_x(x_0) = \frac{a^{x_0}e^{-a}}{x_0!} \qquad x_0 = 0, 1, 2, \ldots$$

$$a > 0$$
$$E(x) = a \qquad \sigma_x^2 = a$$
$$p_x^T(z) = e^{a(z-1)}$$

Rayleigh PDF

$$f_x(x_0) = \begin{cases} a^2 x_0 e^{-a^2 x_0^2/2} & x_0 > 0 \\ 0 & \text{otherwise} \end{cases}$$

$$a > 0$$

$$E(x) = \sqrt{\pi/2}\, a^{-1} \qquad \sigma_x^2 = (2 - \pi/2)a^{-2}$$

Uniform PDF

$$f_x(x_0) = \begin{cases} \dfrac{1}{b-a} & a < x_0 < b \\ 0 & \text{otherwise} \end{cases}$$

$$-\infty < a < b < \infty$$
$$E(x) = (a+b)/2 \qquad \sigma_x^2 = (b-a)^2/12$$
$$f_x^T(s) = (e^{-as} - e^{-bs})[s(b-a)]^{-1}$$

Weibull PDF

$$f_x(x_0) = \begin{cases} abx_0^{b-1} e^{-ax_0^b} & x_0 > 0 \\ 0 & \text{otherwise} \end{cases}$$

$$a > 0 \qquad b > 0$$

$$E(x) = \left(\frac{1}{a}\right)^{1/b} \Gamma(1 + b^{-1})$$

$$\sigma_x^2 = \left(\frac{1}{a}\right)^{2/b} \{\Gamma(1 + 2b^{-1}) - [\Gamma(1 + b^{-1})]^2\}$$

$$\Gamma(c) \equiv \int_0^\infty x^{c-1} e^{-x}\, dx$$

INDEX